Y0-AAF-605

MATH.
STAT.
LIBRARY

Lectures on Ordinary Differential Equations

DR. JOHN H. BARRETT
December 20, 1922–January 21, 1969

Lectures on Ordinary Differential Equations

Edited by

Robert McKelvey

Department of Mathematics
University of Colorado
Boulder, Colorado

Academic Press 1970 *New York and London*

ACADEMIC PRESS, INC.
111 Fifth Avenue, New York, New York 10003

United Kingdom Edition published by
ACADEMIC PRESS, INC. (LONDON) LTD.
Berkeley Square House, London W1X 6BA

LIBRARY OF CONGRESS CATALOG CARD NUMBER: 70-117103

PRINTED IN THE UNITED STATES OF AMERICA

Originally published in *Advances in Mathematics* 1968, 1969

CONTRIBUTORS

JOHN H. BARRETT,*
Department of Mathematics, University of Tennessee, Knoxville, Tennessee

LLOYD K. JACKSON,
Department of Mathematics, University of Nebraska, Lincoln, Nebraska

ROBERT E. O'MALLEY, JR.,
Courant Institute of Mathematical Sciences, New York University, New York, New York

D. WILLETT,
Department of Mathematics, University of Utah, Salt Lake City, Utah

* *(Deceased)*

CONTENTS

Topics in Singular Perturbations

Robert E. O'Malley, Jr.

Classification of Second Order Linear Differential Equations with Respect to Oscillation

D. Willett

PREFACE

The lectures in this volume survey recent important tendencies in some of the active fields of research in ordinary differential equations. Taken together, they provide a comprehensive introduction to the subject, suitable for study in graduate seminars. They also provide a valuable resource for the active researchers in the field.

The paper of Lloyd Jackson surveys the subfunction method in two-point boundary value problems. Introduced originally by O. Perron and F. Riesz in studies of the Dirichlet problem, this method has emerged in the hands of Jackson, his students, and other workers as a comprehensive and systematic approach to a wide class of nonlinear, one-dimensional problems. In the lectures Jackson has reworked much of the subject, creating a unified and esthetically satisfying body of theory.

Both John Barrett and Douglas Willett deal with oscillation theory, though in quite different aspects. Willett systematizes and unifies a considerable body of recent results about the second order equation, often calling on ingenious technical devices to deal with highly complicated situations. Barrett's notes represent a pedigogical approach which he gradually refined in advanced seminars at the University of Utah and later at the University of Tennessee. By treating a series of prototype examples, Barrett gradually develops his reader's powers, finally approaching the difficulties inherent in the oscillation theory for equations of higher order than second.

It is the essence of singular perturbation theory that the perturbed boundary problem is a different kind of mathematical object from the unperturbed, a phenomenon associated with a change in order of the differential equation and an accompanying loss of boundary conditions. This intriguing aspect, combined with the pervasiveness in nature of asymptotically singular phenomena, has produced a voluminous literature, much of it having appeared since Wolfgang Wasow's well-known and now classical study, Robert O'Malley's article surveys the recent developments in this rapidly growing subject.

The articles in this volume are all related in one way or another to the symposium on ordinary differential equations held in Boulder in the summer of 1967. That symposium, the first of the Rocky Mountain Summer Seminars, was organized under the auspices of the Associated Rocky Mountain Universities, and financed by a grant from the Graduate Education Division of the National Science Foundation. The lectures of Barrett and Jackson were actually delivered at the symposium, in much the same form as here. Willett's survey represents an elaboration of an hour talk given at that time. O'Malley's notes, which had appeared earlier in mimeographed form, were the basis for one of several informal seminars held during the symposium.

It is especially fortunate that John Barrett retained the energy to develop through his symposium lectures and in these notes a definitive treatment of a subject which now bears his permanent mark. It was a privilege for me to have known this intelligent, considerate, and courageous man.

Lectures on Ordinary Differential Equations

Reprinted from *Advances in Mathematics* 3, No. 4, 415–509, (1969).

Oscillation Theory of Ordinary Linear Differential Equations

JOHN H. BARRETT*

The University of Tennessee, Knoxville, Tennessee 37916

Preface

The purpose of this article [a compilation of lectures originally presented at the Associated Western Universities Differential Equations Symposium, Boulder, Colorado in the summer of 1967] is to give motivating examples and ideas which have influenced the author in his studies of oscillation properties of solutions of linear ordinary differential equations. This is a subject where the mathematical tools needed are relatively elementary but where it is easy to state an unsolved problem. For example, the familiar second-order equation $y'' + q(x)y = 0$ is still a valid subject for research, although it has a voluminous literature.

As far as oscillation theory is concerned, most texts in Differential Equations, both elementary and advanced, deal only with second-order equations. A few deal with self-adjoint fourth-order equations and, perhaps, those of arbitrary even order and systems of first or second-order

* We regret to report that Dr. Barrett died on January 21, 1969, and therefore that this paper is being published posthumously. We are grateful to Drs. John Bradley, William J. Coles, and John W. Heidel for reading the proofs.

equations. Any discussion of oscillatory properties of third-order equations or other nonself-adjoint equations is hard to find and that is the lowest order where truly nonself-adjoint equations occur.

In this article an attempt is made to give a self-contained inductive development from equations of one order to the next. Most of the discussion will deal with equations of second, third, and fourth orders, with linear systems of second-order equations and with generalizations of those results to equations of higher orders. No attempt is made to survey all of the oscillation theory of equations of orders higher than four. Considerable attention is devoted to equations of order three (Section II) and this is in line with the increased recent interest in these equations, as the Bibliography will show.

Instead of the usual format where proofs follow the respective theorems, motivating examples and developments of the ideas are given first with statements of theorems following as summaries of what has been established in the discussion.

I. Second-Order Equations

Much of the material in this introductory section is contained in introductory texts on Ordinary Differential Equations. Only those topics are included which are pertinent to the succeeding discussion of equations of higher order. For further oscillation theory of second-order equations see Chapters IV and XI of Hartman's recent advanced text (*47*).

1.1. *Basic Properties*

The real linear second-order equation

$$(1.1)\qquad \begin{gathered} l_2[y] \equiv y'' + A(x)\,y' + B(x)\,y = 0; \\ A \text{ and } B \in C(I), \qquad I = [a, b), \qquad a < b \leqslant \infty, \end{gathered}$$

is equivalent to a special case of the *canonical self-adjoint* form:

$$(E_2)\qquad L_2[y] = (ry')' + qy = 0; \qquad r > 0, \qquad r \,\&\, q \in C(I),$$

where $r = \exp(\int A)$ and $q = rB$. A function y is said to be *admissible for* the operator L_2 on an interval I provided y and $ry' \in C[I]$. Note that y'' need not exist when r is not differentiable. A *solution* of (E_2) is an admissible function for L_2 satisfying $L_2[y] = 0$ on I. Existence and uniqueness theorems for (E_2) may be obtained easily from the equivalent vector-matrix form

$$(V_2) \qquad \begin{pmatrix} y \\ D_1 y \end{pmatrix}' = \begin{pmatrix} 0 & \frac{1}{r} \\ -q & 0 \end{pmatrix} \begin{pmatrix} y \\ D_1 y \end{pmatrix}; \qquad D_1 y = ry'.$$

Lemma 1.1. *For any pair of numbers (c_0, c_1) and each $c \in I$ there exists a unique solution of (E_2) satisfying*

$$y(c) = c_0, \qquad (ry')(c) = c_1.$$

There are several ways to transform (E_2) back into the form (1.1).

Lemma 1.2. (a) *If $r \in C'(I)$ then (E_2) can be put into the form* (1.1).
(b) *If r is not differentiable then the change of independent variable:*

$$(1.2) \qquad t = \int_a^x (1/r) \Rightarrow x = x(t), \qquad Y(t) = y[x(t)]$$

yields the form (1.1)*—with respect to t—without a middle term,*

$$\ddot{Y} + QY = 0, \qquad Q(t) = (qr)[x(t)].$$

Note that Lemma 1.2(b) provides a method for removing the middle term of (1.1) and differentiation of coefficients is not required, as it is in the standard variation of parameters substitution

$$y = uv, \qquad v' = (A/2)v.$$

Integration-by-parts of $zl_2[y]$ (or $zL_2[y]$), where z is an arbitrary function with the exhibited derivatives, yields useful *Lagrange Identities*.

Lemma 1.3. (a) *If $y, z \in C''(I)$ then*

$$zl_2[y] = \{zy' - yz' + Ayz\}' + yl_2^+[z],$$

where $l_2^+[z] = (z' - Az)' + Bz$, an adjoint operator of l_2.
(b) *If z and y are admissible functions for L_2 then*

$$zL_2[y] = [r(zy' - yz')]' + yL_2[z].$$

Note that $l_2^+ = l_2$ if $A \equiv 0$, and in this special case l_2 and (1.1) are *self-adjoint*. Since L_2 serves as its own adjoint operator, L_2 and the corresponding equation (E_2) are said to be *self-adjoint*.

1.2. *Factoring and Disconjugacy*

The Polya-Mammana (*73*, *79*) factored form of L_2 is easy to establish and is actually a variation of standard Wronskian properties.

Theorem 1.1. *If* $L_2[v] = 0$ *and* $v \neq 0$ *on* $I' \subset I$ *then*

(a) $$vL_2[y] = [vD_1y - yD_1v]'$$

and

$$L_2[y] = (1/v)[rv^2(y/v)']' \tag{1.4}$$

for each L_2*-admissible* y *on* I' *and*

(b) *no nontrivial solution of* (E_2) *has two zeros on* I'.

Definition 1.1. A *second-order* linear operator (e.g., L_2) and the corresponding homogeneous equation (e.g., E_2) are said to be *disconjugate* on an interval I provided that *no nontrivial solution* of the equation *has two zeros* on I.

Let $v_1 = v_1(x, a)$ be the solution of (E_2) defined by the initial conditions:

$$y(a) = 0 \quad \text{and} \quad D_1y(a) = (ry')(a) = 1. \tag{1.3}$$

This solution is called the *principal* solution of (E_2) at $x = a$ and oscillation properties of (E_2) can be given in terms of $v_1(x, a)$.

The Polya–Mammana factored form (1.4) may be used to prove the following results which form a version of the Sturm comparison theorem.

Lemma 1.4. (a) *If* $a < b \leqslant \infty$ *then the equation* (E_2) *is disconjugate on* $I = [a, b)$ *if, and only if, the principal solution* $v_1(x, a) > 0$ *on* $I^0 = (a, b)$.

(b) *If* (E_2) *is disconjugate on an interval* I *then*

$$(ry')' + q_1y = 0; \qquad q_1 \in C(I) \ \text{and} \ q_1 \leqslant q \ \text{on} \ I \tag{E_2'}$$

is disconjugate on I.

(c) *If, in addition,* $r \leqslant r_1$ *and* $r_1 \in C(I)$ *then*

$$(r_1y')' + q_1y = 0 \tag{E_2''}$$

is disconjugate on I.

W. Leighton and Z. Nehari have isolated a crucial part of the standard proof of the Sturm Separation Theorem as a *Fundamental Lemma* and applied it to linear differential equations of orders greater than two (*70*).

Lemma 1.5. *If* $u(x)$ *and* $v(x)$ *are differentiable functions on* $[a, c]$, $a < c$, $u(a) = u(c) = 0$ *and* $v(x) \neq 0$ *on* $[a, c]$, *then*

(a) $vu' - uv' = v^2(u/v)'$ *and their Wronskian* $uv' - vu'$ *has a zero on* (a, c) *and*

(b) *there is a linear combination* $z = u - kv$ *which has a double zero on* (a, c) (*i.e., at* $x = \xi \in (a, c)$ *where* $z(\xi) = z'(\xi) = 0$).

Note that if u and v are also solutions of (E_2) and $a < c < b$ then they are linearly dependent, which contradicts the original assumptions of Lemma 1.5, thus yielding the Sturm separation theorem.

Lemma 1.4(a) provides a method for proving oscillation theorems, by use of nonlinear Riccati equations.

Lemma 1.6. *Let* $y(x)$ *be a solution of* (E_2) *on* $I = [a, \infty)$, *such that* $y(a) = 0$ *and* $y'(a) > 0$. *If* $y(x) > 0$ *on* (a, ∞) *then*

(a) $h = -ry'/y$ *satisfies a Riccati equation*

$$h' = q + (1/r)h^2 \quad \text{on} \quad (a, \infty). \tag{1.5}$$

(b) *If, in addition,* $y'(x) > 0$ *on* $[a, \infty)$ *then* $\int_a^x q$ *is bounded above on* $[a, \infty)$. *On the other hand, if* $\int_a^\infty q \equiv \infty$ *then* $y'(x)$ *has a zero on* (a, ∞) *and* $y(x)$ *is bounded on* $[a, \infty)$.

Once we have a zero of $y'(x)$ we proceed to force a subsequent zero of $y(x)$. Hille (*58*) seems to have been the first to note the following property, which was of considerable use to Nehari (*75*) and the author (*8*), for establishing necessary conditions for disconjugacy of (E_2).

Lemma 1.7. *Let* $y(x)$ *be a solution of* (E_2) *on* $I = [a, \infty)$ *such that* $y(a) > 0$, $y'(a) = 0$. *If* $y(x) > 0$ *and* $q(x) \geqslant 0$, *but* $\not\equiv 0$ *for large* x, *then* $D_1 y(x) = (r(x)\, y'(x)) \leqslant 0$ *on* (a, ∞) *and* $\int_a^\infty (1/r) < \infty$.

Theorem 1.2. *If* $\int_a^\infty (1/r) = \infty$, $q \geqslant 0$ *but* $\not\equiv 0$ *for large* x, (E_2) *is disconjugate on* $[a, \infty)$ *and* $y(x)$ *is any nontrivial solution of* (E_2) *with* $y(a) = 0$, *then* $y(x)\, y'(x) > 0$ *on* (a, ∞).

It is often useful to know that disconjugacy allows the construction of a nonzero solution on closed or open intervals (68). Recall that the converse (Lemma 1.4) is also true.

The case of the finite closed interval is easy to prove. On the other hand, if $I = [a, b)$, $a < b \leqslant \infty$, let $x_n \in (a, b)$ and $\{x_n\} \uparrow b$ and $y_n(x)$ be the unique solution of (E_2) such that

$$y_n(x_n) = 0, \qquad y_n(x) > 0 \quad \text{on} \quad [a, x_n) \quad \text{and} \quad (c_1{}^n)^2 + (c_2{}^n)^2 = 1,$$

where $y_n(x) = c_1{}^n u(x) + c_2{}^n v(x)$ and u, v is a given fundamental set of solutions of (E_2). There is a subsequence $\{n_j\}$ of $\{n\}$, such that $\{c_1^{n_j}\}$ and $\{c_2^{n_j}\}$ both converge, and these limits define a positive solution on $I^0 = (a, b)$, which may or may not be zero at $x = a$. Similarly, a positive solution may be found for open intervals I.

Theorem 1.3. *If (E_2) is disconjugate on an interval I then there exists a positive solution of (E_2) on I for* (a) $I = [a, b]$, $a < b < \infty$ *and* (b) $I = (a, b)$, $-\infty \leqslant a < b \leqslant \infty$. (c) *If $I = [a, b)$ then there exists a positive solution on $I^0 = (a, b)$.*

1.3. *Oscillation*

By combining Lemma 1.6 with Theorem 1.2 we have:

Lemma 1.8. *If $\int_a^\infty 1/r = \infty$, $q \geqslant 0$ and $\int_a^\infty q = \infty$ then every solution of (E_2) has infinitely many zeros on $[a, b)$.*

This is a weak form of the Leighton–Wintner oscillation theorem. About 1949, both Leighton (*67*) and Wintner (*108*) eliminated the nonnegative condition, $q(x) \geqslant 0$. (See Theorem 1.2 below). However, the nonnegative coefficient case is of special interest and is more readily generalized to certain equations of higher order.

Definition 1.2. *A second-order operator L_2, or the corresponding equation (E_2), is said to be oscillatory {nonoscillatory} on an interval I provided that every solution of (E_2) has infinitely many {at most a finite number of} zeros on I.*

Although disconjugacy is an extreme case of nonoscillation they are essentially equivalent for second-order equations.

Lemma 1.9. *If L_2, or (E_2), is nonoscillatory on $[a, \infty)$ then there is a number $c \in (a, \infty)$ such that L_2, or (E_2), is disconjugate on $[c, \infty)$.*

Although it is well known that nonoscillation on $[a, \infty)$ implies disconjugacy for large x for second-order equations it was noted recently by Nehari (77) that the analogous statement for equations of orders greater than two is known to be true only for special cases. A useful *example* and comparison equation is *Euler's Equation*

$$x^2y'' + ky = 0, \qquad k = \text{constant}, \tag{1.6}$$

which

(i) has solutions of the form x^α where

$$\alpha \text{ is real if } k \leqslant \tfrac{1}{4},$$
$$\alpha \text{ is complex if } k > \tfrac{1}{4},$$

(ii) is disconjugate if $k \leqslant \frac{1}{4}$ and oscillatory if $k > \frac{1}{4}$ on $[1, \infty)$.

However, when (1.6) is put into the form (E_2), with $r = 1$ and $q = k/x^2$, it *does not satisfy the hypothesis of Lemma 1.8* on $[1, \infty)$ since $\int_1^\infty q < \infty$ for all values of k, although we have oscillation for $k > \frac{1}{4}$. The following is well-known.

Lemma 1.10. *If $\int_a^\infty (1/r) < \infty$ and $\int_a^\infty | q | < \infty$ then equation (E_2) is nonoscillatory on $[a, \infty)$.*

Suppose that (E_2) is nonoscillatory, i.e., there exists a solution $u(x)$ and a number $b \in (a, \infty)$ such that $u(x) > 0$ on $[b, \infty)$. Therefore, $h = -ru'/u$ satisfies the *Riccati Equation*

$$h' = q + \frac{1}{r} h^2 \quad \text{on} \quad [b, \infty). \tag{1.5}$$

Let $\int_a^\infty q = \infty$ and $a < b < \infty$, then there is a number $c \in (b, \infty)$ such that $h(b) + \int_b^x q > 0$ on $[c, \infty)$ and, hence,

$$h(x) > g(x) = \int_b^x (1/r)\, h^2 > 0 \quad \text{on} \quad [c, \infty).$$

Consequently,

$$g' > (1/r)\, g^2 \qquad \text{and} \qquad \int_c^x (1/r) < \frac{1}{g(c)} < \infty.$$

This proof for the following Leighton–Wintner Oscillation Theorem was suggested by W. J. Coles.

Theorem 1.4. *If* $\int_a^\infty 1/r = \infty$ *and* $\int_a^\infty q = \infty$ *then the equation* (E_2) *is oscillatory on* $[a, \infty)$.

This theorem is the *principal motivation* for the discussion in the subsequent chapters where higher-order analogs are established. Willet (*107*) has recently pointed out that if $\int_a^\infty (1/r) < \infty$ then the transformation

$$t = \left[\int_x^\infty (1/r)\right]^{-1} \tag{1.6}$$

yields the following corollary of Theorem 1.4.

Corollary 1.4.1. *If* $\int_a^\infty (1/r) < \infty$ *but*

$$\int_a^\infty q(s) \left[\int_s^\infty (1/r)\right]^2 ds = \infty \tag{1.7}$$

then (E_2) *is oscillatory on* $[a, \infty)$.

In the case of (E_2), with $r = 1$ and $q \geqslant 0$, Hille (*58*) achieved better results than Lemma 1.8. Let $\int_a^\infty (1/r) = \infty$, $q \geqslant 0$ but $\neq 0$ for large x and (E_2) be disconjugate on $[a, \infty)$ and let $y(x)$ be a *positive solution* (E_2) on (a, ∞). Then, by Theorem 1.2, $h = -ry'/y < 0$ on (a, ∞). Since h satisfies the Riccati equation (1.5), it follows that

$$|h(x)| = -h(x) \geqslant \int_x^\infty q \qquad \text{and} \qquad \int_a^x (1/r) \leqslant -1/h(x).$$

Theorem 1.5. *If* $\int_a^\infty (1/r) = \infty$, $q(x) \geqslant 0$ *but* $\neq 0$ *for large* x *on* $[a, \infty)$ *then a necessary condition for disconjugacy of* (E_2) *on* $[a, \infty)$ *is*

$$\int_a^x (1/r) \int_x^\infty q \leqslant 1. \tag{1.8}$$

Hence, $\limsup_{x\to\infty} \int_a^x (1/r) \int_x^\infty q > 1$ *is sufficient for oscillation of* (E_2).

1.4. *The Prüfer Transformation*

The change of variables to polar coordinates of a nontrivial solution $y(x)$ of Eq. (E_2) in its phase plane

$$\text{(1.9)} \qquad \begin{cases} y(x) = \rho(x) \sin \theta(x), \\ D_1 y(x) = \rho(x) \cos \theta(x); \qquad D_1 y = ry', \end{cases}$$

is called a *Prüfer Transformation*, after H. Prüfer who introduced the idea in 1926 (*80*). The approach given here is due to W. T. Reid (*87*) and it is convenient for generalization to higher order self-adjoint systems.

Suppose that $y(x)$ is a nontrivial solution of (E_2) and let

$$\text{(1.10)} \qquad \rho(x) = \sqrt{y^2(x) + (D_1 y(x))^2} > 0.$$

Next normalize y and $D_1 y$ by letting

$$\text{(1.11)} \qquad s(x) = y(x)/\rho(x) \quad \text{and} \quad s_1(x) = D_1 y(x)/\rho(x).$$

By differentiating (1.10) and (1.11) we have

$$\text{(1.12)} \qquad \rho'/\rho = (1/r - q)\, s s_1$$

and

$$\text{(1.13)} \qquad \begin{pmatrix} s \\ s_1 \end{pmatrix}' = \begin{pmatrix} 0 & b(x) \\ -b(x) & 0 \end{pmatrix} \begin{pmatrix} s \\ s_1 \end{pmatrix}; \qquad b(x) = s_1^2(x)/r(x) + q(x)\, s^2(x).$$

Therefore, if $\theta'(x) = b(x)$ then (1.9) is fulfilled and the Prüfer differential equations for (E_2) are

$$\text{(1.14)} \qquad \begin{cases} \text{(a)} \quad \rho' = (1/2)[(1/r - q) \sin 2\theta]\rho, \\ \text{(b)} \quad \theta' = (1/r) \cos^2 \theta + q \sin^2 \theta \\ \qquad\quad = (1/2)(1/r + q) + (1/2)(1/r - q) \cos 2\theta. \end{cases}$$

An alternate approach is to show that the nonlinear θ-equation of (1.14) has a unique solution for each given initial value of θ at some $a \in I$ and then to solve the other (linear) equation for ρ. Also, (1.14) may be derived by differentiating (1.9).

Theorem 1.6. *Each solution of equation* (E_2) *may be expressed by* (1.9) *whose polar components satisfy* (1.14).

Observe that if $\theta(x)$ is a solution of (1.14b) and $\theta(b) = k\pi$ then

$$\theta'(b) = 1/r(b) > 0.$$

Therefore, even though $\theta(x)$ may not be monotone it is always increasing at multiples of π, which has an important bearing on oscillation properties.

Theorem 1.7. *In order for a nontrivial solution $y(x)$ of (E_2) to be oscillatory on $I = [a, \infty)$ it is necessary and sufficient that for any $\theta(x) \in C'[a, \infty)$, which satisfies* (1.9),

$$\lim_{x\to\infty} \theta(x) = \infty.$$

A classical example of the usefulness of Theorem 1.7 is its application to the Bessel differential equation. For simplicity let us take the order zero and note that $u = J_0(x)$ satisfies

$$x^2u'' + xu' + x^2u = 0 \qquad \text{on} \qquad (0, \infty).$$

Eliminating the middle term by

$$y = \sqrt{x}\, J_0(x)$$

we have

$$y'' + (1 + \tfrac{1}{4}x^2)\, y = 0 \qquad \text{on} \qquad (0, \infty).$$

Now (1.14b) yields

$$\theta' = 1 + (\tfrac{1}{4}x^2) \sin^2 \theta$$

and, hence, $\theta(x) \to \infty$ as $x \to \infty$, so that $J_0(x)$ is oscillatory on $(0, \infty)$.

The other Prüfer equation (1.14b) also gives the useful information that, for

$$\rho(x) = \sqrt{y^2(x) + [y'(x)]^2},$$

$$\rho' = [(\tfrac{1}{8}x^2) \sin 2\theta]\, \rho$$

and, hence, for $0 < 1 \leqslant x < \infty$,

$$0 < \rho(x) < \rho(1) \exp(\tfrac{1}{8}) < \infty.$$

Consequently $y = \sqrt{x}\, J_0(x)$ is bounded on $[1, \infty)$ and

$$J_0(x) \leqslant M/\sqrt{x} \qquad \text{on} \qquad [1, \infty), \quad M < \infty.$$

A boundedness theorem for equation (E_2) may be found in the same way from (1.14a). However, stronger results can be obtained by modifying the transformation (1.9) into

$$y(x) = \rho(x) \sin \theta(x), \qquad D_1 y(x) = w(x)\, \rho(x) \cos \theta(x), \tag{1.9$'$}$$

where $w(x)$ is an arbitrary positive function of class $C'(I)$. In this case (see (4)) the Prüfer equations are

$$\begin{cases} \rho' = [(1/2)(w/r - q/w) \sin 2\theta - (w'/w) \cos^2 \theta] \rho, \\ \theta' = (1/2)(w/r + q/w) + (1/2)(w/r - q/w) \cos 2\theta + (w'/2w) \sin \theta. \end{cases} \tag{1.14'}$$

The special case where $w(x) = k > 0$, a constant, yields immediately a bound on solutions of (E_2).

Theorem 1.8. *If k is a positive number such that*

$$\int_a^\infty | k/r - q/k | < \infty$$

then all solutions of (E_2) *are bounded on* $I = [a, \infty)$. *Furthermore, if* $y(x)$ *is any nontrivial solution of* (E_2) *and*

$$\beta(x) = (1/2) \int_a^x (k/r + q/k)$$

then there is a positive number A and a number α *such that*

$$\lim_{x\to\infty} [y(x) - A \sin[(\beta(x) + \alpha)] = 0.$$

1.5. *Quadratic Functionals*

Let $y(x)$ be any L_2-admissible function on $[a, b]$, $a < b < \infty$, then

$$yL_2[y] = \{ryy'\}' - r(y')^2 + qy^2. \tag{1.15}$$

If, in addition,

$$L_2[y] = 0, \qquad y(a)\, y'(a) = 0, \qquad \text{and} \qquad y(b)\, y'(b) = 0,$$

then the quadratic functional

$$I[y; a, b] \equiv \int_a^b [r(y')^2 - qy^2] = 0. \tag{1.16}$$

Suppose that Eq. (E_2) is disconjugate on $[a, \infty)$ and $v(x)$ is a positive solution of (E_2) on $[a, b]$, then the factored form (1.4) yields

$$yL_2[y] = \{ryv(y/v)'\}' + rv^2[(y/v)']^2 \qquad \text{on} \qquad [a, b]. \tag{1.15'}$$

By combining (1.15) and (1.17) and integrating over $[a, b]$ we have

$$(1.17) \qquad I[y; a, b] \equiv \int_a^b [r(y')^2 - qy^2] = \int_a^b rv^2[(y/v')]^2 + [\{ry^2v'/v\}]_a^b.$$

If further conditions are imposed on $q(x)$ then Theorem 1.2 gives the sign of $v'(x)$.

Theorem 1.9. *If (E_2) is disconjugate on $[a, \infty)$, $\int_a^\infty (1/r) = \infty$ and $q(x) \geqslant 0$ but $\neq 0$ for large x then*

$$(1.18) \qquad I[y; a, b] \equiv \int_a^b [r(y')^2 - qy^2] > 0, \qquad a < b < \infty,$$

for all nontrivial L_2-admissible functions $y(x)$ on $[a, b]$ for which $y(a) = 0$.
W. T. Reid (*84*) established and utilized a more general criterion in terms of *focal points*, i.e., $x = a$ is said to be a focal point of $x = b$ provided there is a nontrivial solution $y(x)$ of (E_2) such that $y(a) = 0 = y'(b)$. Note that (1.17) holds for more general $y(x)$.

Lemma 1.11. *In order for (1.18) to hold for all nontrivial $y(x)$ of class D' (i.e., piecewise continuous derivative) satisfying $y(a) = 0$, it is necessary and sufficient that if $L_2[y] = 0$, $y(b) = 1$, $y'(b) = 0$ then $y(x) > 0$ on $[a, b]$, i.e., $[a, b)$ contains no (left) focal point of $x = b$.*

It is a simple matter to check that the positivity of the quadratic functional (1.18) is equivalent to Nehari's criterion (*75*) which follows in a slightly more general form given by the author (*8*).

Theorem 1.10. *If λ_b denotes the least eigenvalue of*

$$(ry')' + \lambda qy = 0, \qquad y(a) = y'(b) = 0, \qquad -\infty < a < b < \infty,$$

$\int^\infty (1/r) = \infty$, $q(x) \geqslant 0$ and $q(x) \not\equiv 0$ on any subinterval of $[a, \infty)$, then (E_2) is disconjugate on $[a, \infty)$ if, and only if, $\lambda_b > 1$ for all $b \in (a, \infty)$.

Various choices of functions of class D' in Lemma 1.11 yield *necessary conditions for disconjugacy*, e.g., let $0 \leqslant \alpha < 1 < \beta < \infty$ and

$$y(x) = \begin{cases} \left[\int_a^x (1/r) \Big/ \int_a^t (1/r)\right]^{\beta/2} & \text{for } a \leqslant x \leqslant t < b, \\ \left[\int_a^x (1/r) \Big/ \int_a^t (1/r)\right]^{\alpha/2} & \text{for } a < t \leqslant x \leqslant b < \infty, \end{cases}$$

and the positivity criterion (1.18), together with $\int_a^\infty 1/r = \infty$, yields

$$\Big(\int_a^x 1/r\Big)^{1-\beta} \int_a^x q(t) \Big(\int_a^t 1/r\Big) dt + \Big(\int_a^x 1/r\Big)^{1-\alpha} \int_x^\infty q(t) \Big(\int_a^t (1/r)\Big)^\alpha dt$$

$$\leqslant \frac{\beta - \alpha}{4}\left[1 + \frac{1}{(\beta - 1)(1 - \alpha)}\right] < \infty.$$

Note that (1.8) of Theorem (1.5) is the special case when $\alpha = 0$ and $\beta = 2$. For other cases and for the discussion when $\int_a^\infty q = \infty$ see (*8*).

If we have the general case where we do not know the sign of $r'(x)$ we have the same conclusion (1.18) when the admissible functions $y(x)$ have zeros at $x = a$ and $x = b$. Recall, also, that (1.17) holds for functions in class D' on $[a, b]$, i.e., $y \in D'[a, b]$.

Lemma 1.12. *If* (E_2) *is disconjugate on* $[a, b]$, $a < b < \infty$, *then* (1.18) *holds for each nontrivial function* $y(x) \in D'[a, b]$ *satisfying the zero end conditions*

$$y(a) = 0 = y(b).$$

Conversely, if there is a nontrivial solution $u(x)$ of (E_2) on $[a, b]$ with two zero, say,

$$u(x_1) = 0 = u(x_2), \qquad a \leqslant x_1 < x_2 \leqslant b,$$

then

$$y(x) = \begin{Bmatrix} u(x) & \text{on} \quad [x_1, x_2] \\ 0, & \text{elsewhere on } [a, b] \end{Bmatrix} \in D'[a, b];$$

$y(a) = 0 = y(b)$ and $I[y; a, b] = 0$. Hence, (1.18) is a sufficient condition for disconjugacy.

The results of the preceding discussion are summarized in the concise set of equivalent statements set down by Reid (*83*).

Theorem 1.11. *The following statements are equivalent.*

(i) *Equation* (E_2) *is disconjugate on* $[a, b]$.

(ii) *If* $L_2[u_1] = 0$, $u_1(a) = 0$ *and* $u_1'(a) \neq 0$ *then* $u_1 \neq 0$ *on* $(a, b]$.

(iii) *If* $L_2[u_2] = 0$, $u_2(b) = 0$ *and* $u_2'(b) \neq 0$ *then* $u_2 \neq 0$ *on* $[a, b)$.

(iv) *There is a nonzero solution* $u(x)$ *of* (E_2) *on* $[a, b]$.

(v) *There is a continuous function* $u(x)$ *such that*

(v) *There is a continuous function* $u(x)$ *such that*

$$ru' \in C[a, b] \quad \text{and} \quad uL_2[u] \leqslant 0 \quad \text{on} \quad [a, b].$$

(vi) $I[y; a, b] > 0$ *for all nontrivial functions* $y(x)$ *of class* D' *on* $[a, b]$ *for which* $y(a) = y(b) = 0$.

More general admissible classes of functions have been used by Reid (*85–89*).

Recently, Leighton (*69*) has used the positivity of the quadratic functional to establish a very useful comparison theorem.

Theorem 1.12. *If the coefficients of Eq.* (E_2) *and another equation of the same type,*

$$(\bar{E}_2) \qquad \bar{L}_2[y] = (\bar{r}(x)\, y')' + \bar{q}(x)\, y = 0,$$

satisfy the inequality

$$(1.19) \qquad \int_a^b [(\bar{r} - r)\, u'^2 + (q - \bar{q})\, u^2] > \int_a^b [\bar{r}u'^2 - qu^2]$$

for some L_2*-admissible and* $\bar{L}_2$*-admissible function* $u(x)$ *on* $[a, b]$ *satisfying* $u(a) = 0 = u(b)$, *then any solution* $y(x)$ *of* (E_2) *such that* $y(a) = 0$ *has a zero on* (a, b).

A particularly useful special case is the following.

Corollary 1.12.1. *If in Theorem 1.11,* $r(x) \equiv \bar{r}(x)$ *on* $[a, b]$ *then* (1.19) *implies*

$$\int_a^b (q - \bar{q})\, u^2 > 0.$$

1.6. *Asymptotic values at* ∞

Hille (*50*) solved the integral equation

$$Y(x) = 1 - \int_x^\infty (t - x)\, q(t)\, Y(t)\, dt$$

to establish the following.

Theorem 1.13. *If $r \equiv 1$ and $\int_a^\infty x\,|q(x)|\,dx < \infty$ and*

$$\text{(a)} \qquad Y_0(x) = 1, \qquad Y_k(x) = 1 - \int_x^\infty (t - x)\, q(t)\, Y_{k-1}(t)\, dt$$

then

$$|Y_k(x) - Y_{k-1}(x)| \leqslant \left(\int_x^\infty t\,|q(t)|\,dt\right) k/k!\,,$$

(b) *There exists a solution $Y(x)$ of the integral equation and, furthermore,*

$$Y'' + q(x)\,Y = 0 \qquad \text{and} \qquad \lim_{x\to\infty} Y(x) = 1.$$

Theorem 1.14. *If $r \equiv 1$ and $q(x) \geqslant 0$ on $[a, \infty)$ then in order for (E_2) to have a solution approaching 1 as $x < \infty$ it is necessary and sufficient that $\int_a^\infty xq(x)\,dx < \infty$.*

T. G. Hallam (*44*) has extended these theorems to equations of general *n*th order.

1.7. *Complex Equations with a Real Independent Variable*

If complex coefficients and solutions are admitted in the second-order equation (E_2) some properties of real equations carry over but some do not. For example, $y = e^{ix}$ is a complex solution of $y'' + y = 0$, which has no zeros, although all real solutions are oscillatory. In general, if $u(x)$ and $v(x)$ are linearly independent solutions of Eq. (E_2) then

$$u(x) + iv(x)$$

is a nonzero solution. Hence, the factoring (1.4) can be accomplished for any operator L_2 if complex factors are allowed.

In (*7*) the author introduced a class of equations

$$(1.20) \qquad [y'/q(x)]' + \bar{q}(x)\,y = 0, \qquad q(x) = q_1(x) + iq_2(x) \neq 0, \qquad q_k \in C[a, \infty),$$

which for $q(x)$ *real* has the familiar solutions

$$\sin \int^x q \qquad \text{and} \qquad \cos \int^x q.$$

With this in mind, for any complex continuous function $q(x)$ define two complex functions

$$s(x) = s[a, x; q] \quad \text{and} \quad c(x) = c[a, x; q] \tag{1.21}$$

to form the unique solution pair of

$$s' = q\bar{c}, \qquad c' = -q\bar{s} \tag{1.22}$$

subject to the initial conditions

$$s(a) = 0, \qquad c(a) = 1. \tag{1.123}$$

It is not difficult to verify the following identities, which shows that (1.20), or (1.22) retains the boundedness property for complex $q(x)$.

Lemma 1.13. $|s|^2 + |c|^2 = 1.$

The odd and even properties of the real case of (1.21) take on an interesting form.

Lemma 1.14. *If k is a complex number such that $|k| = 1$ then*

$$s[a, x; kq] = ks[a, x; q]$$

and

$$c[a, x; kq] = c[a, x; q] \qquad \text{on} \quad [a, \infty).$$

By use of the complex polar form, i.e.,

$$s[a, x; q] = h(x) \exp\{i\alpha(x)\}, \qquad q(x) = r(x) \exp\{i\theta(x)\},$$

various oscillation and nonoscillation theorems can be established. In particular, the system (1.22) becomes

$$\begin{cases} (h'/r)' + \{(1 + b^2) - (\alpha'/r - b)^2\} rh = 0 \\ h^2\alpha'/r = 2 \int_a^x bhh', \qquad b = \theta'/2r. \end{cases} \tag{1.24}$$

Theorem 1.15. *If the real second-order*

$$(y'/r)' + (1 + b^2) ry = 0$$

is nonoscillatory then $s[a, x; q]$ has at most a finite number of zeros on $[a, \infty)$.

Theorem 1.16. *If $r \in C[a, \infty)$, θ, h and $\alpha \in C'[a, \infty)$ and*

$$b(x) = \theta'(x)/2r(x) \equiv \text{constant on } [a, \infty)$$

then $s[a, x; q]$ has infinitely many zeros on $[a, \infty)$ if and only if $\int^{\infty} |r| = \infty$. Furthermore, if $b \neq 0$ then $C[a, x; q]$ has no zeros on $[a, \infty)$.

The real Prüfer transformation of Section 1.4 can be paralleled (7).

Theorem 1.17. *If $y(x)$ is a nontrivial solution of (E_2) with continuous complex coefficients*

$$r(x) = r_1(x) + ir_2(x), \qquad q(x) = q_1(x) + iq_2(x), \qquad r_k \text{ and } q_k \in C[a, \infty),$$

such that $y(a) = 0$, then there exist a nonzero function $\rho(x) \in C'[a, \infty)$ and a function $Q(x) \in C[a, \infty)$ such that

$$\begin{cases} y(x) = \rho(x)\, s[a, x; Q], \\ r(x)\, y'(x) = \bar{\rho}(x)\, \bar{c}[a, x; Q]. \end{cases} \tag{1.25}$$

Furthermore,

$$\begin{aligned} \rho' &= \overline{\rho s c}(1/r - \bar{q}), \\ q &= (\bar{\rho}/\rho)(|c|^2/r + \bar{q}\,|s|^2). \end{aligned} \tag{1.26}$$

A slight modification of the first equation of (1.26) yields a boundedness theorem.

Theorem 1.18. *If k is a real positive constant and $y(x)$ is a solution of (E_2) with complex coefficients then there is a positive constant M so that*

$$|y(x)| \leq M \exp(1/2) \int_a^x \{|kr_1/|r|^2 - q_1/k| + |kr_2/|r|^2 - q_2/k|\}.$$

A real matrix formulation of the complex scalar equation (E_2) is noted in Section V.

II. A. Third-Order Equations

Although differential equations of second and fourth orders have been studied extensively, it is only recently that third-order equations have been given serious study. Of course, there is the classical paper of Birkhoff (*16*) of 1908 where he applied methods of projective geometry. While Birkhoff's paper is a necessary reference fo any paper on third-order equations, his results or methods are seldom quoted. In 1948 Sansone (*91*) gave a summary of results known to that date, as well as a number of new results. The current interest in third-order equations was kindled by Hanan's 1961 paper (*45*), the 1960 paper of Azbelev and Caljuk (*3*) and the papers of Gregus since 1955 (*30–44*). Other important papers are those of Kondrat'ev (*62, 63*), Svec (*101*) and Lazer (*66*)

2.1. *Examples*

If one solves and examines the oscillatory and nonoscillatory properties. of solutions of

$$\text{(2.1)} \qquad \begin{array}{ll} \text{(a)}\quad y''' + y = 0, & \text{(b)}\quad y''' - y = 0, \\ \text{(c)}\quad y''' + y' = 0, & \text{(d)}\quad y''' - y' = 0, \end{array}$$

on half-line intervals $[a, \infty)$, then he is armed with suitable motivating examples for most of this chapter. For a long time it was thought that the following was a typical property of all third-order equations.

Lemma 2.1. *Every third-order equation with real constant coefficients,*

$$\text{(2.2)} \qquad y''' + c_2 y'' + c_1 y' + c_0 y = 0$$

has a nonzero solution on $[a, \infty)$.

However, Sansone (*91*) and Gregus (*30*) have exhibited equations of the form

$$\text{(2.2)} \qquad [y'' + P(x)y]' = y''' + P(x)y' + P'(x)y = 0; \qquad P(x) \in C'[a, \infty),$$

for which every solution has infinitely many zeros on $[a, \infty)$. Lazer (*66*) has recently used the following as motivating examples.

Lemma 2.2.

(a) *If the constants* $c_1 < 0$ *and* $c_0 > 0$ *then*

$$y''' + c_1 y' + c_0 y = 0 \quad 0 \qquad on \quad [a, \infty)$$

has a nontrivial solution with infinitely many zeros if, and only if,

$$c_0 - \frac{2}{3\sqrt{3}} | c_1 |^{3/2} > 0.$$

Also, all solutions have infinitely many zeros, except for nonvanishing solutions.

(b) *If* $c_1 < 0$ *and* $c_0 < 0$ *then the above criterion is replaced by*

$$| c_0 | - \frac{2}{3\sqrt{3}} | c_1 |^{3/2} > 0.$$

Also, there exist two independent solutions with infinitely many zeros.

(c) *If* $c_0 > 0$ *and* $c_1 > 0$ *all solutions of* (2.1) *have infinitely many zeros, except for constant multiples of one nonvanishing solution.*

Hanan (*45*) has used the third-order Euler Equation

$$y''' + (a/x^2)y' + (b/x^3)y = 0; \qquad a, b = \text{constants,} \tag{2.3}$$

as his basic comparison equation.

Lemma 2.2′. (a) *If* $a \geqslant 1$ *then there is a nontrivial oscillatory solution of* (2.3) (*i.e., one with infinitely many zeros*) *on* $[1, \infty)$.

(b) *If* $a < 1$ *and* $a + b > 0$ *then there is a nontrivial oscillatory solution of* (2.3) *if and only if*

$$a + b - 2\left(\frac{1-a}{3}\right)^{3/2} > 0.$$

(c) *If* $a < 1$ *and* $a + b < 0$ *then there is a nontrivial oscillatory solution of* (2.3) *if and only if*

$$a + b + 2\left(\frac{1-a}{3}\right)^{3/2} < 0.$$

The author (*13*) used the following product principle of Appel (*1*) to predict his general third-order Canonical Equation [see (E_3) below]. This property is well known to those who work with Special Functions.

Lemma 2.3. *If w and z are solutions of* (E_2) $(ry')' + qy = 0$ *then their product* $y = wz$ *is a solution of the third-order equation*

$$\{r[(ry')' + 2qy]\}' + 2q(ry') = 0; \qquad r > 0, \quad r \,\&\, q \in C(I). \tag{2.4}$$

Let $v_0 = v_0(x, a)$ and $v_1 = v_1(x, a)$ be fundamental solutions of (E_2) with $v_0(a) = 1$, $D_1v_0(a) = (rv_0')(a) = 0$ and $v_1(a) = 0$, $D_1v_1(a) = (rv_1')(a) = 1$, then a fundamental set of solutions of (2.4) is

(2.5)

$$\left.\begin{array}{l} y = v_1^2/2 : y(a) = D_1y(a) = 0, \quad D_2y(a) = 1 \\ y = v_0v_1 : y(a) = 0, \ D_1y(a) = 1, \ D_2y(a) = 0 \\ y = v_0^2 : y(a) = 1, \quad D_1y(a) = 0, \ D_2y(a) = 0 \end{array}\right\} \text{ where } \left\{\begin{array}{l} D_1y = ry', \\ \\ D_2y = r(D_1y + 2qy)'. \end{array}\right.$$

Note that Eq. (2.4) has a nontrivial solution with three zeros if, and only if, equation (E_2) has a nontrivial solution with two zeros. Of course, two arbitrary zeros can be specified for any linear third-order equation.

Definition 2.1. *A third-order linear differential operator* $L[y]$ *(and its corresponding homogeneous equation* $L[y] = 0$*) is said to be disconjugate on an interval* I *provided that no nontrivial solution has three zeros on* I.

Hence, the third-order equation (2.4) is disconjugate if, and only if, the second-order equation (E_2) is disconjugate.

A well-known special case is that when $r \equiv 1$ and $q \in C'$, for which (2.4) becomes

$$y''' + 4qy' + 2q'y = 0, \tag{2.6}$$

whose solutions are linear combinations of products of two solutions of $y'' + qy = 0$. An interesting example of (2.6) is

$$y''' + 4y' = 0, \tag{2.6'}$$

whose solutions are linear combinations of $\sin^2 x$, $\cos^2 x$ and $\sin x \cos x$.

Hanan (*45*) failed to take into account such examples which contradict his Fundamental Lemma. Fortunately, such cases did not enter into his subsequent applications. Note that $y = \sin^2 x$ and $y = \sin x \cos x$ are independent nontrivial solutions with at least three zeros on $[0, \pi]$. In fact, $y = \sin^2 x$ has infinitely many double zeros on $[0, \infty)$. A corrected version is given in Theorem 2.9.

2.2. *A Canonical Form*

The equation (2.4), which was generated by products of solutions of (E_2), suggested to the author (*13*) the *third-order canonical form*

$$(E_3) \qquad L_3[y] = \{r_2(r_1y')' + q_1y]\}' + q_2(r_1y') = 0,$$

where $r_i > 0$, r_i and $q_i \in C(I)$, $I = [a, b)$, $a < b \leqslant \infty$. Let $D_1y = r_1y'$, $D_2y = r_2L_2[y] = r_2[(r_1y')' + q_1y]$; then ($E_3$) becomes

$$L_3[y] = (D_2y)' + q_2D_1y = 0.$$

It may be that $y''' + p(x)y = 0$ is not a typical equation of third-order but, instead, that either

$$(2.7) \qquad \text{(a)}\ \ (y'' + py)' = 0 \qquad \text{or} \qquad \text{(b)}\ \ y''' + py' = 0$$

is more typical for predicting oscillation properties. It should be noted that (E_3) contains Shinn's quasi-differential operator (*99*). A function y is L_3-*admissible* provided that y, D_1y and $D_2y \in C'(I)$, and a solution of (E_3) is an L_3-admissible function which satisfies (E_3).

The equation (E_3) is equivalent to a special case of the *first-order vector-matrix* equation:

$$(V_3) \qquad \alpha' = B(x)\alpha; \qquad B(x) = (b_{ij}(x)), \qquad b_{ij} \in C(I),$$

where

$$B(x) = \begin{pmatrix} 0 & 1/r_1 & 0 \\ -q_1 & 0 & 1/r_2 \\ 0 & -q_2 & 0 \end{pmatrix},$$

in the sense that if y is a solution of (E_3) then

$$\alpha = \begin{pmatrix} y \\ y_1 \\ y_2 \end{pmatrix}$$

is a solution of (V_3) and, conversely, if $\alpha = (a_i)$ is a (vector) solution of (2.1) then $y = a_1$ is a solution of (E_3).

Again, the standard uniqueness and existence properties follow easily by use of the system (V_3).

Lemma 2.4. *Any classical third-order equation*

$$l_3[y] = y''' + p_2(x)\, y'' + p_1(x) y' + p_0(x) y = 0; \qquad p_i \in C(I), \tag{2.8}$$

can be expressed in the canonical form (E_3) *since it is equivalent to*

$$(sy'')' + sp_1 y' + sp_0 y = 0; \qquad s = \exp\left(\int p_2\right),$$

and

$$\left[sy'' + \left(\int sp_0\right) y\right]' + \left[sp_1 - \int sp_0\right] y' = 0.$$

Obviously, the canonical form (E_3) is not unique.

Suppose that y is admissible for L_3 and z has the desired derivatives; then the following adjoint operator and *Lagrange Identity* are easily derived by integrations-by-parts.

Lemma 2.5. *If y is L_3-admissible and z is $L_3{}^+$-admissible, where the adjoint operator is*

$$L_3{}^+[y] = \{r_1[(r_2 y')' + q_2 y]\}' + q_1(r_2 y'), \tag{2.9}$$

then y and z satisfy the Lagrange Identity

$$zL_3[y] = \{zD_2 y - D_1{}^+ z D_1 y + y D_2{}^+ z\}' - y L_3{}^+[z], \tag{2.10}$$

where $D_1{}^+ y = r_2 y'$ and $D_2{}^+ y = r_1[(D_1{}^+ y)' + q_2 y]$.

The corresponding *adjoint equation* is

$$L_3{}^+[y] = (D_2{}^+ y)' + q_1 D_1{}^+ y = 0. \tag{$E_3{}^+$}$$

Note that the adjoint operators L^+ and $D_i{}^+$ are obtained from L and D_i, respectively, by *interchanging* r_1 with r_2 and q_1 with q_2. If $r_1 \equiv r_2$ and $q_1 \equiv q_2$ then $L^+ = L$ and (E_3) is said to be *self-adjoint* (sometimes called *anti-self-adjoint* for the *odd* order 3). Observe that Eq. (2.4) and its classical special case (2.6) are self-adjoint third-order equations. Also, the equations (2.7) are adjoints of each other. *Define certain fundamental solutions* of (E_3) by the *initial conditions*:

$$u_2 = u_2(x, a) : y(a) = 0 = D_1 y(a), \qquad D_2 y(a) = 1,$$

(*First principal solution*)

$$(2.11)\qquad u_1 = u_1(x, a) : y(a) = 0, \quad D_1y(a) = 1, \quad D_2y(a) = 0; \quad (\textit{Second principal solution})$$

$$u_0 = u_0(x, a) : y(a) = 1, \quad D_1y(a) = 0 = D_2y(a).$$

Define $u_i^+ = u_i^+(x, a)$ to be the corresponding fundamental solutions for the adjoint equation (E_3^+). Recall that these fundamental solutions of the self-adjoint equation (2.4) are given by (2.5).

Lemma 2.6. (a) *Any solution of* (E_3) *with a zero at* $x = a$ *has the form*

$$y(x) = c_1u_1(x, a) + c_2u_2(x, a).$$

(b) *In order for there to exist a nontrivial solution of* (E_3) *with a zero at* $x = a$ *and a double zero at* $x = b \neq a$ *it is necessary and sufficient that the "Wronskian" of the principal solutions* $u_2(x, a)$ *and* $u_1(x, a)$,

$$(2.12)\qquad \sigma(x) = W[u_2, u_1] = u_2D_1u_1 - u_1D_1u_2 = r_1(u_2u_1' - u_1u_2'),$$

vanish at $x = b$.

The Leighton–Nehari Fundamental Lemma 1.5 is very useful for showing when (2.12) has zeros. Apparently, Birkhoff (*16*) first noted that the Wronskian of two solutions satisfies an adjoint equation.

Lemma 2.7. (a) *If* y *and* z *are solutions of* (E_3) *then their Wronskian* $w = yD_1z - zD_1y$ *(note inclusion of* r_1*) is a solution of the adjoint equation* (E_3^+).

(b) *The special Wronskian* (2.12) *satisfies the initial conditions:* $\sigma(a) = 0 = \sigma'(a)$, $D_2^+\sigma(a) = 1$ *and, hence, is the first principal solution of* (E_3^+), *i.e.*,

$$(2.13)\qquad W[u_2, u_1](x, a) = u_2^+(x, a).$$

2.3. *Self-adjoint Equations*

We have already noted the self-adjoint third-order examples (2.4) and (2.6) but under what conditions is the classical general equation (2.8) equivalent to a self-adjoint equation? Recently Giuliano (*29*) gave sufficient conditions, which were later simplified by the author (*13*).

Theorem 2.1. (a) *The third-order operator*

$$l_3[y] = y''' + p_2(x)y'' + p_1(x)y' + p_0(x)y = 0; \qquad p_i \in C^{(i)}(I),$$

satisfies the identity,

$$r^2 l_3[y] = \left\{r\left[(ry')' + (1/r)\left(\int r^2 p_0\right) y\right]\right\}' + (1/r)\left[r^2 p_1 - (r^2)''/2 - \int r^2 p_0\right](ry'),$$

with $r = \exp\{(1/3) \int p_2\}$.

(b) *The equation* (2.8), $l_3[y] = 0$, *is equivalent to a self-adjoint equation if* $z = r^2 = \exp\{(2/3) \int p_2\}$ *satisfies*

$$(2.14) \qquad (z'' - 2p_1 z)' + 4p_0 z = 0.$$

For $p_2 = 0$ *the conditions* (2.14) *simplify to the special case* (2.6), *i.e.*,

$$(2.14') \qquad p_1' = p_0/2.$$

Since the canonical form (E_3) is not unique the self-adjointness conditions $r_1 = r_2$ and $q_1 = q_2$ can be improved slightly.

Theorem 2.2. *The canonical third-order equation* (E_3) *is equivalent to a self-adjoint equation on an interval* I *if there exist constants* $k \neq 0$ *and* m *such that*

$$(2.15) \qquad r_2(x) = kr_1(x) \quad \textit{and} \quad q_2(x) = kq_1(x) + m/r_1(x) \quad \textit{on} \quad I.$$

2.4. *The Lagrange Bracket*

The Lagrange Identity (2.10) contains the bilinear functional

$$\{z; y\} = zD_2 y - D_1{}^+ z D_1 y + (D_2{}^+ z) y$$

which is sometimes known as the *concomittant* or the *Lagrange Bracket* of the operator L_3. Observe that if $L_3[y] = 0$ and $L_3{}^+[z] = 0$ on an interval I then

$$\{z; y\} \equiv \text{constant on } I,$$

and this constant is determined by initial values at any particular point $x = a$. Also, if z is given then $\{z; y\}$ is a third-order differential operator in y.

Lemma 2.8. (a) $\{u_2{}^+; u_i\} = 0$ *for* $i = 1, 2$.

(b) *If* $z \neq 0$ *on an interval* I *then*

$$(2.16) \qquad \{z; y\} = r_2 z^2 \left[\left(\frac{r_1 y'}{z}\right)' + \left(\frac{D_2{}^+ z + r_2 q_1 z}{r_2 z^2}\right) y\right] \quad \textit{on} \quad I.$$

(c) *In the self-adjoint case*

$$\{y; y\} = 2yD_2y - (D_1y)^2. \tag{2.17}$$

It is well-known that, for all $x, t \in I$,

$$u_2(x, t) \equiv u_2^+(t, x) \tag{2.18}$$

and recently, Dolan (*24*) has noted a more general identity.

Lemma 2.9. *The fundamental solutions* (2.11) *of* (E_3) *and its adjoint* (E_3^+) *satisfy*

$$D_\alpha u_\beta(x, t) \equiv (-1)^{\alpha+\beta} D_{2-\beta} u^+_{2-\alpha}(t, x) \qquad \textit{for} \quad \alpha, \beta = 0, 1, 2. \tag{2.19}$$

Note that (2.18) is the special case where $\alpha = 0$ and $\beta = 2$.

2.5. *Factoring and Disconjugacy*

If $v^+(x)$ is a nonzero solution of the adjoint equation (E^+) on an interval I, i.e.,

$$L_3^+[v^+] = 0 \qquad \text{and} \qquad v^+ \not\equiv 0, \tag{i}$$

then the Lagrange Identity (2.10) yields

$$L_3[y] = (1/v^+)\{v^+; y\}' \tag{2.20}$$

for all L_3-admissible functions $y(x)$ on I. Now the Lagrange Bracket $\{v^+; y\}$ is a second-order differential operator (2.16) and is constant if $L_2[y] = 0$. Any solution v of (E_3) for which

$$\{v^+; v\} = 0 \tag{ii}$$

at some point of I is by Lemma 2.8 a solution of the differential equation (ii) on the whole interval I. If

$$L_3[v] = 0 \qquad \text{and} \qquad v \not\equiv 0 \tag{iii}$$

and (ii) is satisfied then the Polya–Mammana factoring (1.4) may be applied to $\{v^+; y\} = 0$. Actually, such factoring was first done by Frobenius (*28*) and later utilized by Polya (*79*) and Mammana (*73*).

Lemma 2.10. *Under the conditions* (i), (ii), *and* (iii) *on an interval I, the third-order canonical operator L_3 factors into*

$$L_3[y] = \frac{1}{v^+}\left(\frac{r_2 v^{+2}}{v}\left[\frac{r_1 v^2}{v^+}(y/v)'\right]'\right)' \tag{2.21}$$

for all L_3-admissible functions $y(x)$ on I.

We have already noted that, if v_1 and v_2 are solutions of (E_3), then their Wronskian

$$w = W[v_1, v_2] = v_1 D_1 v_2 - v_2 D_1 v_1$$

is a solution of the adjoint equation (E^+). Furthermore, it turns out that

$$\{w; v_i\} = 0 \qquad \text{for} \quad i = 1, 2.$$

Therefore, we can obtain the Polya–Mammana factored form (2.21) by letting

$$v^+ = W[v_2, v_1] \qquad \text{and} \qquad v = v_i \qquad \text{for} \quad i = 1 \text{ or } 2.$$

Note that (2.21) implies that L_3 is disconjugate on I by Definition 2.1. If we choose two principal solutions from (2.11),

$$v_1 = u_1(x, a) \qquad \text{and} \qquad v_2 = u_2(x, a)$$

then $v^+ = W[u_2, u_1] = u_2^+$ and it takes only a slight extension of the argument for Lemma 2.10 to yield a criterion for disconjugacy.

Theorem 2.3. *The operator L_3 [and the equation (E_3)] is disconjugate on the half-closed interval $[a, b)$, $a < b \leqslant \infty$, if and only if the principal solutions of (E_3) satisfy*

$$u_2(x, a) > 0 \qquad \text{and} \qquad u_2^+(x, a) > 0 \qquad \text{on} \quad (a, b). \tag{2.22}$$

As a consequence of the symmetry property (2.18) Theorem 2.4 becomes

Corollary 2.3.1. *The criterion (2.22) for disconjugacy on $[a, b)$ or $(a, b]$ is equivalent to*

$$u_2(x, t) \neq 0 \qquad \text{for} \quad x, t \in (a, b), \quad x \neq t. \tag{2.22'}$$

Because of the dual role of L_3 and its adjoint, L_3^+, we have

Corollary 2.3.2. *An operator* L_3 *{equation* (E_3)*} is disconjugate on* $[a, b)$, $a < b \leqslant \infty$, *if and only if its adjoint* $L_3^+\{E_3^+\}$ *is disconjugate on* $[a, b)$.

By combining two adjacent factors of L_3 in (2.21) with $v^+ = w$, we have another factored form which Mammana (*73*) noted for the classical equation (2.8).

Lemma 2.11. *If* v *is a nonzero solution of* (E_3) *on* I *then*

$$L_3[y] = l_2(l_1[y]), \tag{2.18$'$}$$

where

$$l_1[y] = r_1 v^2 (y/v)' = r_1 v \left[y' - \frac{v'}{v} y \right]$$

and

$$l_2[y] = \left(\frac{D_1^+ y}{v} \right)' + \left(\frac{D_2 v + r_1 q_2 v}{r_1 v^2} \right) y,$$

for all L_3*-admissible functions on* I.

2.6. *First Conjugate Points*

Suppose that there is a nontrivial solution $y(x)$ of (E_3) with three zeros (counting any double zero as two zeros) $x_1 \leqslant x_2 \leqslant x_3$, $x_1 < x_3$ on $I = [a, \infty)$; then the disconjugacy criterion (2.19) asserts that either $u_2(x, a)$ or $u_2^+(x, a)$ has a zero on $(a, x_3]$. If $u_2(x, a)$ has a zero at $x = b$ then $y = u_2(x, a)$ satisfies the two-point boundary conditions

$$y(a) = D_1 y(a) = 0 = y(b). \tag{2.23}$$

On the other hand, if $u_2^+(x, a)$ has a zero at $x = b$ then there is a nontrivial solution $y(x)$ of (E_3) satisfying

$$y(a) = 0 = y(b) = D_1 y(b). \tag{2.24}$$

Definition 2.2. *Let* $z_{21}(a)\{z_{12}(a)\}$ *be the minimum number* $b \in (a, \infty)$ *such that the boundary conditions* (2.23) {(2.24)} *are satisfied by a nontrivial solution of* (E_3) *and* $z_{ij}(a) = \infty$ *if the respective boundary conditions are not satisfied.*

Note that for the examples (2.1),

(a): $z_{21}(a) < \infty$, $z_{12}(a) = \infty$,
(b): $z_{21}(a) = \infty$, $z_{12}(a) < \infty$.

Definition 2.3. *If* (E_3) *has a nontrivial solution with three zeros on* $[a, \infty)$ *then the first conjugate point* $\eta_1(a)$ *of* $x = a$ *is defined to be*

$$\eta_1(a) = \inf\{x_3 \,;\, a \leqslant x_1 \leqslant x_2 \leqslant x_3 \,,\, y(x_i) = 0, y \not\equiv 0, L_3[y] = 0\}. \tag{2.25}$$

If (E_3) *is disconjugate we write* $\eta_1(a) = \infty$.

The disconjugacy Theorem 2.3 may now be restated.

Theorem 2.4. $\eta_1(a) = \min\{z_{21}(a), z_{12}(a)\}$.

Azbelev and Caljuk (*3*) were the first to arrive at an equivalent result. Note that in the *self-adjoint* case (2.4),

$$z_{21}(a) = z_{12}(a) = z_{22}(a).$$

For the example (2.1),

(c): $z_{12}(a) = z_{21}(a) = z_{22}(a) = a + 2\pi$,
(d): $z_{12}(a) = z_{21}(a) = z_{22}(a) = \infty$.

Definition 2.2′. *The number* $z_{22}(a)$ *is the smallest* $b \in (a, \infty)$ *such that*

$$y(a) = D_1 y(a) = 0 = y(b) = D_1 y(b) \tag{2.26}$$

is satisfied by a nontrivial solution of (E_3). *Otherwise, write* $z_{22}(a) = \infty$.

Hanan's special classes of classical third-order equations (2.8) may be described simply as

$$C_{\mathrm{I}} : z_{12}(t) = \infty \qquad \text{for all} \quad t \in [a, \infty)$$

and

$$C_{\mathrm{II}} : z_{21}(t) = \infty \qquad \text{for all} \quad t \in [a, \infty).$$

Motivating examples in these classes are given by

$$y''' + p(x)\, y = 0, \qquad p \in C[a, \infty)$$

which belongs to C_{II} if $p < 0$ and to C_{I} if $p > 0$ on $[a, \infty)$. The author

(*13*) has extended Hanan's classes to the canonical equation (E_3) by examining the behavior of the functional

$$\Phi[y] = 2yD_2y - (D_1y)^2.$$

Theorem 2.5. *If* (r_2/r_1) *and* $(r_2q_1 - r_1q_2) \in C'[a, \infty)$ *and*

$$(r_2/r_1)' \leqslant 0\{\geqslant 0\}, \qquad (r_2q_1 - r_1q_2)' \leqslant 0\{\geqslant 0\} \tag{2.27}$$

but $\neq 0$ *on any subinterval of* $[a, \infty)$, *then* $z_{21}(t) = \infty$ $\{z_{12}(t) = \infty\}$ *for all* $t \in [a, \infty)$.

For the equation considered by Hanan,

$$L_3[y] = y''' + p_1(x)y' + p_0(x)y = 0; \qquad p_i \in C^{(i)}[a, \infty), \tag{2.28}$$

the conditions (2.24) Theorem 2.6 reduces to:

$$L_3 \in C_{\mathrm{I}}\{C_{\mathrm{II}}\}, \quad \text{if} \quad 2p_0 - p_1' \geqslant 0\{\leqslant 0\}$$

but $\neq 0$ on any subinterval.

2.7. *Disconjugacy Numbers*

While Hanan discussed the cases where either z_{12} or $z_{21} = \infty$, Azbelev and Caljuk (*3*) considered the more general case where both z_{12} and z_{21} might both be finite. As a motivating example, they constructed a third-order differential equation of the type

$$(ry'')' + qy' = 0, \qquad r > 0, \quad q \geqslant 0, \qquad r \,\&\, q \in C[0, 2\pi], \tag{2.29}$$

which has as a fundamental set of solutions

$$1, \ \sin x + 0.001x^3, \qquad 1 - \cos x$$

and for which (in our notation)

$$a = 0 < z_{12}(0) = 5.7 < z_{21}(0) = 2\pi.$$

Azbelev and Caljuk defined and used the following for the classical equation (2.3).

Definition 2.3′. (a) *An adjacent pair of zeros x_1, x_2 of a nontrivial solution of (E_3) is said to be an (i, j)-adjacent pair provided the multiplicity at $x = x_1$ is at least i and at x_2, at least j.*

(b) *For each pair (i, j) of positive integers*

$$r_{ij}(a) = \sup\{t\colon \textit{no } (i,j)\textit{-adjacent pair of zeros in } [a, t)\}. \tag{2.30}$$

(c) *The maximum interval of disconjugacy $[a, r(a))$ is defined by*

$$r(a) = \sup\{t\colon (E_3) \textit{ is disconjugate on } [a, t)\}. \tag{2.31}$$

Using these concepts, they proved that

(i) $r(a) = \min[r_{12}(a), r_{21}(a)]$,

and purported to prove that

(ii) $r_{22}(a) = \max[r_{12}(a), r_{21}(a)]$.

The first assertion is equivalent to Theorem 2.4 since

$$r(a) = \eta_1(a) = \min[z_{12}(a), z_{21}(a)].$$

However, the assertion follows only for certain cases, as we will now see.

After these lectures were given, T. L. Sherman pointed out the following example.

Example 2.1. The three linearly independent functions

$$u = (1/2)\sin^2 x \cos x, \qquad v = \cos^2 x \sin x, \qquad w = 1$$

satisfy the third-order equation

$$(y''/\alpha_{12})' + (\alpha_{23}/\alpha_{12}^2)y' = 0 \quad \text{on} \quad [0, \infty), \tag{2.32}$$

where

$$\alpha_{ij} = \begin{vmatrix} v^{(i)} & u^{(i)} \\ v^{(j)} & u^{(j)} \end{vmatrix} \quad \text{for} \quad i, j = 0, 1, 2, 3.$$

Since $\alpha_{12} = 1 + (1/2)\sin^2 x \geqslant 1 > 0$, then (2.32) is nonsingular. It is easy to calculate that

$$\eta_1(0) = z_{12}(0) = z_{21}(0) = \pi/2 = r_{12}(0) = r_{21}(0) < z_{22}(0) = r_{22}(0) = \pi$$

and, since $\alpha_{01} = u_2^+(x, 0) = (1/8)\sin^2(2x)$,

$$\eta_1^+(0) = z_{12}^+(0) = z_{21}^+(0) = \pi/2 = z_{22}^+(0) = r_{22}^+(0).$$

Therefore, the assertion (ii) does not hold for Eq. (2.32) but it does hold for its adjoint equation.

With respect to the results of Azbelev and Caluk, a 2–2 adjacent pair of zeros of a third-order equation (E_3) does not imply such a pair for the adjoint equation (E_3^+). Azbelev and Caljuk denoted

$$R(a) = \max[r_{12}(a), r_{21}(a)]$$

and Dolan (*24*) introduced the number

$$Z(a) = \max[z_{12}(a), z_{21}(a)]. \tag{2.33}$$

Let η_1^+, r_{ij}^+, z_{ij}^+, r^+, R^+, and Z^+ be the respective numbers for the adjoint equation (E_3^+) corresponding to the previously defined numbers for (E_3). There are some immediate observations.

Theorem 2.6. (a) $r_{ij}(a) = r_{ji}^+(a)$ *for* $i + j = 3$,

(b) $\eta_1(a) = \eta_1^+(a) = r(a) = r^+(a)$,

and

(c) $\eta_1(t)$ *is an increasing function of* t.

Also, by definition

$$a < r(a) = \eta_1(a) \leqslant R(a) \leqslant Z(a) \leqslant \infty.$$

Since $r_{ij}(a) = \inf\{z_{ij}(t), t \geqslant a\}$ for $i + j = 3$ then for each $\epsilon > 0$ there exists a nontrivial solution $y_\epsilon(x)$ which has an (i, j)-adjacent pair of zeros on $[a, r_{ij}(a) + \epsilon)$ and which is positive between these zeros. Let $r_{ij}(a) < \infty$, $\epsilon > 0$ and $\tau_\epsilon \in [a, r_{ij}(a) + \epsilon)$ such that $z_{ij}(\tau_\epsilon, r_{ij}(a) + \epsilon)$. Furthermore, let us normalize $y_\epsilon(x)$ so that

$$y_\epsilon(x) = c_0^\epsilon u_0(x, a) + c_1^\epsilon u_1(x, a) + c_2^\epsilon u_2(x, a), \qquad (c_0^\epsilon)^2 + (c_1^\epsilon)^2 + (c_2^\epsilon)^2 = 1.$$

Now there exists a sequence $\{\epsilon_n\}$ of positive numbers approaching zero such that for $i = 0, 1$ and 2 the sequence $\{c_i^{\epsilon_n}\}$ and $\{\tau_{\epsilon_n}\}$ converge, say, to

c_i and τ_{ij}, respectively. Let

$$Y(x) = c_0u_0(x, a) + c_1u_1(x, a) + c_2u_2(x, a);$$

then $\{y_{\epsilon_n}(x)\}$ *converges uniformly* to $Y(x)$ on $[a, r_{ij}(a) + \epsilon_1]$. Hence $Y(r_{ij}) = Y'(r_{ij}) = 0$ and there exists $\tau_{ij} = \tau_{ij}(a) \in [a, r_{ij}(a))$ such that $Y(\tau_{ij}) = 0$, i.e.,

$$r_{ij}(a) = z_{ij}(\tau_{ij}). \tag{2.34}$$

Lemma 2.12. *For $i + j = 3$, there exists $\tau_{ij} \in [a, r_{ij}(a))$ such that $r_{ij}(a) = z_{ij}(\tau_{ij})$ and $r_{ij}(a) \neq z_{ij}(t)$ for $t > \tau_{ij}$.*

Of course, we know that if $\eta_1(a) = r_{ij}(a)$ then $\tau_{ij}(a) = a$.

On the other hand, suppose that

$$\eta_1(a) = z_{21}(a) < z_{12}(a) < \infty. \tag{2.35}$$

We note an improved form of the Leighton–Nehari Fundamental Lemma 1.5.

Lemma 2.13. *If $u(x)$ and $v(x)$ are differentiable functions on $[a, b]$, $a < b < \infty$, such that*

$$v(x) \neq 0 \quad \text{on} \quad (a, b) \quad \text{and} \quad (u/v)(a^+) = 0 = (u/v)(b^-)$$

then there is a number λ and a number $\xi \in (a, b)$ such that $u(x) + \lambda v(x)$ has a double zero at $x = \xi$.

Therefore, letting $u(x) = u_2(x, a)$ and $v(x) = u_2(x, z_{12})$ we see that the latter has at least one zero c on (a, z_{12}). Furthermore,

$$D_1u_2(c, z_{12}) \neq 0.$$

By an argument of the type used for Lemma 2.12, or by recalling that solutions of (E_3) are continuous functions of initial conditions, we have:

Lemma 2.14. *If $X \in (a, \infty)$ and $b \in (a, X)$ then $\{u_2(x, b - \epsilon)\}$ approaches $u_2(x, b)$ uniformly on $[a, X]$.*

Hence, there is a positive number ϵ such that $u_2(x, z_{12}(a) - \epsilon)$ has a zero on $(a, z_{12}(a) - \epsilon)$ and, of course, a double zero at $z_{12}(a) - \epsilon > a$, i.e.,

$$a < r_{12}(a) < z_{12}(a).$$

Putting this together with Lemma 2.12 we see that $r_{12}(a) = z_{12}(\tau_{12})$, $\tau_{12} \in (a, r_{12}(a))$, and $u_2(x, r_{12}) > 0$ on (τ_{12}, r_{12}). By Lemma 2.14, $D_1u_2(\tau_{12}, r_{12}) \neq 0$ contradicts the definition of $r_{12} = r_{12}(a)$. Consequently,

$$D_1u_2(\tau_{12}, r_{12}) = 0 = u_2(\tau_{12}, r_{12})$$

and

$$u_2(x, r_{12}) \equiv ku_2(x, \tau_{12}) > 0,$$

for some positive constant k. Furthermore, $u_2{}^+(x, \tau_{12}) > 0$ on (τ_{12}, r_{12}) so that $\eta_1(\tau_{12}) = r_{12}(a) = z_{22}(\tau_{12})$.

Theorem 2.7. *If* $z_{21}(a) < z_{12}(a) < \infty$ *then* $r_{12}(a) \in (z_{21}(a), z_{12}(a))$ *and there is a number* $\tau_{12}(a) \in (a, r_{12}(a))$ *such that* $\tau_{12}(a)$, $r_{12}(a)$ *form a* (2, 2)-*adjacent pair of zeros for* (E_3).

For the other case,

$$\eta_1(a) = z_{12}(a) < z_{21}(a) < \infty, \tag{2.36}$$

we recall that $z_{ij} = z_{ji}^+$ and $r_{ij} = r_{ji}^+$ and, hence, Theorem 2.7 applies to the adjoint equation $(E_3{}^+)$. Consequently,

$$R^+(a) = r_{12}^+(a) = r_{22}^+(a) < z_{12}^+(a)$$

and, hence,

$$R(a) = r_{21}(a) < z_{21}(a).$$

Because of Example 2.1, it does not follow that $R(a) = z_{22}(a)$ and we now summarize the valid portions of the conclusions of Azbelev and Caljuk.

Theorem 2.8. *If* $\eta_1(a) < R(a) < \infty$ *then there is a number* $\tau \in (a, R(a))$ *such that*

$$R(a) = \eta_1(\tau)$$

and τ, $R(a)$ *form a 2–2 adjacent pair of zeros of*

$$(E_3), \qquad \textit{if} \quad \eta_1(a) = z_{21}(a) \qquad (\textit{e.g., case } 2.35),$$

or

$$(E_3{}^+), \qquad \textit{if} \quad \eta_1(a) = z_{12}(a) \qquad (\textit{e.g., case } 2.36).$$

We can draw a few more conclusions about the pair τ, $R(a)$. Returning to the case (2.35), let $R = R(a)$ and τ be as given by Theorem 2.8, $c \in (\tau, R)$ and define the solution

$$(2.37) \qquad w^+(x) = u_2(x, c)\, D_1u_2(x, \tau) - u_2(x, \tau)\, D_1u_2(x, c)$$

of the adjoint equation $(E_3{}^+)$. Then

$$w^+(\tau) = w^+(c) = w^+(R) = 0$$

and, furthermore, these zeros are simple zeros. Finally, we note that

$$R = z_{22}(\tau) = z_{21}(\tau) = z_{12}(\tau) = z_{12}^+(\tau) = z_{21}^+(\tau).$$

Corollary 2.8.1. *If* $\eta_1(a) = z_{21}(a) < r_{12}(a) = R(a) < \infty$ *and* τ *is the number guaranteed by Theorem 2.8 then* τ, $R(a)$ *form both a* 2–1 *and a* 1–2 *adjacent pair of zeros of the adjoint equation* $(E_3{}^+)$ *and, furthermore, if* $c \in (\tau, R(a))$ *then* τ, c *and* $R(a)$ *are simple zeros of a solution of* $(E_3{}^+)$.

2.8. *Subsequent Conjugate Points*

For a third-order equation, having a nontrivial solution with $\nu + 2$ zeros, Hanan (*45*) defined the νth conjugate point.

Definition 2.4. *If* ν *is a positive integer and there is a nontrivial solution* $y(x)$ *of* (E_3) *with* $\nu + 2$ *zeros,* $a = x_1 \leqslant x_2 \leqslant \cdots \leqslant x_{\nu+2}$, *then*

$$\eta_\nu(a) = \inf_y \{x_{\nu+2}\}.$$

If no solution has $\nu + 2$ *zeros, let* $\eta_\nu(a) = \infty$.

For $\nu = 1$ this definition differs from $\eta_1(a)$ in Definition 2.3 by requiring that $x = a$ is the first zero, i.e., $y(a) = 0$, but, it is readily seen that the two definitions are equivalent. For $\nu > 1$ it is a different matter as noted by Example 2.1, except when $R(a) = \infty$ for which Hanan established the equivalence.

Suppose that $n \geqslant 3$ and (E_3) has a nontrivial solution $y(x)$ with n zeros, $a = x_1 \leqslant x_2 \leqslant x_3 \leqslant \cdots \leqslant x_n < \infty$. For $n = 3$ we have already seen the possible nontrivial solutions of (E_3) which have three zeros on

$[a, \eta_1]$. Note that in each such case one of the solutions has a double zero, although, as in Example 2.1, another solution may have only simple zeros.

For general ν, where $\eta_\nu(a) < \infty$, Hanan established the existence of a nontrivial solution of the classical equation (2.8), having $\nu + 2$ zeros on $[a, \eta_\nu(a)]$. Suppose this is not so and let $\{x_{\nu+2}^{(n)}\}$ be a decreasing sequence of numbers approaching $\eta_\nu(a)$ such that for each n, $x_{\nu+2}^{(n)}$ is the $(\nu + 2)$th zero on $[a, \infty)$ of the normalized solution

$$y(x) = c_1 u_1(x, a) + c_2 u_2(x, a), \qquad c_1^2 + c_2^2 = 1, \tag{2.38}$$

which have $\nu + 2$ zeros on $[a, \infty)$. Therefore, there is a solution

$$Y(x) = C_1 u_1(x, a) + C_2 u_2(x, a)$$

which is the uniform limit of a subsequence of solutions from the set (2.42). Thus $Y(x)$ has $\nu + 2$ zeros,

$$x_1 \leqslant x_2 \leqslant x_3 \leqslant \cdots \leqslant x_{\nu+2} \in [a, \infty)$$

and $x_{\nu+2} = \eta_\nu(a)$. Such a solution $Y(x)$ is called an extremal solution.

Definition 2.5. *If $\eta_\nu(a) < \infty$ then any nontrivial solution of (E_3) which has $\nu + 2$ zeros on $[a, \eta_\nu(a)]$ is called an extremal solution for $\eta_\nu(a)$.*

Therefore the minimum in the definition of (a) can be replaced by the minimum value of $x_{\nu+2}$. Note that if $y(x)$ is an extremal solution for $\eta_\nu(a)$ then

$$y(a) = 0 = y(\eta_\nu(a)).$$

Suppose that $Y(x)$ is an extremal solution of (E_3) for $\eta_\nu(a) < \infty$ and that all $n + 2$ zeros x_i of $Y(x)$ on $[a, \eta_\nu(a)]$ are simple, i.e., $Y'(x_i) \neq 0$ and

$$a = x_1 < x_2 < \cdots < x_{\nu+1} < x_{\nu+2} = \eta_\nu(a) = \eta_\nu .$$

Define a set of solutions of (E_3) for $\epsilon > 0$ by

$$\begin{cases} y_\epsilon(a) = 0 = y_\epsilon(\eta_\nu - \epsilon), \\ D_1 y_\epsilon(a) = D_1 Y(a) \neq 0. \end{cases}$$

Then for sufficiently small $\epsilon > 0$,

$$y_\epsilon(x) = D_1 Y(a) \left[u_1(x, a) - \frac{u_1(\eta_\nu - \epsilon, a)}{u_2(\eta_\nu - \epsilon, a)} u_2(x, a) \right], \tag{2.39}$$

if

$$u_2(\eta_\nu(a), a) \neq 0.$$

Consequently,

$$y_\epsilon(x) \to Y(x), \qquad \text{uniformly on } [a, \eta_\nu], \qquad \text{as } \epsilon \to 0$$

and, since the n zeros of $Y(x)$ on (a, η_ν) are simple, there is a number $\epsilon > 0$ such that $y_\epsilon(x)$ has n zeros on $(a, \eta_\nu - \epsilon)$. But this gives a total of $n + 2$ zeros on $[a, \eta_\nu - \epsilon]$, which contradicts the definition of $\eta_\nu = \eta_\nu(a)$, so one of the zeros of $Y(x)$ is not simple. A similar argument can be given if $u_2(a, \eta_\nu(a)) = u_2^+(\eta_\nu(a), a) \neq 0$. In this way Dolan (*24*) proved a corrected version of Hanan's Fundamental Lemma (*45*).

Theorem 2.9. *If* $\eta_\nu(a) < \infty$ *and either*

$$u_2(\eta_\nu(a), a) \neq 0 \qquad \text{or} \qquad u_2(a, \eta_\nu(a)) = u_2^+(\eta_\nu(a), a), \neq 0 \tag{2.40}$$

then for every extremal solution $Y(x)$ *of* (E_3) *for* $\eta_\nu(a)$ *there is a number* $\tau \in [a, \eta_\nu(a)]$ *and a nonzero number* k *such that*

$$Y(x) = ku_2(x, \tau), \qquad x \in [a, \infty).$$

Note that for $\nu = 1, \tau = a$, or $\eta_\nu(a)$. In fact, Hanan showed that a double zero of an extremal solution occurs at $x = a$ or $\eta_\nu(a)$ if $R(a) = \infty$. His arguments also hold on $[a, R(a))$, which establishes the following characterization.

Theorem 2.10. *If* $\eta_\nu(a) < R(a) \leqslant \infty$ *and* $R(a) = r_{12}(a)\{r_{21}(a)\}$ *then* $u_2(x, a)\{u_2(x, \eta_\nu(a))\}$ *is an extremal solution of* (E_3) *for* $\eta_\nu(a)$, *is a simple zero of* $u_2(x, a)$, *and* $\eta_\nu(a)$ *is a simple zero of* $u_2(x, a)\{u_2^+(x, a)\}$. *Furthermore,*

$$\eta_\nu(a) = \eta_\nu^+(a). \tag{2.41}$$

We note that, in Example 2.1, Eq. (2.32) has the property that

$$R(a) = \pi/2 = \eta_1(0) = \eta_1^+(0) = \eta_2^+(0) < \eta_2(0) = \pi$$

and, hence, (2.41) does not always hold outside the interval $[a, R(a))$. Another case where (2.41) obviously holds is that of self-adjoint equations. Note that the hypothesis (2.40) eliminates self-adjoint equations from consideration. But this case is not difficult (*24*).

Theorem 2.11. *If* (E_3) *is self-adjoint on* $[a, \infty)$ *and* $\eta_\nu(a) < \infty$, *then* $u_2(x, a)$ *is an extremal solution for* $\eta_\nu(a)$ *having double zeros at each of the conjugate points* $\eta_1(a), \ldots, \eta_\nu(a)$ *and is positive elsewhere on* $(a, \eta_\nu(a)]$.

Dolan (*24*) also showed that $\{\eta_\nu(a)\}$ is a nondecreasing sequence of numbers such that if $\eta_{\nu+2}(a) < \infty$ then either

(i) $\eta_{\nu+2}(a) > \eta_{\nu+1}(a) \geqslant \eta_\nu(a)$

or

(ii) $\eta_{\nu+2}(a) \geqslant \eta_{\nu+1}(a) > \eta_\nu(a)$.

A constructive characterization of all conjugate points $\eta_\nu(a)$ for $\nu > 1$ and for the general equation (E_3) is still an open question.

2.9. *Oscillation of* (E_3) *and its Adjoint*

Another basic question, which remains unanswered except for a few special cases, is:

> *If* (E_3) *is oscillatory on* $[a, \infty)$ *then is its adjoint equation* $(E_3{}^+)$ *also oscillatory*?

Hanan (*45*) and Svec (*102*) have answered this question in the affirmative for (E_3) in Hanan's classes C_{I} or C_{II}. But the answer is not known in the general situation.

Dolan (*24*) has given the following partial answers.

Lemma 2.15. *If* (E_3) *is oscillatory on* $[a, \infty)$ *then*

$$\eta_\nu(a) < \infty \qquad \textit{for} \quad \nu = 1, 2, 3, \ldots.$$

Lemma 2.16. *If* $\eta_\nu(a) < \infty$ *for each positive integer* ν *then*

$$\eta_\nu(a) \to \infty, \qquad \textit{as} \quad \nu \to \infty.$$

Theorem 2.12. *If* $\eta_\nu(a) < \infty$ *for each positive integer* ν *then either* (E_3) *or* $(E_3{}^+)$ *is oscillatory on* $[a, \infty)$.

Corollary 2.12.1. *If* (E_3) *is nonoscillatory on* $[a, \infty)$ *then either*

(i) *at most a finite number of conjugate points* $\eta_\nu(a) < \infty$ *exist or*

(ii) $(E_3{}^+)$ *is oscillatory.*

Theorem 2.13. *If* (E_3) *is oscillatory and* $(E_3{}^+)$ *is nonoscillatory on* $[a, \infty)$ *then every solution of* (E_3) *is oscillatory on* $[a, \infty)$.

Sansone (*91*) has given examples of equations (E_3) for which every solution is oscillatory on $[a, \infty)$.

Corollary 2.13.1. *If* (E_3) *is oscillatory but has a nonoscillatory solution on* $[a, \infty)$ *then* $(E_3{}^+)$ *is oscillatory.*

Note that in Hanan's classes C_{I} and C_{II}, either (E_3) or $(E_3{}^+)$ has a nonoscillatory solution.

Theorem 2.14. *If* (E_3) *has an oscillatory nontrivial solution* $y(x)$ *on* $[a, \infty)$ *such that* $y(a) = 0$, $(E_3{}^+)$ *is nonoscillatory on* $[a, \infty)$, *and* λ *is the largest zero of* $u_2{}^+(x, a)$, *then every solution* $y(x)$ *of* (E_3) *for which* $y(a) = 0$ *satisfies the second-order equation*

$$\left(\frac{r_1}{z}y'\right)' + \frac{1}{z^2 r_2}[D_2{}^+z + q_1 r_2 z]y = 0 \qquad \text{on} \quad (\lambda, \infty), \tag{2.42}$$

where $z = u_2{}^+(x, a)$.

Corollary 2.14.1. *If* (E_3) *is oscillatory on* $[a, \infty)$, $r_2 q_1 \in C'[a, \infty)$ *and* $(r_2 q_1)'$ *has constant sign on* $[a, \infty)$, *then* $(E_3{}^+)$ *is oscillatory.*

Theorem 2.15. *If* (E_3) *and* $(E_3{}^+)$ *are both nonoscillatory on* $[a, \infty)$ *then they are both disconjugate for large* x *on* $[a, \infty)$.

II. B. Nonnegative Coefficients

2.10. *The Classical Equation*

Here we will be concerned with

$$(\mathcal{E}_3) \qquad \mathcal{L}_3[y] = y''' + P(x)\,y' + Q(x)\,y = 0; \qquad P \text{ and } Q \in C[a, \infty),$$

whose coefficients satisfy

$$(\mathrm{H}_1) \qquad P(x) \geqslant 0 \quad \text{and} \quad Q(x) \geqslant 0 \quad \text{but} \quad P(x) + Q(x) \not\equiv 0 \text{ on any subinterval of } [a, \infty).$$

These are recent results of the author (*14*).

We recall from Section I that second-order equations with non-negative coefficients have special properties, e.g., if $P(x) \geqslant 0$ but $\not\equiv 0$ for large x, and the equation

$$(\mathcal{E}_2) \qquad \mathcal{L}_2[y] = y'' + P(x) = 0; \qquad P \in C[a, \infty)$$

is disconjugate on $[a, \infty)$ then

(i) there is a positive solution $v(x)$ of $(\mathcal{E}_2)$ on $[a, b]$ for each $b \in (a, \infty)$,

(ii) any nontrivial solution $y(x)$ of $(\mathcal{E}_2)$ for which $y(a) = 0$ satisfies

$$y'(x) \neq 0 \quad \text{on} \quad [a, \infty),$$

and

(iii) $\int_x^\infty P \leqslant 1/(x-a) \quad \text{for} \quad x \in (a, \infty).$

Hanaṇ (*45*), Lazer (*66*), and Gregus (*44, 45*) have recently studied oscillatory properties for various cases of $(\mathcal{E}_3)$. Waltman (*105*) and Heidel (*49*) have replaced the last term of $(\mathcal{E}_3)$ with the possibly nonlinear term $Q(x)y^\gamma$. Most of these discussions require one or both of the additional assumptions:

$$(\mathrm{H}_2) \qquad \mathcal{L}_2[y] = y'' + P(x)y = 0 \quad \text{disconjugate on} \quad [a, \infty),$$

$$(\mathrm{H}_3) \qquad P \in C'[a, \infty) \quad \text{and} \quad 2Q - P' \geqslant 0 \quad \text{but} \quad \not\equiv 0 \quad \text{on any subinterval of } [a, \infty).$$

Under the assumption (H_2), the positive solution $v(x)$, asserted in (i), can be used to factor the first two terms of $(\mathcal{E}_3)$ which then becomes

$$\frac{1}{v}\left[v^2\left(\frac{y'}{v}\right)'\right]' + Qy = 0 \quad \text{on} \quad [a, b]. \tag{2.43}$$

By means of the change of variables

$$t = \int_a^x v, \qquad Y(t) = y(x), \tag{2.44}$$

Eq. (E_3) becomes

$$(R(t)\ddot{Y})^{\cdot} + T(t)Y = 0; \qquad R(t) = v^3(x), \qquad T(t) = Q(x), \tag{2.45}$$

thus eliminating the middle term without changing signs of the coefficients. Therefore, as Hanan(*45*) noted, Eq. (2.45) is in class C_{I}, i.e., $r_{12}(a) = \infty$.

Also, Hanan showed that the hypothesis (H_3) restricts Eq. ($\mathcal{E}_3$) to class C_{I}. Another way of showing this property will lead to some other results. Rewrite Eq. ($\mathcal{E}_3$) as

$$\mathfrak{L}_3[y] = y''' + Py' + (P'/2)y + (Q - P'/2)y = 0 \tag{2.46}$$

and note that the first three terms form a self-adjoint operator,

$$\mathfrak{L}_3{}^*[y] = y''' + Py' + (P'/2)y = [y'' + (P/2)y]' + (P/2)y'. \tag{2.47}$$

Therefore, we can express ($\mathcal{E}_3$) in the form

$$\mathfrak{L}_3[y] = \mathfrak{L}_3{}^*[y] + xm(x)y = 0, \qquad m \in C[a, \infty), \tag{2.48}$$

which Gregus studied extensively (*30–42*). For any nontrivial solution $y(x)$ of ($\mathcal{E}_3$), $y\mathfrak{L}_3[y] = 0$, which yields

$$\{y[y'' + (P/2)y] - y'^2/2\}' = -(Q - P'/2)y^2. \tag{2.49}$$

Hence, the bracketed quantity is decreasing so that again $(\mathcal{E}_3) \in C_{\mathrm{I}}$. As we pointed out in Subsection 2.11, Hanan has shown that

Theorem 2.16. *If* $R(a) = \infty$ *and* ($\mathcal{E}_3$) *is nonoscillatory on* $[a, \infty)$ *then* ($\mathcal{E}_3$) *is disconjugate for large* x.

If $R(a) = \infty$ is replaced by (H_1) we have not been able to answer this question but have been able to prove many of the other results of Hanan, Lazer and Waltman for ($\mathcal{E}_3$) and show that neither (H_2) nor (H_3) is necessary. We will be primarily concerned with necessary conditions for disconjugacy of ($\mathcal{E}_3$).

2.11. *The Lagrange Identity and the Adjoint Equation*

The Lagrange Identity for $\mathfrak{L}_3[y]$ is

$$u\mathfrak{L}_3[v] = \{u; v\}' - v\mathfrak{L}_3{}^+[u], \tag{2.50}$$

where

$$\{u, v\} = uv'' - u'v' + v\mathfrak{D}_2u, \quad \mathfrak{D}_2y = l_2[y] = y'' + Py, \tag{2.51}$$

and the adjoint operator is

$$\mathcal{L}_3^+[y] = (\mathcal{D}_2 y)' - Qy.$$

Note that if $P \in C'[a, \infty)$ then the adjoint operator can be written in the more familiar forms

$$\mathcal{L}_3^+[y] = y''' + (Py)' - Qy = y''' + Py' - (Q - P')y.$$

The quantity $y\mathcal{L}_3^+[y] = 0$ may be integrated by parts to obtain

Lemma 2.17. *If $y(x)$ is a solution of the adjoint equation*

$$(\mathcal{E}_3^+) \qquad \mathcal{L}_3^+[y] = [y'' + Py]' - Qy = 0; \qquad P \text{ and } Q \in C[a, \infty),$$

then for $x \in [a, \infty)$,

$$(2.52) \qquad y\mathcal{D}_2 y - y'^2/2]_a^x = \int_a^x Qy^2 + \int_a^x Pyy'.$$

If we can find a solution $y(x)$ of $(\mathcal{E}_3^+)$ whose coefficients satisfy (H_1), *for which*

$$(2.53) \qquad y(x) > 0, \quad y'(x) > 0 \quad \text{and} \quad \mathcal{D}_2 y(x) < 0 \quad \text{on} \quad (a, \infty$$

then (2.52) *implies the inequalities*

$$(2.54) \qquad \int_a^\infty Qy^2 < \infty, \qquad \int_a^\infty Pyy' < \infty.$$

2.12. *Properties of Principal Solutions*

Assume that equation $(\mathcal{E}_3)$, whose coefficients satisfying (H_1), is disconjugate on $[a, \infty)$, i.e.,

$$(2.55) \qquad u_2(x, a) > 0 \quad \text{and} \quad u_2^+(x, a) > 0 \quad \text{on} \quad [a, \infty).$$

Consider quotients of $u_i = u_i(x, a)$ and $u_i^+ = u_i^+(x, a)$ and their derivatives, using $\mathcal{D}_2 y = y'' + Py$ in place of y'' for u_i^+,

$$(2.56) \qquad \lambda_0 = u_1/u_2, \qquad \lambda_1 = u_1'/u_2', \qquad \lambda_2 = u_1''/u_2'',$$

(2.56⁺)

$$\lambda_0^+ = u_1^+/u_2^+, \qquad \lambda_1^+ = u_1^{+\prime}/u_2^{+\prime}, \qquad \lambda_2^+ = \mathcal{D}_2 u_1^+/\mathcal{D}_2 u_2^+; \qquad \mathcal{D}_2 y = y'' + Py.$$

The identity (2.13) applied to $(\mathcal{E}_3)$ gives

$$
(2.57)\quad \begin{cases} u_2^+ = u_1 u_2' - u_2 u_1', & u_2 = u_1^+ u_2^{+\prime} - u_2^+ u_1^{+\prime}, \\ u_2^{+\prime} = u_1 u_2'' - u_2 u_1'', & u_2' = u_1^+ \mathfrak{D}_2 u_2^+ - u_2^+ \mathfrak{D}_2 u_1^+, \\ \mathfrak{D}_2 u_2^+ = u_1' u_2'' - u_2' u_1'', & \mathfrak{D}_2 u_2 = u_1^{+\prime} u_2^{+\prime\prime} - u_2^{+\prime} u_1^{+\prime\prime}, \\ & u_2'' = u_1^{+\prime} \mathfrak{D}_2 u_2^+ - u_2^{+\prime} \mathfrak{D}_2 u_1^+. \end{cases}
$$

The derivatives of the quotients (2.56) and (2.56^+) are

(2.56′)

$$\lambda_0' = -u_2^+/u_2^2, \qquad \lambda_1' = -\mathfrak{D}_2 u_2^+/(u_2')^2, \qquad \lambda_2' = -(P\mathfrak{D}_2 u_2^+ + Q u_2^+)/(u_2'')^2,$$

$$(2.56^{+\prime})\qquad \lambda_0^{+\prime} = -u_2/(u_2^+)^2, \qquad \lambda_1^{+\prime} = -\mathfrak{D}_2 u_2/(u_2^{+\prime})^2, \qquad \lambda_2^{+\prime} = Q u_2'/(\mathfrak{D}_2 u_2^+)^2.$$

and the differences are

(2.56^-)

$$\lambda_0 - \lambda_1 = u_2^+/u_2 u_2', \qquad \lambda_1 - \lambda_2 = \mathfrak{D}_2 u_2^+/u_2' u_2'', \qquad \lambda_0 - \lambda_2 = u_2^{+\prime}/u_2 u_2'',$$

$(2.56^\pm)$

$$\lambda_0^+ - \lambda_1^+ = u_2/u_2^+ u_2^{+\prime}, \quad \lambda_1^+ - \lambda_2^+ = u_2''/u_2^{+\prime}\mathfrak{D}_2 u_2^+, \quad \lambda_0^+ - \lambda_2^+ = u_2'/u_2^+ \mathfrak{D}_2 u_2^+.$$

Observe that if, in addition to (2.55),

$$(2.58)\qquad u_2' > 0, \qquad u_2^{+\prime} > 0, \qquad u_2'' > 0 \quad \text{and} \quad \mathfrak{D}_2 u_2^+ > 0 \quad \text{on} \quad (a, \infty),$$

then

(2.59)

$$\lambda_0^+ > \lambda_1^+ > \lambda_2^+ > 0, \qquad \lambda_0^{+\prime} < 0, \qquad \lambda_1^{+\prime} < 0 \quad \text{and} \quad \lambda_2^{+\prime} > 0 \quad \text{on} \quad (a, \infty).$$

Hence, there exists a positive constant λ such that

$$\lambda_0^+(x) > \lambda_1^+(x) > \lambda > \lambda_2^+(x) \quad \text{on} \quad (a, \infty)$$

or, in other words, the solution

$$(2.60)\qquad Y(x) = u_1^+(x, a) - \lambda u_2^+(x, a)$$

of the adjoint equation $(\mathcal{E}_3^+)$ satisfies the inequalities

$$(2.61)\qquad Y > 0, \qquad Y' > 0 \quad \text{and} \quad \mathfrak{D}_2 Y = Y'' + PY < 0 \quad \text{on} \quad (a, \infty).$$

Consequently, (2.54) holds for $y = Y$.

We will now show that (2.58) follows from (2.55) and hypothesis (H_1). Similar results for special cases have been obtained by Waltman (*105*) and Lazer (*66*).

For the special cases where $P \equiv 0$ or $Q \equiv 0$ it is a simple matter to show that (2.55) also implies that

$$u_2''(x, a) > 0 \quad \text{on} \quad [a, \infty).$$

But the general case is not so easy and will now be dealt with.

Suppose there exists a zero of $u_2''(x, a)$ on (a, ∞) and let

$$u_2''(\mu_1, a) = 0, \qquad u_2''(x, a) > 0 \quad \text{on} \quad [a, \mu_1]. \tag{2.62}$$

By Lemma 2.24, $u_2'(x, a)$ has a zero on (μ_1, ∞) and let

$$u_2'(\xi_1, a) = 0, \qquad u_2'(x, a) > 0 \quad \text{on} \quad (a, \xi_1).$$

Since $u_2''' = -Pu_2' - Qu_2 \leqslant 0$, as long as $u_2' > 0$ and since $P + Q \not\equiv 0$ on any subinterval, then $u_2''(x, a)$ has a second zero, i.e.,

$$u_2''(\mu_2, a) = 0, \qquad u_2''(x, a) \neq 0 \quad \text{on} \quad (\mu_1, \mu_2), \quad \mu_2 \in (\xi_1, \infty).$$

However, if $u_2'(x, a)$ has a second zero ξ_2 then

$$u_2'(\xi_2, a) = 0, \qquad u_2'(x, a) < 0 \quad \text{on} \quad (\xi_1, \xi_2), \quad \xi_2 \in (\mu_2, \infty),$$

and the first identity of (2.57) yields $u_1'(\xi_2, a) < 0$ so that

$$\lambda_1(x) \to +\infty \qquad \text{as} \qquad x \to \xi_2 \quad \text{on} \quad (\xi_1, \xi_2).$$

But this contradicts

$$\lambda_1'(x) = -\mathfrak{D}_2 u_2^+/(u_2')^2 < 0 \quad \text{on} \quad (\xi_1, \xi_2).$$

Lemma 2.18. *Under hypotheses* (H_1), (2.57) *and* (2.62),

(a) $u_2''(x, a)$ *has a second zero* $\mu_2 \in (\mu_1, \infty)$ *and*

(b) $u_2'(x, a)$ *has only one zero* ξ_1 *on* (a, ∞), $\xi_1 \in (\mu_1, \mu_2)$ *and* $u_2'(x, a) < 0$ *on* (ξ_1, ∞).

The hypothesis (H_1) implies that $u_2''(x, a) > 0$ immediately to the right of $x = \mu_2$. If there is a third zero μ_3,

$$u_2''(\mu_3, a) = 0, \qquad u_2''(x, a) > 0 \quad \text{on} \quad (\mu_2, \mu_3),$$

then the third identity of (2.57) implies that $u_1''(\mu_3, a) > 0$ and

$$\lambda_2(x) \to +\infty \qquad \text{as} \qquad x \to \mu_3 \quad \text{on} \quad (\mu_2, \mu_3).$$

But this contradicts

$$\lambda_2' = -[P\mathfrak{D}_2 u_2^+ + Q u_2^+]/(u_2'')^2 \leqslant 0 \quad \text{on} \quad (\mu_2, \mu_3).$$

Furthermore, the Lagrange identity (2.51) yields

$$u_2^{+\prime} u_2' = u_2^+ u_2'' + u_2 \mathfrak{D}_2 u_2^+.$$

Lemma 2.19. (a) *If* $u''(x, a) > 0$ *on* (a, ∞) *then* $u_2'(x, a) > 0$ *and* $u_2(x, a) > 0$ *on* $[a, \infty)$.

(b) *If* $u_2^+(x, a) > 0$ *on* (a, ∞) *then* $\mathfrak{D}_2 u_2^+(x, a) > 1 > 0$ *on* (a, ∞).

Next suppose that $u_2'(x, a) > 0$ on (a, ∞); then $z = u_2'$ is a positive solution of a second-order equation with nonnegative coefficients,

$$z'' + [P + Q u_2/u_2']z = 0 \quad \text{on} \quad (a, \infty),$$

and

$$z(a) = 0, \qquad z'(a) = 1.$$

As we noted in Section 2.12 (i),

$$z'(x) = u_2''(x, a) > 0 \quad \text{on} \quad [a, \infty)$$

which proves the following.

Lemma 2.20. *If* $u_2'(x, a) > 0$ *and* (H_1) *is assumed then*

$$u_2''(x, a) > 0 \quad \textit{on} \quad [a, \infty).$$

Conversely, suppose that $u_2^{+\prime}(x, a)$ has a zero on (a, ∞) and let

$$u_2^{+\prime}(t_1, a) = 0, \qquad u_2^{+\prime}(x, a) > 0 \quad \text{on} \quad (a, t_1).$$

From (2.55) and (2.57) it follows that $u_1^{+\prime}(t_1, a) < 0$ and, hence,

$$\lambda_1^+(x) \to -\infty, \qquad \text{as} \qquad x \to t_1 \quad \text{on} \quad (a, t_1).$$

On the other hand, if $u_2''(x, a) > 0$ on (a, ∞) then

$$\lambda_1^+(x) > \lambda_2^+(x) = \mathfrak{D}_2 u_1^+(x, a)/\mathfrak{D}_2 u_2^+(x, a) > -\infty \quad \text{on} \quad [a, \infty).$$

Lemma 2.21. *If, in addition to* (H_1) *and* $(\mathcal{E}_3)$ *disconjugate, it is assumed that* $u''(x, a) > 0$ *and* $u_2^+(x, a) > 0$ *then*

$$u_2^{+'}(x, a) > 0 \quad on \quad [a, \infty).$$

Lemma 2.22. *Under hypotheses* (H_1), (2.57), *and* (2.62),

(a) $u_2''(x, a) > 0$ *on* (μ_2, ∞),
(b) $u_2^{+'}(x, a) < 0$ *on* (μ_2, ∞).

We have already noted that $\mathfrak{D}_2 u_2^+ \geqslant 1$ on $[a, \infty)$. Therefore,

$$u_2^{+'}(x, a) - u_2^{+'}(c, a) + \int_c^x P u_2^+ \geqslant x - c, \qquad \mu_2 \leqslant c \leqslant x < \infty,$$

and $\int_c^\infty P = \infty$ so that the Leighton–Wintner oscillation theorem, applied to $\mathfrak{L}_2[y] = y'' + Py = 0$, yields a contradiction.

Lemma 2.23. *Let* $(\mathcal{E}_3)$ *be disconjugate and its coefficients satisfy* (H_1). *In order for* $u_2''(x, a)$ *to have a zero on* (a, ∞) *it is necessary and sufficient that*

$$\int_a^\infty P = \infty.$$

As a result of Lemmas 2.25, 2.26, and 2.27, the quotients $\lambda_0(x)$, $\lambda_1(x)$, and $\lambda_2(x)$ are eventually decreasing. Therefore there is a number α and a number $c \in (\mu_2, \infty)$ such that

$$\lambda_i(x) < \alpha \qquad \text{for} \quad i = 1, 2, 3 \qquad \text{and} \qquad x \in (c, \infty).$$

Consequently, the solution $z(x) = \alpha u_2(x, a) - u_1(x, a)$ of $(\mathcal{E}_3)$ satisfies

$$z > 0, \qquad z' > 0 \qquad \text{and} \qquad z'' > 0 \quad \text{on} \quad (c, \infty).$$

But, as Lazer pointed out, these conditions provide $w = z'$ a positive solution of

$$w'' + [P(x) + Q(x)\, z(x)/z'(x)]w = 0 \quad \text{on} \quad (c, \infty)$$

which is disconjugate on (c, ∞); this contradicts Lemma 2.27 by the Leighton–Wintner Oscillation Theorem.

Therefore, all of the inequalities of (2.58) follow from (2.57).

Theorem 2.17. *If* ($\mathcal{E}_3$) *is disconjugate and its coefficients satisfy* (H_1) *then there is a solution*

$$Y(x) = u_1^+(x, a) - \lambda u_2^+(x, a), \qquad \lambda > 0,$$

of (E_3^+) *such that* $Y > 0$, $Y' > 0$, $\mathfrak{D}_2 Y < 0$ *on* (a, ∞) *and*

$$\int_a^\infty QY^2 < \infty, \qquad \int_a^\infty PYY' < \infty. \tag{2.63}$$

2.13. *Necessary Conditions for Disconjugacy*

The results of the preceeding section are necessary conditions for disconjugacy of ($\mathcal{E}_3$) under hypothesis (H_1) on its coefficients. We continue to make these assumptions and Theorem 2.17 has some immediate corollaries.

Note that $y = Y(x)$ is a positive solution of

$$y'' + [P + | \mathfrak{D}_2 Y |/Y]y = 0 \quad \text{on} \quad (a, \infty)$$

and, consequently,

$$\mathfrak{L}_2[y] = y'' + Py = 0 \quad \text{is disconjugate on} \quad [a, \infty). \tag{2.64}$$

Therefore, by Hille's Theorem [i.e., (iii) of Subsection 2.10],

$$\int_x^\infty P < 1/(x - a) \quad \text{on} \quad (a, \infty). \tag{2.65}$$

It is easy to see that

Lemma 2.24. *If Eq.* ($\mathcal{E}_3$), *under hypothesis* (H_1), *is disconjugate on* $[a, \infty)$ *then*

$$\lim_{x\to\infty} \mathfrak{D}_2 Y(x) = 0, \qquad \int_a^\infty (QY) = \lambda$$

and

$$\mathfrak{D}_2 Y(x) = Y''(x) + P(x)\, Y(x) = -\int_x^\infty QY \quad \text{on} \quad (a, \infty). \tag{2.66}$$

Furthermore, (2.63) implies that $\int_a^\infty Q < \infty$ and since $Y(x)$ is increasing

$$Y'' + \left[P(x) + \int_x^\infty Q\right] Y \leqslant 0.$$

Lemma 2.25. *If Eq.* ($\mathcal{E}_3$), *under hypothesis* (H_1), *is disconjugate on* $[a, \infty)$ *then* $\int_a^\infty Q < \infty$ *and the second-order equation*

$$y'' + \left[P(x) + \int_x^\infty Q\right] y = 0 \tag{2.67}$$

is disconjugate on $[a, \infty)$. *Furthermore,*

$$\int_x^\infty P + \left(\int_x^\infty\right)^2 Q < 1/(x-a) \quad on \quad (a, \infty). \tag{2.67'}$$

Note that

$$\left(\int_x^\infty\right)^2 Q = \int_x^\infty (t-x)\, Q(t)\, dt < \infty, \qquad x \in (a, \infty)$$

if and only if

$$\int_a^\infty tQ(t)\, dt < \infty. \tag{2.68}$$

One of Hanan's results follows immediately.

Lemma 2.26. *If* ($\mathcal{E}_3$) *is disconjugate,* $Q(x) \geqslant 0$, $P(x) \in C'[a, \infty)$ *and* $2Q(x) - P'(x) \geqslant 0$ *on* $[a, \infty)$, *then*

$$\int_a^\infty t[2Q(t) - P'(t)]\, dt < \infty.$$

Recently, Lazer (*66*) (in the course of proving his main oscillation theorem) has shown, under hypotheses (H_1) and (H_3), that if (E_3) is disconjugate then so is the second-order equation

$$y'' + [P(x) + mxQ(x)]y = 0 \tag{2.67*}$$

for each number $m < 1/2$.

In a private communication, *J.* H. E. Cohn has provided examples showing that oscillation of (2.67) or (2.67*) does not imply oscillation of the other.

Now recall that Y' is positive and decreasing and suppose that

$$(2.69) \qquad Y'(x) > K > 0, \qquad \text{i.e.,} \qquad Y(x) > K(x-a) \quad \text{on} \quad (a, \infty).$$

Then (2.63) implies

$$(2.70) \qquad \int_a^\infty t^2 Q(t)\, dt < \infty \qquad \text{and} \qquad \int_a^\infty tP(t)\, dt < \infty.$$

But, according to a special case of the nth-order asymptotic results of Hallam (*44*),

$$(2.70') \qquad \int_a^\infty t^2 \,|\, Q(t)|\, dt < \infty \qquad \text{and} \qquad \int_a^\infty t \,|\, P(t)|\, dt < \infty$$

implies disconjugacy of $(\mathcal{E}_3)$ for large x and, hence, nonoscillation of $(\mathcal{E}_3)$ on $[a, \infty)$.

On the other hand, if either

$$(2.71) \qquad \int_a^\infty t^2 Q(t)\, dt = \infty \qquad \text{or} \qquad \int_a^\infty tP(t)\, dt = \infty,$$

then $Y'(x)$ decreases to zero as $x \to \infty$ and integration of (2.66) yields

$$(2.72) \qquad Y'(x) = \left(\int_x^\infty\right)^2 (QY) + \int_x^\infty (PY) \quad \text{on} \quad [a, \infty).$$

Using, again, the fact that $Y'(x)$ is decreasing,

$$(2.73) \qquad Y'(x) \geqslant H(x)\, Y(x); \qquad H(x) = \left(\int_x^\infty\right)^2 Q + \int_x^\infty P.$$

Note that (2.67′) of Lemma 2.25 implies that

$$H(x) < 1/(x-a) \quad \text{on} \quad (a, \infty).$$

Also, (2.71) implies that

$$\int_a^\infty H = \infty.$$

The differential inequality (2.73) may be integrated to yield

$$Y(x) \geqslant Y(b) \exp\left(\int_b^x H\right), \qquad a < b \leqslant x < \infty, \tag{2.74}$$

which may be substituted into (2.63).

Theorem 2.18. *If Eq.* ($\mathcal{E}_3$), *with coefficients satisfying* (H_1), *is disconjugate then*

$$\int_a^\infty Q(t) \exp\left(2 \int_a^t H\right) dt < \infty \quad \textit{and} \quad \int_a^\infty P(t)\, H(t) \exp\left(2 \int_a^t H\right) dt < \infty, \tag{2.75}$$

where

$$H(x) = \left(\int_x^\infty\right)^2 Q + \int_x^\infty P = \int_x^\infty [(t-x)\,Q(t) + P(t)]\, dt.$$

2.14. *Nonoscillation Theorems*

Here we reprove an important nonoscillation theorem of Lazer. For simplicity let us first consider the special case of ($\mathcal{E}_3$) with $P(x) \equiv 0$,

$$\mathcal{L}_3{}^0[y] = y''' + Q(x)\, y = 0, \qquad Q(x) \geqslant 0, \qquad Q \in C[a, \infty), \qquad 0 < \int_a^\infty Q \leqslant \infty. \tag{2.76}$$

Note that Eq. (2.76) $\in C_{\mathrm{I}}$; i.e.,

$$u_2(x, b) > 0 \qquad \text{for} \quad a \leqslant x < b < \infty.$$

In fact, (2.76) has stronger properties, e.g.,

$$u_2'(x, b) < 0 \qquad \text{for} \quad a \leqslant x < b < \infty. \tag{2.77}$$

Suppose that (2.76) is oscillatory on $[a, \infty)$ and $y(x)$ is any nontrivial solution so that for some $c \in [a, \infty)$, $y(a) = 0$. Since (2.76) $\in C_{\mathrm{I}}$ and is oscillatory then $u_2(x, c)$ is oscillatory. But $y(x)$ and $u_2(x, a)$ are both solutions of the nonsingular second-order equation

$$\{u_2{}^+; y\} = 0 \quad \text{on} \quad (a, \infty),$$

where $u_2{}^+ = u_2{}^+(x, a) > 0$ on (a, ∞), and, hence, $y(x)$ is oscillatory.

Since this conclusion does not depend on the particular equation (2.76) we have a theorem first proved by Hanan.

Lemma 2.27. *If* $(\mathcal{E}_3) \in C_{\mathrm{I}}$ *and is oscillatory on* $[a, \infty)$ *then any solution which vanishes once is also oscillatory.*

Next suppose, not only that (2.76) is oscillatory, but that there is a nonoscillatory solution $u(x)$. Then by Lemma 2.30, $u(x) \neq 0$ on $[a, \infty)$. Let

$$u(x) > 0 \quad \text{on} \quad [a, \infty).$$

Since, $u_2(x, a)$ is oscillatory let b be its first zero, i.e.,

$$u_2(b, a) = 0, \qquad u_2(x, a) > 0 \quad \text{on} \quad (a, b), \qquad b \in (a, \infty).$$

Then by Lemma 2.13 there is a number λ such that $u_2(x, a) - \lambda u(x)$ has a double zero at some $x = \alpha \in (a, b)$. Hence,

$$\lambda u(x) - u_2(x, a) = k u_2(x, \alpha); \qquad \lambda > 0, \quad k > 0.$$

Therefore, $\lambda u'(x) = u_2'(x, a) + k u_2'(x, \alpha)$ and from (2.77),

$$u'(a) = (k/\lambda)\, u_2'(a, \alpha) < 0.$$

Lemma 2.28. *If* $(\mathcal{E}_3) \in C_{\mathrm{I}}$ *and is oscillatory but has a nonoscillatory solution* $u(x)$ *on* $[a, \infty)$ *then* $u(x) \neq 0$ *on* $[a, \infty)$. *Furthermore, if* $(\mathcal{E}_3)$ *satisfies the condition* (2.77), *then* $|\, u(x)|' < 0$ *on* $[a, \infty)$.

Suppose that $y(x)$ is an oscillatory solution of (2.76); then $z = (y/u)'$ is an oscillatory solution of the second-order equation

$$z'' + (3u'/u)z' + (3u''/u)z = 0 \tag{2.79}$$

and $w = u^{3/2}z$ is an oscillatory solution of

$$w'' + (\tfrac{3}{2})[u''/u - (\tfrac{1}{2})(u'/u)^2]w = 0 \quad \text{on} \quad [a, \infty). \tag{2.80}$$

Since the nonoscillatory solution $u(x)$ satisfies

$$u(x) > 0, \qquad u'(x) < 0 \qquad \text{and} \qquad u''(x) > 0,$$

it is not difficult to show that $u(x)$ is bounded and

$$\lim_{x\to\infty} u'(x) = \lim_{x\to\infty} u''(x) = 0. \tag{2.81}$$

Furthermore, by multiplying $u''' + Pu = 0$ by $u(x)$ and integrating we have

$$\{uu'' - u'^2/2\}' = -Qu^2 \leqslant 0 \tag{2.82}$$

and by (2.81),

$$\{uu'' - u'^2/2\}(x) = \int_x^\infty Qu^2 > 0 \quad \text{on} \quad [a, \infty). \tag{2.83}$$

If, in addition, $\int_a^\infty Q < \infty$, then

$$\{uu'' - u'^2/2\}(x) \leqslant u^2(x) \int_x^\infty Q \tag{2.84}$$

and comparison with (2.80) implies oscillation of

$$y'' + (\tfrac{3}{2})\left(\int_x^\infty Q\right) y = 0 \quad \text{on} \quad [a, \infty). \tag{2.85}$$

Theorem 2.19. *If* (2.85) *is nonoscillatory on* $[a, \infty)$ *then* $y''' + Qy = 0$, $Q \geqslant 0$, *is nonoscillatory on* $[a, \infty)$.

We will now extend the preceeding discussion to Lazer's case of $(\mathcal{E}_3)$, whose coefficients satisfy hypotheses (H_1) and (H_3). Proceeding as before, if $y(x)$ is an oscillatory solution and $u(x)$ is a nonzero solution of (E_3) then $w = u^{3/2}(y/u)'$ is an oscillatory solution of

$$w'' + \{P + (3/2)[u''/u - (\tfrac{1}{2})(u'/u)^2]\}w = 0. \tag{2.86}$$

Because of (H_3), Eq. $(\mathcal{E}_3)$ can be rewritten as

$$[y'' + (P/2)y]' + (P/2)y' + (Q - P'/2)y = 0, \tag{$\mathcal{E}_3'$}$$

which is the form studied by Gregus (*34–42*). Then for any solution $y(x)$ of $(\mathcal{E}_3)$,

$$\{y\mathfrak{D}_2 y - y'^2/2\}' = -(Q - P'/2)y^2 \leqslant 0; \qquad \mathfrak{D}_2 y = y'' + (P/2)y. \tag{2.87}$$

Note that (2.87) yields immediately that Eq. $(\mathcal{E}_3) \in C_{\mathrm{I}}$; i.e.,

$$u_2(x, t) > 0, \qquad a \leqslant x < t < \infty.$$

For third-order equations in C_{I}, Gregus (*40, 41*) and Svec (*102*) have constructed nonoscillatory solutions with properties needed here.

For each positive integer $n > a$ let $y_n(x)$ be the solution of (E_3) such that

$$y_n(n) = y_n{}'(n) = 0, \qquad y_n''(n) > 0$$

which is normalized so that

$$y_n(x) = c_0{}^n u_0(x, a) + c_1{}^n u_1(x, a) + c_2{}^n u_2(x, a); \qquad (c_0{}^n)^2 + (c_1{}^n)^2 + (c_2{}^n)^2 = 1.$$

Therefore, there exists a sequence $\{n_i\}$ of positive integers $\to \infty$ such that $\{c_j^{n_i}\}$ converges for $j = 0, 1, 2$, say,

$$\{c_j^{n_i}\} \to c_j\,; \qquad c_0{}^2 + c_1{}^2 + c_2{}^2 = 1.$$

Hence the sequence $\{y_{n_i}(x)\}$ converges to the nontrivial solution

$$Y(x) = c_0 u_0(x, a) + c_1 u_1(x, a) + c_2 u_2(x, a) \quad \text{on} \quad [a, \infty)$$

and uniformly on closed finite subintervals $[a, X]$, $a < X < \infty$.

From (2.87) it follows that

$$\{y_{n_i} \mathfrak{D}_2 y_{n_i} - y_{n_i}'^2/2\}(x) = \int_x^{n_i} (Q - P'/2)\, y_{n_i}^2.$$

Let $a < X < \infty$. Then, for $n_i > X$ and $x \in [a, X]$,

$$\{y_{n_i} \mathfrak{D}_2 y_{n_i} - y_{n_i}'^2\}(x) \geqslant \int_x^{X} (Q - P'/2)\, y_{n_i}^2,$$

from which we obtain

$$\{Y \mathfrak{D}_2 Y - Y'^2/2\}(x) \geqslant \int_x^{\infty} (Q - P'/2)\, Y^2.$$

Now that the right-hand integral exists it is easy to complete the proof of the following result due to Gregus (*42*).

Lemma 2.29. *If the coefficients of* $(\mathcal{E}_3)$ *satisfy* (H_3) *then there exists a positive solution* $Y(x)$ *on* $[a, \infty)$ *such that*

$$\{Y \mathfrak{D}_2 Y - Y'^2/2\}(x) = \int_x^{\infty} (Q - P'/2)\, Y^2 > 0 \quad \textit{on} \quad [a, \infty). \tag{2.88}$$

If in addition $Y'(x) < 0$ and $\int_a^\infty Q < \infty$, then

$$\{Y\mathfrak{D}_2 Y - Y'^2/2\}(x) \leqslant Y^2(x) \int_x^\infty (Q - P'/2) \leqslant Y^2(x) \left[P(x)/2 + \int_x^\infty Q\right] \tag{2.89}$$

and comparison with (2.86), where $u = Y$, yields that

$$y'' + \left[P(x) + (3/2) \int_x^\infty Q\right] y = 0 \tag{2.90}$$

is oscillatory on $[a, \infty)$.

Suppose that $(\mathcal{E}_3)$ is oscillatory but that (2.90) is nonoscillatory on $[a, \infty)$; then

$$Q \geqslant 0 \quad \text{and} \quad y'' + Py = 0 \quad \text{is nonoscillatory on} \quad [a, \infty). \tag{H_4}$$

Hanan (*45*) asserted that, under these assumptions, any nonoscillatory solution is decreasing in absolute value. However, his proof appears to be incorrect.

As we have already noted, (H_4) implies that $y'' + Py = 0$ is disconjugate for large x, i.e., there is a number $c \in [a, \infty)$ such that $y'' + Py = 0$ is disconjugate on $[c, \infty)$. Consequently, if $d \in (c, \infty)$ there is a positive solution $v(x)$ of $y'' + Py = 0$ on $[c, d]$ and $(\mathcal{E}_3)$ can be rewritten in the two-term form

$$[v^2(y'/v)']' + Qvy = 0 \quad \text{on} \quad [c, d]. \tag{$\hat{\mathcal{E}}_3$}$$

From (2.19) we recall that

$$u_2'(c, \alpha) = -u_1^+(\alpha, c)$$

and, since $u_1^+(x, c)$ satisfies

$$[(1/v)(v^2y')']' - Qvy = 0 \quad \text{on} \quad [c, \infty), \tag{$\hat{\mathcal{E}}_3^+$}$$

it is easy to show that

$$u_1^+(x, c) > 0 \qquad \text{and} \qquad u_2'(c, x) < 0 \qquad \text{for} \quad c < x < \infty.$$

Now that property (2.77) is satisfied, we have Lemma 2.28 on $[c, \infty)$. Since $Y(x)$ of Lemma 2.29 is a positive nonoscillatory solution then

$$Y'(x) < 0 \quad \text{on} \quad [c, \infty),$$

and (2.90) is oscillatory on $[c, \infty)$, a contradiction. Note that Lazer's additional assumption that $P(\infty) = 0$ is unnecessary.

Theorem 2.20. *If the coefficients of* $(\mathcal{E}_3)$ *satisfy* (H_1) *and* (H_3), $\int_a^\infty Q < \infty$, *and the second-order equation*

$$y'' + \left[P(x) + (3/2)\int_x^\infty Q\right] y = 0$$

is nonoscillatory on $[a, \infty)$, *then so is the third-order equation* $(\mathcal{E}_3)$.

2.15. *A Third-order Prüfer Transformation*

Recalling the second-order Prüfer Transformation (1.9) of Section I, let $y(x)$ be any nontrivial solution of the third-order equation (E_3) and define an *amplitude function*

$$\rho(x) = \sqrt{y^2(x) + (D_1 y)^2(x) + (D_2 y)^2(x)} > 0. \tag{2.91}$$

Next, normalize the phase components of $y(x)$ by letting

$$s(x) = y(x)/\rho(x), \qquad s_1(x) = D_1 y(x)/\rho(x), \qquad s_2(x) = D_2 y(x)/\rho(x). \tag{2.92}$$

Differentiating (2.91) we have

$$\rho' = [(1/r_1 - q_1)\, s s_1 + (1/r_2 - q_2)\, s_1 s_2]\rho \tag{2.93}$$

and differentiating (2.92) yields the vector-matrix equation

$$\begin{pmatrix} s \\ s_1 \\ s_2 \end{pmatrix}' = \begin{pmatrix} 0 & b_1(x) & 0 \\ -b_1(x) & 0 & b_2(x) \\ 0 & -b_2(x) & 0 \end{pmatrix} \begin{pmatrix} s \\ s_1 \\ s_2 \end{pmatrix}; \tag{2.94}$$

$$b_1 = 1/r_1 - (1/r_1 - q_1)\, s^2 - (1/r_2 - q_2)\, s s_2\,,$$
$$b_2 = q_2 + (1/r_1 - q_1)\, s s_2 + (1/r_2 - q_2)\, s_2^2.$$

This differential equation for the direction cosines may be rewritten by use of the vector cross-product, i.e.,

$$\sigma' = -\beta \times \sigma = \sigma \times \beta \tag{2.94'}$$

where $\sigma = (s, s_1, s_2)$, $\beta = (b_2, 0, b_1)$. Note that the tangent vector σ' is orthogonal to the plane of β and σ and the locus of $\sigma(x)$ is a path on the unit sphere. For the case where b_1 and b_2 are linearly dependent the locus of it is a circle. This interesting equation warrants further study and should shed further light on third-oder oscillation properties.

By altering (2.92) slightly we let

$$y = \rho s, \quad D_1 y = k_1 \rho s_1, \quad D_2 y = k_1 k_2 \rho s_2; \quad k_1 \,\&\, k_2 = \text{constants} \neq 0 \tag{2.92'}$$

and obtain

$$\rho' = [(k_1/r_1 - q_1/k_1)\, s s_1 + (k_2/r_2 - q_2/k_2)\, s_1 s_2]\rho, \tag{2.93'}$$

which immediately yields bounds for solutions of (E_3) (*13*).

Theorem 2.21. *If there exist nonzero constants k_1 and k_2 such that $\int_a^\infty |\, k_1/r_1 - q_1/k_1 \,| < \infty$ and $\int_a^\infty |\, k_2/r_2 - q_2/k_2 \,| < \infty$ then every solution of (E_3) is bounded on $I = [a, \infty)$.*

Other boundedness theorems for third-order equations have been reported by Dobrohotova (*23*) and Rab (*81*).

2.18. *Gregus' Asymptotic Theorems.*

In his numerous papers (*34–42*) since 1955, Gregus has studied the equation

$$y''' + 2Ay' + (A' + b)y = 0; \qquad A, A' \,\&\, b \in C(I). \tag{2.95}$$

He observed that the classical equation

$$l_3[y] = y''' + p_2 y'' + p_1 y' + p_0 y = 0; \qquad p_i \in C^i(I),$$

can be transformed into the form (2.95) by means of the substitution

$$\hat{y} = y \exp\left\{(1/3) \int^x p_2\right\}$$

and the subsequent equation $(\mathcal{E}_3)$ rewritten as $(\mathcal{E}_3')$, as in the preceding section. In the form $(\mathcal{E}_3')$, Eq. (2.95) becomes

$$[y'' + Ay]' + Ay' + by = 0; \qquad A \,\&\, b \in C(I), \tag{2.95'}$$

and the coefficient $A(x)$ need not be differentiable. Note that the first three terms form a self-adjoint operator. Let $v_i(x, a)$ be the principal solutions of

$$y'' + (A/2)y = 0$$

defined by the initial conditions

$$y = v_0(x, a), \qquad y(a) = 1, \qquad y'(a) = 0$$

$$y = v_1(x, a); \qquad y(a) = 0, \qquad y'(a) = 1.$$

If we rewrite (2.95′) in the vector-matrix form

$$\begin{pmatrix} y \\ y' \\ \mathfrak{D}_2 y \end{pmatrix}' + \begin{pmatrix} 0 & -1 & 0 \\ A & 0 & -1 \\ 0 & A & 0 \end{pmatrix} \begin{pmatrix} y \\ y' \\ \mathfrak{D}_2 y \end{pmatrix} = \begin{pmatrix} 0 \\ 0 \\ -by \end{pmatrix},$$

then a fundamental matrix solution of the homogeneous part is

$$Y = \begin{pmatrix} v_0^2 & v_0 v_2 & v_1^2/2 \\ 2v_0 v_0' & (v_0 v_1)' & v_1 v_1' \\ 2(v_0')^2 & 2v_0' v_1' & (v_1')^2 \end{pmatrix}, \qquad \text{where} \quad v_i = v_i(x) = v_i(x, a),$$

and solutions of the complete nonhomogeneous equation are solutions of the system of integral equations

$$(2.96) \quad \begin{cases} y(x) = y(a)\, v_0^2(x) + y'(a)\, v_0(x)\, v_1(x) + D_2 y(a)\, v_1^2(x)/2 \\ \qquad - \int_a^x K(x, t)\, b(t)\, y(t)\, dt, \\ y'(x) = 2y(a)\, v_0(x)\, v_0'(x) + y'(a)(v_0 v_1)'(x) + D_2 y(a)\, v_1(x)\, v_1'(x) \\ \qquad - \int_a^x K_1(x, t)\, b(t)\, y(t)\, dt, \\ \mathfrak{D}_2 y(x) = 2y(a)[v_0'(x)]^2 + 2y'(a)\, v_0'(x)\, v_1'(x) + D_2 y(a)[v_1'(x)]^2 \\ \qquad - \int_a^x K_2(x, t)\, b(t)\, y(t)\, dt, \end{cases}$$

where $K(x, t) = [v_0(x)\, v_1(t) - v_1(x)\, v_0(t)]^2/2 = v_1^2(x, t)/2,$

$$K_1(x, t) = K'(x, t) = v_1(x, t)\, v_1'(x, t),$$

and

$$K_2(x, t) = [v_1'(x, t)]^2.$$

Lemma 2.30. (a) *If* $b(x) \leqslant 0$ *on* $I = [a, \infty)$ *then* $u_2(x, a) > 0$ *and* $\mathfrak{D}_2 u_2(x, a) > 0$ *on* $[a, \infty)$.

(b) *If, in addition,* $A(x) \leqslant 0$ *on* $[a, \infty)$ *then*

$$\lim_{x\to\infty} u_2(x, a) = \lim_{x\to\infty} u_2'(x, a) = \infty$$

and $\mathfrak{D}_2 u_2(x, a)$ *is an increasing function on* $[a, \infty)$.

Since $\mathfrak{D}_2 u_2 \geqslant |b| u_2 + |A| u_2' \geqslant 0$ on (a, ∞) then

$$\mathfrak{D}_2 u_2(x, a) \geqslant \begin{cases} 1 + (1/2)\int_a^x (t-a)^2 |b(t)|\, dt \\ 1 + \int_a^x (t-a) |A(t)|\, dt. \end{cases}$$

Lemma 2.31. *If, in addition to the hypotheses of Lemma* 2.32, *either*

(i) $$\int_a^\infty t^2 |b(t)|\, dt = \infty$$

or

(ii) $$\int_a^\infty t\, |A(t)|\, dt = \infty,$$

then

$$\lim_{x\to\infty} \mathfrak{D}_2 u_2(x, a) = \infty = \lim_{x\to\infty} u_2''(x, a).$$

Also, the latter limit is assured if

(iii) $$\int_a^\infty t^2 |A(t)|\, dt = \infty.$$

Note that the sufficient condition $\int_a^\infty t^2[A'(t) - b(t)]\, dt = \infty$ of Gregus (*42*) is implied by (i) or (ii) but Gregus' conclusion that $u_2 + 2Au_2 \to \infty$ is slightly stronger.

Suppose with Gregus that

(2.97) $\quad A(x) \leqslant 0 \quad$ and $\quad b(x) \geqslant 0$ but $\not\equiv 0$ on any interval of $[a, \infty)$.

Then, as in the preceeding Subsection, Eq. (2.95) $\in C_{\mathrm{I}}$ and

$$(y'' + Ay)' + Ay'$$

is a disconjugate operator and, hence, there exists a solution $Y(x)$ such that

$$Y(x) > 0 \quad \text{and} \quad Y'(x) < 0 \quad \text{on} \quad [a, \infty).$$

Also, the adjoint equation satisfies Lemma 2.32. The Lagrange identity

$$u_2^+ \mathfrak{D}_2 Y - u_2^{+\prime} Y' + Y \mathfrak{D}_2 u_2^+ = Y(a)$$

then yields an improvement of Gregus' theorem (*42*).

Theorem 2.22. *Under hypothesis* (2.97), *there exists a solution* $Y(x)$ *of* (2.95) *such that*

$$Y(x) > 0, \qquad Y'(x) < 0 \qquad \text{and} \qquad Y''(x) > 0,$$

and $\lim_{x\to\infty} Y'(x) = \lim_{x\to\infty} \mathfrak{D}_2 Y(x) = 0$. *If, in addition, either* (i) *or* (ii) *of Lemma* 2.33 *is assumed then*

$$\lim_{x\to\infty} Y(x) = 0.$$

III. Fourth-Order Equations

3.1. *Examples*

Motivating examples, which illustrate the oscillatory behavior with which we are concerned, are

$$\text{(3.1)} \quad \begin{array}{lll} \text{(a)}\ \ y^{\text{iv}} - y = 0, & \text{(c)}\ \ y^{\text{iv}} + y' = 0, & \text{(e)}\ \ y^{\text{iv}} + y'' = 0, \\ \text{(b)}\ \ y^{\text{iv}} + y = 0, & \text{(d)}\ \ y^{\text{iv}} - y' = 0, & \text{(f)}\ \ y^{\text{iv}} - y'' = 0, \end{array}$$

The first two examples (3.1a) and (3.1b) are special cases of

$$\text{(3.2)} \quad \begin{array}{ll} \text{(a)}\ \ (r(x)y'')'' - p(x)y = 0, & \\ \text{(b)}\ \ (r(x)y'')'' + p(x)y = 0; & r, p > 0, \quad r, p \in C[a, \infty), \end{array}$$

for which Nehari and Leighton (*70*) made a rather complete study. Other authors (*9–11, 54, 88–90*) have extended certain parts of the theory developed by Nehari and Leighton to

$$\text{(3.3)} \qquad [(Ry'')' + Qy']' + Py = 0.$$

Whyburn (*106*) and Kondrat'ev (*62*) have also contributed basic discussions of Eqs. (3.2). Kamke (*60*) has listed other examples and Handelman and Tu (*46*) studied a phsical application of (3.3), i.e., vibrations of a beam with a compressive stress distribution.

Somewhat different examples are generated by recalling that the product of two solutions of a second-order equation is a solution of a third-order (self-adjoint) equation and by taking the *product of three solutions*, which turns out to satisfy a fourth-order equation.

Lemma 3.1. *Recall that* $L_2[y] = (ry')' + qy = 0$ *and let* $L_2[v_i] = 0$ *for* $i = 1, 2, 3$. *Then* $y = v_1v_2v_3$ *satisfies a fourth-order equation*

$$[r(r\{r[(ry')' + 3qy]\}' + 4qry')]' + 3qr[(ry')' + 3qy] = 0. \tag{3.4}$$

An interesting special case of (3.4) is that when $r = q = 1$ and (3.4) becomes

$$y^{\text{iv}} + 10y'' + 9y = 0 \tag{3.4'}$$

which has the fundamental set of solutions

$$y = \sin^3 x, \qquad \cos^3 x, \qquad \cos^2 x \sin x, \qquad \sin^2 x \cos x.$$

An obvious way to generate fourth-order examples is to iterate operators of lower order, e.g.,

$$L[y] = \lambda_4\{w_3\lambda_3(w_2\lambda_2[w_1\lambda_1(y)])\}; \tag{3.5}$$

$$\lambda_i[y] = y' + p_i y; \qquad w_i > 0, \qquad w_i, p_i \in C(I)$$

and

$$L[y] = \Lambda_2(w\Lambda_1[y]); \tag{3.6}$$

where

$$\Lambda_i[y] = (r_i y')' + q_i y, \qquad r_i > 0; \qquad w, r_i; p_i \in C(I).$$

Note that for the operator of (3.5) every nontrivial solution of the equation $L[y] = 0$ has at most three zeros in the interval I.

Definition 3.1. *A fourth-order operator* $L[y]$ *and the corresponding equation* $L[y] = 0$ *is said to be disconjugate on an interval* I *if no nontrivial solution has four zeros on* I.

Later on we will show that any disconjugate fourth-order operator can be factored into the form (3.5). An interesting special case of (3.6) is the constant coefficient case,

$$L[y] = (D^2 + m^2)(D^2 + n^2)y = y^{\text{iv}} + (m^2 + n^2)y'' + m^2n^2y \qquad [m, n = \text{arbitrary constants}],$$

which discourages general separation theorem conjectures although such theorems have been established for such special cases as (3.2) (*63*, *70*).

Although the sum of two solutions of a single second-order equation yields nothing of interest here, the sum (in fact, any linear combination) of solutions of different second-order equations satisfies a fourth-order equation.

Lemma 3.2. *If* $(Rv_i')' + Q_iv_i = 0$ *for* $i = 1, 2$ *and* $Q_1 \neq Q_2$ *then* $y = v_1 + v_2$ *satisfies*

$$\left\{R\left(\frac{2}{Q_1 - Q_2}\left[(Ry')' + \left(\frac{Q_1 + Q_2}{2}\right)y\right]\right)'\right\}' + \frac{Q_1 + Q_2}{Q_1 - Q_2}\left[(Ry')' + \left(\frac{Q_1 + Q_2}{2}\right)y\right] + \left(\frac{Q_1 - Q_2}{2}\right)y = 0. \tag{3.7}$$

The case $Q_1 = -Q_2 \neq 0$ and $R = 1$, where (3.7) becomes

$$\left(\frac{y''}{Q_1}\right)'' + Q_1y = 0,$$

has already been noted (*9*); and the constant coefficien tcases, with $R = 1$, $Q_1 = K_1$, $Q_2 = K_2$ [K_i = arbitrary constants], become

$$y^{\text{iv}} + (K_1 + K_2)\,y'' + \left(\frac{K_1^2 + K_2^2}{2}\right)y = 0.$$

The product of solutions of different second-order equations is also of interest. W. Hahn (*43*) has made use of this idea in his study of orthogonal functions.

Lemma 3.3. *For solutions* v_1 *and* v_2 *of Lemma 3.2,* $y = v_1v_2$ *satisfies*

$$\left[\frac{1}{Q_2 - Q_1}\{R[(Ry')' + (Q_1 + Q_2)y]\}' + (Q_1 + Q_2)\,Ry'\right]' + (Q_2 - Q_1)y = 0. \tag{3.8}$$

Finally, if $y(x)$ is a solution of $L_3[y] = h$ where $h \equiv 0$, $\{h \in C(I)$ and $h \neq 0\}$ on I and L_3 is the canonical third-order operator, then $y(x)$ satisfies the fourth-order equation

$$(L_3(y))' = 0 \left\{\left[\frac{1}{h} L_3(y)\right]' = 0\right\}.$$

3.2. *A Fourth-Order Canonical Form*

The preceding suggest the following general fourth-order equation (*12*)

$$(E_4) \qquad L_4[y] = (D_3 y)' + q_3 D_2 y + q_4 y = 0,$$

where $D_3 y = r_3 L_3[y]$, L_3 is the third-order canonical operator of Subsection 2.2, which was defined in terms of the quasi-derivatives $D_1 y$ and $D_2 y$, and the coefficients satisfy

$$r_i > 0; \qquad r_i, q_i \in C(I).$$

It is easy to see that (3.2), (3.3), (3.5), and (3.6)–(3.8) are special cases of (E_4). Shinn's quasi-differential operator (*98, 99*) is also equivalent to (E_4). Furthermore, the classical fourth-order equation

$$l_4[y] = y^{\text{iv}} + p_3 y''' + p_2 y'' + p_1 y' + p_0 y = 0 \tag{3.9}$$

is equivalent to a special case of (E_4), since it is easily obtained from

$$\left[y''' \exp \int p_3 + \left(\int p_1 \exp \int p_3\right) y'\right]' + \left(p_2 \exp \int p_3 - \int p_1 \exp \int p_3\right) y'' + \left(p_0 \exp \int p_3\right) y = 0. \tag{3.10}$$

The corresponding phase-vector form of (E_4) is

$$\begin{pmatrix} y \\ D_1 y \\ D_2 y \\ D_3 y \end{pmatrix}' = \begin{pmatrix} 0 & 1/r_1 & 0 & 0 \\ -q_1 & 0 & 1/r_2 & 0 \\ 0 & -q_2 & 0 & 1/r_3 \\ -q_4 & 0 & -q_3 & 0 \end{pmatrix} \begin{pmatrix} y \\ D_1 y \\ D_2 y \\ D_3 y \end{pmatrix} \tag{3.11}$$

from which appropriate existence and uniqueness properties are easily obtained. The preceding formulation is a special case of the nth-order canonical form given by Zettl (*110*) who used it and the third-order case, L_3, as motivating examples.

The usual well-known methods provide the *Lagrange Formula*

$$zL_4[y] - yL_4^+[z] = \{z; y\}', \tag{3.12}$$

$$\{z; y\} = zD_3y - D_1^+zD_2y + D_2^+zD_1y - (D_3^+z)y,$$

where the *adjoint operators* L_4^+, D_1^+, D_2^+, D_3^+ are obtained from L_4, D_1, D_2, D_3, respectively, by interchanging the coefficients:

$$r_1 \text{ with } r_3 \quad \text{and} \quad q_1 \text{ with } q_3 .$$

Note that if $L_3[y] = 0$ and $L_3^+[z] = 0$ then $\{z; y\}$ is a constant.

3.3. *Self-Adjoint Equations*

Equation (E_4) with

$$r_1 = r_3 \quad \text{and} \quad q_1 = q_3$$

represents a large class of self-adjoint equations of order four; i.e.,

$(E_4{}^*)$

$$L_4{}^*[y] = \{r_1[(r_2[(r_1y')' + q_1y])' + q_2r_1y']\}' + q_1r_2[(r_1y')' + q_1y] + q_4y = 0,$$

where $D_i = D_i^+$ for $i = 1, 2, 3$.

Note that (3.2)–(3.4) and (3.7) are self-adjoint; (3.5) is self-adjoint if $\lambda_1 = \lambda_3$ and $w_1 = w_3$; (3.6) is self-adjoint if $\Lambda_1 = \Lambda_2$; and (3.10) is equivalent to a self-adjoint equation if $p_3 = 0$ and $p_2' = p_1$. Also, if $r = \exp(\int p_3)$, and $p_i \in C^{(i-1)}(I)$ for $i = 1, 2, 3$ then

$$rl_4[y] = (ry'')'' + (rp_2 - r'')\, y'' + rp_1y' + rp_0y, \tag{3.13}$$

and $rl_4[y]$ is self-adjoint if $r = \exp(\int p_3)$ satisfies the

$$(r'' - p_2r)' + p_1r = 0. \tag{3.14}$$

Next we note that certain nonself-adjoint third-order equations can be imbedded in self-adjoint fourth-order equations.

Lemma 3.4. *If $z(x)$ is a nonvanishing solution over an interval I of the* *canonical third-order equation* $L_3[y] = 0$, *then* $(1/z)(r_1 z^2 L_3[y])'$ *is a self-adjoint fourth-order operator. In particular,*

$$(3.15)\quad (1/z)(r_1 z^2 L_3[y])' = \{r_1(D_2 y)' - [D_2(D_1 z)' + (r_2 q_1 - r_1 q_2)]\, D_1 y\}' + q_1 D_2 y + (q_1 D_2 z) y.$$

3.4. *Factoring Fourth-Order Operators*

Here we proceed as we did when factoring the third-order operator L_3. Suppose that the adjoint equation $(E_4{}^+)$ has a nonvanishing solution on an interval I, i.e.,

$$\text{(i)}\qquad L_4{}^+[v^+] = 0, \qquad v^+ \neq 0 \quad \text{on} \quad I.$$

Then, by the Lagrange Formula (3.12),

$$L_4[y] = \frac{1}{v^+}\{v^+; y\}'$$

and $\{v^+; y\}$ is a third-order operator of the form $L_3[y]$; i.e.,

$$(3.16)\qquad \{v^+; y\} = r_3 v^{+2}\left\{\frac{r_2}{v^+}\left[(D_1 y)' + \left(q_1 - \frac{v^+}{r_2}\int \frac{D_3{}^+ v^+}{r_2 v^{+2}}\right) y\right]\right\}' + \left[\frac{r_3 q_2 v^+ + D_2{}^+ v^+}{r_3 v^{+2}} + \frac{1}{r_1}\int \frac{D_3{}^+ v^+}{r_3 v^{+2}}\right] D_1 y$$

Next, assume there are two solutions v_1 and v_2 of (E_4) such that

$$\text{(ii)}\qquad \{v^+; v_i\} = 0 \qquad \text{for} \quad i = 1, 2 \quad \text{on} \quad I,$$

$$\text{(iii)}\qquad v_1 \neq 0,$$

and

$$\text{(iv)}\qquad w_2 = v_1 D_1 v_2 - v_2 D_1 v_1 \neq 0 \quad \text{on} \quad I.$$

By Lemma 2.10 the third-order operator $\{v^+; y\}$ can be factored yielding

$$(3.17)\qquad L_4[y] = \frac{1}{v^+}\left\{\frac{r_3 v^{+2}}{w_2}\left(\frac{r_2 w_2{}^2}{v_1 v^+}\left[\frac{r_1 v_1{}^2}{w_2}(y/v_1)'\right]'\right)'\right\}'.$$

In addition, suppose that v_3 is another solution of (E_4) and consider the Wronskian of the three solutions v_1, v_2, v_3,

$$(3.18)\qquad w_3 = W[v_1, v_2, v_3] = \alpha_{012},$$

where

$$\alpha_{ijk} = \begin{vmatrix} D_i v_1 & D_i v_2 & D_i v_3 \\ D_j v_1 & D_j v_2 & D_j v_3 \\ D_k v_1 & D_k v_2 & D_k v_3 \end{vmatrix}.$$

Successive differentiations of the determinant α_{012} yield

$$(3.19) \qquad D_1{}^+ w_3 = \alpha_{013}\,, \qquad D_2{}^+ w_3 = \alpha_{023}\,, \qquad D_3{}^+ w_3 = \alpha_{123}$$

and, finally, that w_3 satisfies the adjoint equation $(E_4{}^+)$. Furthermore, it is easy to show that

$$(3.20) \qquad \{w_3\,;\, v_i\} = 0 \qquad \text{for} \quad i = 1, 2, 3.$$

Consequently, w_3 may be substituted for v^+ in (iii) and (3.17), the latter becoming the Polya–Mammana factorization and a special case of (3.5). Note that Lemma 3.4 is suggested by (3.17). By combining pairs of factors in (3.17) we obtain a special case of (3.6), namely,

$$(3.21) \qquad L_4[y] = \mathscr{L}_2(r_2 w_2{}^2 L_2[y]),$$

where

$$L_2[y] = \left(\frac{r_1 y'}{w_2}\right)' + \left[\frac{r_1(r_2 w_2{}')' + (r_1 q_2 + 2r_2 q_1)\, w_2}{2r_2 w_2{}^2}\right] y$$

and $\mathscr{L}_2[y]$ is obtained from $L_2[y]$ by replacing r_1 and q_1 by r_3 and q_3, respectively. In the self-adjoint case, (3.21) becomes

$$(3.22) \qquad L_4{}^*[y] = L_2(r_2 w_2{}^2 L_2[y]) = L_2{}^+(L_2[y]).$$

Only the assumption (iv), $w_2 \neq 0$, is needed for (3.21) and (3.22) to hold.
Since $L_2[v_i] = 0$ for $i = 1$ or 2, the Wronskian w_2 satisfies

$$(3.23) \qquad l_2[y] = \left(\frac{r_2 y'}{v_i{}^2}\right)' + \left(\frac{2D_2 v_i + q_2 v_i}{v_i{}^3}\right) y = 0.$$

The special case for Eq. (3.3) with $Q = 0$,

$$l_2[y] = \left(\frac{Ry'}{v_i{}^2}\right)' + \left(\frac{2Rv_i''}{v_i{}^3}\right) y = 0,$$

has been investigated previously (*9, 70*).

Finally, under assumptions (iii) and (iv), i.e., disconjugacy,

$$L_4^*[y] = l_1^+[w_2 l_2(l_1[y])] \tag{3.24}$$

is another factoring of the self-adjoint operator, where

$$l_1[y] = r_1 v_i^2 (y/v_i)' = r_1 v_i \left[y' - \frac{v_i'}{v_i} y \right].$$

The factored expressions (3.22) and (3.24) are equivalent to those previously given by Mammana (*73*) for the special case (3.3).

3.5. *First Conjugate Points*

In the preceding section conditions (i)–(iv) were given which insure that $L_4[y]$ can be factored into first-order real factors (3.17) over an interval I and, hence, that (E_4) is disconjugate on I. Conversely, if (E_4) has a nontrivial solution with four zeros (counting each multiple zero as that many zeros) on I then either v_1 or w_2 or w_3 has a zero on I. This result may be improved by choosing particular solutions of (E_4).

Let $u_i(x, a)$ for $i = 0, 1, 2, 3$ be the fundamental set of solutions of (E_4) on $I = [a, \infty)$ as defined by the initial conditions

$$D_j u_i(a, a) = \delta_{ij}\,; \qquad i, j = 0, 1, 2, 3, \tag{3.25}$$

and $u_i^+(x, a)$ be the corresponding solutions of the adjoint equation (E_4^+). Because of these initial conditions at $x = a$, if $v_1 = u_3(x, a)$, $v_2 = u_2(x, a)$ and $v_3 = u_1(x, a)$ in (3.18), then

$$w_3 = W_3[u_3, u_2, u_1](x, a) = u_3^+(x, a). \tag{3.26}$$

Therefore, if $a < b \leqslant \infty$ and

$$u_3(x, a) > 0, \quad u_3^+(x, a) > 0 \quad \text{and} \quad W_2[u_3, u_2](x, a) > 0 \quad \text{on} \quad (a, b), \tag{3.27}$$

then (E_4) is disconjugate on (a, b).

Suppose $a < b \leqslant \infty$, that (E_4) has a nontrivial solution $y(x)$ with four zeros on $[a, b)$, and that (3.27) holds. Then $y(a) = 0$ and $y(x)$ satisfies the third-order equation

$$\{u_3^+; y\} = 0 \quad \text{on} \quad (a, b), \tag{3.28}$$

which has u_3 and u_2 as particular solutions. Again, because of the first and last inequalities of (3.27), Eq. (3.28) is disconjugate, i.e., $y(x)$ has at most two zeros on (a, b), which determines $y'(a) = 0 = D_1 y(a)$. But since $u_3(x, a) > 0$ and $W_2[u_3, u_2](x, a) > 0$ on (a, b) then $y(x)$ cannot have two zeros on (a, b), so that $y(a) = D_1 y(a) = D_2 y(a) = 0$. But this implies the contradictory conclusion that $y(x) \equiv K u_3(x, a)$.

Theorem 3.1. *If the inequalities* (3.27) *hold on the open interval* (a, b), $a < b \leqslant \infty$, *then the* Eq. (E_4) *is disconjugate on the half-closed interval* $[a, b)$.

On the other hand, if (E_4) has a solution with four zeros on $[a, b)$ then either $u_3(x, a)$ or $u_3^+(x, a)$ or $W_2[u_3, u_2](x, a)$ has a zero in (a, b). If $u_3(c, a) = 0$, $a < c < b$, then $u_3(x, a)$ satisfies the two-point boundary conditions

$$y(a) = D_1 y(a) = D_2 y(a) = 0 = y(c). \tag{3.29}$$

Since $u_3^+ = W_3[u_3, u_2, u_1]$ then $u_3^+(c, a) = 0$ yields a nontrivial solution of (E_4) satisfying

$$y(a) = 0 = y(c) = D_1 y(c) = D_2 y(c). \tag{3.30}$$

Similarly, $W_2[u_3, u_2](c, a) = 0$ yields a solution of (E_4) satisfying

$$y(a) = D_1 y(a) = 0 = y(c) = D_1 y(c). \tag{3.31}$$

Definition 3.2. *For* $i + j = 4$ *let* $z_{ij}(a)$ *be the smallest number* $c > a$ *such that* (E_4) *has a solution satisfying*

$$y(a) = D_1 y(a) = \cdots = D_{i-1} y(a) = 0 = y(c) = D_1 y(c) = \cdots = D_{j-1} y(c) \tag{3.32}$$

[*i.e.*, $y(x)$ *has a zero of multiplicity* i *at* $x = a$ *and* j *at* $x = c = z_{ij}(a)$]. *If no such finite number* c *exists let* $z_{ij}(a) = \infty$.

Note that $c = z_{31}(a)$, $z_{13}(a)$, and $z_{22}(a)$ satisfies (3.29), (3.30) and (3.31), respectively.

Definition 3.3. *If* (E_4) *has a nontrivial solution with at least four zeros on* $[a, \infty)$, *the number*

$$\eta_1(a) = \inf\{x_4 ; x_1 \leqslant x_2 \leqslant x_3 \leqslant x_4 , y(x_i) = 0, y \not\equiv 0, L_4[y] = 0\} \tag{3.33}$$

is called the first conjugate point of $x = a$ *with respect to* (E_4) *in* (a, ∞). *If* (E_4) *is disconjugate let* $\eta_1(a) = \infty$.

Coppel (*21*) has given a more general definition of conjugate points and has established some comparison theorems for "canonical" systems of differential equations, but these will not be treated here.

Let $c = \min\{z_{ij}(a); i + j = 4\}$; then (E_4) is disconjugate on $[a, c)$ and if $c < \infty$ there is a nontrivial solution of (E_4) having four zeros (counting multiplicities) on $[a, c]$.

Theorem 3.2. $\eta_1(a) = \min\{z_{13}(a), z_{22}(a), z_{31}(a)\} \leqslant \infty$.

For the self-adjoint equation $(E_4{}^*)$, since $u_3{}^+(x, a) = u_3(x, a)$, then $z_{13}(a) = z_{31}(a)$ and there is one less alternative.

Corollary 3.2.1. *For the self-adjoint equation* $(E_4{}^*)$,

$$\eta_1(a) = \min\{z_{31}(a), z_{22}(a)\}.$$

It is now clear how the special cases (3.2), treated by Leighton and Nehari (*70*), fit into the whole picture. Indeed, they are the mutually exclusive extreme cases where for Eq. 3.2(a): $z_{31}(a) = \infty$ and $z_{22}(a) = \eta_1(a)$, and for Eq. 3.2(b): $z_{22}(a) = \infty$ and $z_{13}(a) = \eta_1(a) = z_{31}(a)$.

They also pointed out in certain cases in which the middle term of Eq. (3.3) can be removed and we now give a simple explanation of this procedure.

Suppose that the second-order equation

$$L_2[y] = (Ry')' + Qy = 0$$

is disconjugate on $[a, b)$ and $a < b < c \leqslant \infty$. Then there exists a nonzero solution $z(x)$ of $L_2[y] = 0$ on $[a, c]$ and $L_2[y]$ can be factored as in Section I. Therefore (3.3) becomes

$$\{(1/z)[Rz^2(y'/z)']'\}' + Py = 0. \tag{3.34}$$

Let $z > 0$ and the change of variable

$$t = \int_a^x z \Leftrightarrow x = \alpha(t) \tag{3.35}$$

yields the two-term form

$$(\mathcal{R}\ddot{\mathcal{Y}})^{\cdot\cdot} + \mathcal{P}\mathcal{Y} = 0, \tag{3.34'}$$

where $\mathcal{R}(t) = (Rz^3)[\alpha(t)]$, $\mathcal{P}(t) = P[\alpha(t)]$ and $\mathcal{Y}(t) = y[\alpha(t)]$.

The author (*10, 11*), W. T. Reid (*88, 89*), D. Hinton (*52*), and Howard (*54*) established conditions for the existence of a finite $z_{22}(a)$ for the three-term equation (3.3). However, there has been very little on the existence of a finite $z_{31}(a)$ beyond the simple case (3.28b) of Leighton and Nehari. R. W. Hunt (*56, 57*) has extended much of this fourth-order theory to two-term equations of higher orders. Most of the discussions for even orders are direct generalizations of the vector-matrix formulation which is given in the next section.

IV. Self-Adjoint Fourth-Order Equations

4.1. *Vector-Matrix Formulations of Fourth-Order Scalar Equations*

With the use of a standard formulation (*18*), as applied by Sternberg (*100*) and the author (*10*) to (3.3), the self-adjoint equation

$$(E_4{}^*) \qquad L_4[y] = (D_3 y)' + q_1 D_2 y + q_4 y = 0$$

may be expressed in the vector-matrix system

$$(4.1) \qquad \begin{cases} \alpha' = A\alpha + B\hat{\alpha} \\ \hat{\alpha}' = C\alpha - A^{\dagger}\hat{\alpha}, \end{cases}$$

where $A^{\dagger}$ is the transpose of the matrix A and

$$(4.2) \qquad \alpha = \begin{pmatrix} y \\ D_1 y \end{pmatrix}, \qquad \hat{\alpha} = \begin{pmatrix} -D_3 y \\ D_2 y \end{pmatrix},$$

$$A = \begin{pmatrix} 0 & 1/r_1 \\ -q_1 & 0 \end{pmatrix}, \qquad B = \begin{pmatrix} 0 & 0 \\ 0 & 1/r_2 \end{pmatrix}, \qquad C = \begin{pmatrix} q_4 & 0 \\ 0 & -q_2 \end{pmatrix}.$$

By means of matrix integrating factors, the system (4.1) is simplified to

$$(4.3) \qquad \begin{cases} \beta' = E\hat{\beta}, \\ \hat{\beta}' = -F\beta, \end{cases}$$

where

$$(4.4) \qquad \begin{cases} \beta = T^{-1}\alpha \\ \hat{\beta} = T^{\dagger}\hat{\alpha} \end{cases}, \qquad \begin{matrix} E = T^{-1}B(T^{\dagger})^{-1} \\ F = T^{\dagger}CT \end{matrix}, \qquad \begin{matrix} T' = AT \\ \hat{T}' = -\hat{T}A^{\cdot} \end{matrix}$$

Here, it is convenient to take the interval $I = [a, \infty)$ and

$$(4.5) \qquad T = \begin{pmatrix} u & v \\ D_1 u & D_1 v \end{pmatrix}, \qquad \hat{T} = \begin{pmatrix} D_1 v & -D_1 u \\ -D_1 v & u \end{pmatrix} = (T^{-1})^{\dagger},$$

where u and v are the fundamental set of solutions of the self-adjoint second-order equation

$$(r_1y')' + q_1y = 0 \tag{4.6}$$

with $u(a) = 1$, $D_1u(a) = 0$ and $v(a) = 0$, $D_1v(a) = 1$. Therefore, the coefficients of (4.3) become

$$\begin{cases} E = (1/r_2)\,E_1\,, & E_1 = \begin{pmatrix} v^2 & -uv \\ -uv & u^2 \end{pmatrix}, \\ F = -q_4F_1 + q_2F_2\,, & F_1 = \begin{pmatrix} u^2 & uv \\ uv & v^2 \end{pmatrix}, \quad F_2 = \begin{pmatrix} (D_1u)^2 & D_1u\,D_1v \\ D_1u\,D_1v & (D_1v)^2 \end{pmatrix}. \end{cases} \tag{4.7}$$

Note that $\det(E_1) = \det(F_1) = \det(F_2) \equiv 0$ and E_1, F_1, and F_2 are symmetric and positive-semidefinite, which we write

$$E_1 \geqslant 0, \qquad F_1 \geqslant 0, \qquad F_2 \geqslant 0.$$

The special case of Eq. (3.3), where $r_1 = 1$, $r_2 = R$, $q_1 = 0$, $q_2 = Q$, $q_3 = 0$, $q_4 = P$ and $a = 0$ yields the matrix coefficients

$$E_1 = \begin{pmatrix} x^2 & -x \\ -x & 1 \end{pmatrix}, \quad F_1 = \begin{pmatrix} 1 & x \\ x & x^2 \end{pmatrix}, \quad F_2 = \begin{pmatrix} 0 & 0 \\ 0 & 1 \end{pmatrix}, \tag{4.8}$$

has already been studied (*10*). R. W. Hunt (*56*) and R. L. Sternberg (*100*) have investgated the higher-order case

$$(Ry^{(n)})^{(n)} + Py = 0. \tag{4.9}$$

W. T. Reid (*85–89*) has applied variational methods to the vector-matrix system (4.1), including complex coefficients (see Subsection 4.4).

Note that $E \geqslant 0$ and if $F \leqslant 0$, then

$$(\beta^{\dagger}\hat{\beta})' = \beta^{\dagger}(-F)\,\beta + \hat{\beta}^{\dagger}E\hat{\beta} \geqslant 0 \tag{4.10}$$

and hence

$$\beta^{\dagger}\hat{\beta} = \alpha^{\dagger}\hat{\alpha} = D_1y\,D_2y - yD_3y \tag{4.11}$$

is nonincreasing on $[a, \infty)$. Consequently, certain two-point boundary problems are impossible in this case.

Theorem 4.1. *If* $q_4 \geqslant 0$ *but* $\neq 0$ *on any subinterval of* $[a, b]$ *and*

$q_2 \leqslant 0$ on $[a, b]$ then no nontrivial solution of $(E_4{}^*)$ satisfies any of the following two-point boundary conditions:

(3.31) $y(a) = D_1y(a) = 0 = y(b) = D_1y(b),$ *"clamped ends"*;

(4.12) $y(a) = D_1y(a) = 0 = D_2y(b) = D_3y(b),$ *right "free end"*;

(4.13) $y(a) = D_2y(a) = 0 = y(b) = D_2y(b),$ *"supported ends"* [22];

(4.14) $y(a) = D_2y(a) = 0 = D_1y(b) = D_3y(b).$

Another way of stating the first conclusion of Theorem 4.1 is that $z_{22}(a) = \infty$ and one such case (3.2b) of this has been studied by Leighton and Nehari (*70*), where they showed that the conjugate points of $x = a$ for (3.2b) are the zeros of $u_3(x, a)$.

In the following sections we will establish conditions on the coefficients of (E_4) which insure that (3.31) is satisfied nontrivially, i.e., that $z_{22}(a) < \infty$.

Atkinson (*2*) has studied the boundary problem of $(E_4{}^*)$ and (4.13) by use of the vector-matrix system

$$(4.15) \quad \begin{pmatrix} y \\ D_2y \end{pmatrix}' = \begin{pmatrix} 0 & 1/r_1 \\ 1/r_1 & -q_2 \end{pmatrix}\begin{pmatrix} D_3y \\ D_1y \end{pmatrix}, \quad \begin{pmatrix} D_3y \\ D_1y \end{pmatrix}' = \begin{pmatrix} -q_4 & -q_1 \\ -q_1 & 1/r_2 \end{pmatrix}\begin{pmatrix} y \\ D_2y \end{pmatrix}$$

and the author (*9*) has reported on the special case as (4.15) applies to the two-term equation (3.2a), where he investigated relations between solutions satisfying (3.31) and those satisfying (4.13). Coppel (*21*) has considered more general boundary conditions.

4.2. *Subwronskians*

Although the second-order equation (3.23) is sometimes useful (*9, 70*) there are other relations involving 2×2 subwronskians (i.e., subdeterminants of the Wronskian w_3) whose coefficients are those of the original differential equation.

For the fundamental solutions $u_2(x, a)$ and $u_3(x, a)$ of the self-adjoint equation $(E_4{}^*)$ denote the various subwronskians by

$$(4.16) \qquad \alpha_{ij} = \begin{vmatrix} D_iu_2 & D_iu_3 \\ D_ju_2 & D_ju_3 \end{vmatrix} \qquad \text{for } i, j = 0, 1, 2, 3.$$

Note that $\alpha_{ij} = -\alpha_{ji}$, $\alpha_{ii} = 0$ and $\{u_3 ; u_2\} = 0$ is equivalent to

$$(4.17) \qquad \alpha_{03} = \alpha_{12}.$$

Successive differentiations yield the following:

Lemma 4.1. (a) $\alpha_{01}' = \alpha_{02}/r_2$;
(b) $\alpha_{02}' = 2\alpha_{12}/r_1 - q_2\alpha_{01}$;
(c) $\alpha_{12}' = \alpha_{13}/r_1 - q_1\alpha_{02}$;
(d) $\alpha_{13}' = \alpha_{23}/r_2 - 2q_1\alpha_{12} + q_4\alpha_{01}$;
(e) $\alpha_{23}' = -q_2\alpha_{13} + q_4\alpha_{02}$;

and

(f) $\alpha_{01}(a) = 0 = \alpha_{02}(a) = \alpha_{12}(a) = \alpha_{13}(a),$
$\alpha_{13}'(a) = 1/r_2(a), \qquad \alpha_{23}(a) = 1.$

An equivalent result to Lemma 4.1 is that $\sigma = \alpha_{01}$ is the unique solution of the fifth-order self-adjoint equation and initial conditions

$$(4.18) \qquad \begin{cases} L_5[\sigma] = (\mathfrak{D}_4\sigma)' + q_2\mathfrak{D}_3\sigma - q_4\mathfrak{D}_1\sigma = 0, \\ \sigma(a) = \mathfrak{D}_1\sigma(a) = \mathfrak{D}_2\sigma(a) = \mathfrak{D}_3\sigma(a) = 0, \qquad \mathfrak{D}_4\sigma(a) = 1, \end{cases}$$

where $\mathfrak{D}_1\sigma = \alpha_{02}$, $\mathfrak{D}_2\sigma = 2\alpha_{12}$, $\mathfrak{D}_3\sigma = \alpha_{13}$, and $\mathfrak{D}_4\sigma = \alpha_{23}$. Mammana (*73*) noted the fifth-order equation of (4.18) for the special case (3.3).

Lemma 4.2. *There exists a number* $\delta > 0$ *such that*

$$\alpha_{01}(x) = u_2(x, a)\, D_1u_3(x, a) - u_3(x, a)\, D_1u_2(x, a) > 0$$

in $(a, a + \delta)$, i.e., $a < z_{22}(a) \leqslant \infty$.

A routine calculation of the Lagrange Formula for $L_5[\sigma]$ shows that it is self-adjoint. Next we note a *nonlinear* identity among the subwronskians α_{ij} , which is easily verified.

Lemma 4.3. $\alpha_{02}\alpha_{13} = \alpha_{12}^2 + \alpha_{23}\alpha_{01}$.

The following Riccati-type equations are also useful.

Lemma 4.4. (a) *If* $\alpha_{13} \neq 0$ *on an interval* I *then* $h = -\alpha_{23}/\alpha_{13}$ *satisfies*

$$h' = q_2 + \frac{1}{r_2} h^2 - q_4 \left(\frac{\alpha_{12}}{\alpha_{13}}\right)^2 - 2q_1 \left(\frac{\alpha_{12}}{\alpha_{13}}\right) h.$$

(b) *If* $\alpha_{23} \neq 0$ *on* I *then* $k = \alpha_{13}/\alpha_{23}$ *satisfies*

$$k' = \frac{1}{r_2} + q_2k^2 - q_4 \left(\frac{\alpha_{12}}{\alpha_{23}}\right)^2 - 2q_1 \left(\frac{\alpha_{12}}{\alpha_{23}}\right).$$

(c) *If* $\alpha_{01} \neq 0$ *on I then* $m = \alpha_{13}/\alpha_{01}$ *satisfies*

$$m' = q_4 - \frac{1}{r_2}\left(\frac{\alpha_{12}}{\alpha_{02}}\right)^2 - 2q_1\left(\frac{\alpha_{12}}{\alpha_{01}}\right).$$

We now summarize the results obtained by the author (*12*) for the special case (3.3), i.e., where

$$r_1 = 1, \qquad r_2 = R, \qquad q_1 = 0, \qquad q_2 = Q, \qquad \text{and} \qquad q_4 = P,$$

by use of Lemma 4.1–4.4.

Recall that the first zero on (a, ∞) of α_{01} is $z_{22}(a)$ and designate the first zeros on (a, ∞) of the subwronskians α_{23} and α_{13} as $\mu_1(a)$ and $\xi_1(a)$, respectively.

Theorem 4.2. *For Eq.* (3.3) *in* $[a, \infty)$:

(a) *If* $z_{22}(a) < \infty$ *then* $a < \mu_1(a) < \xi_1(a) < z_{22}(a)$.

(b) *If* $P \leqslant 0$ *and the second-order equation* $(Ry')' + Qy = 0$ *has a nontrivial solution with three zeros,* $x_1 < x_2 < x_3$, *then* $z_{22}(a) \leqslant x_3$.

(c) *If* $P \leqslant 0$ *and the second-order equation* $(Ry')' + Qy = 0$ *has a nontrivial solution satisfying* $y(a) = y'(b) = 0$, $a < b < \infty$, *then* $\mu_1(a) \leqslant b$.

(d) *If* $P \leqslant 0$ *and* $Q \geqslant 0$, *but* $|P| + Q \not\equiv 0$ *for large x, and* $\int^{\infty} (1/R) = \infty$ *then in order for* $z_{22}(a) < \infty$ *it is necessary and sufficient that* $\mu_1(a) < \infty$ *or* $\xi_1(a) < \infty$.

4.3. *Matrix Riccati Equations*

Here we apply the arguments, which established certain oscillation results for scalar equations in Section I, to the matrix system

$$\begin{cases} Y' = E\hat{Y}, \\ \hat{Y}' = -FY, \end{cases} \tag{4.19}$$

where E and F are the coefficient matrices (4.7) of the vector-matrix equation (4.3). Note that $\det T(x) \equiv 1$ and

$$Y = T^{-1}\begin{pmatrix} u_2 & u_3 \\ D_1u_2 & D_2u_3 \end{pmatrix}, \qquad \hat{Y} = T^{\dagger}\begin{pmatrix} -D_2u_2 & -D_3u_3 \\ D_2u_2 & D_2u_3 \end{pmatrix}$$

is a solution pair of (4.19) satisfying the initial conditions

$$Y(a) = 0, \qquad \hat{Y}(a) = 2\begin{pmatrix} 0 & -1 \\ 1 & 0 \end{pmatrix}. \tag{4.20}$$

Howard (*55*) has established a theorem involving the minimum eigenvalue of $\int^x F$. Reduced to the special case (4.3) his theorem becomes

Theorem H. If (i) min e.v.$(\int^x F) \to \infty$ *as* $x \to \infty$ *and* (ii) $E(x) \geqslant e(x)I$, $e \geqslant 0$ *with* $\int^\infty e = \infty$, *then* $z_{22}(a) < \infty$.

However, min e.v. $E(x) \equiv 0$ and, hence, no such $e(x)$ exists for (4.3).

Since det $\hat{Y}(a) = 1 > 0$, then $\hat{Y}^{-1}(x)$ exists on $[a, b)$ for some $b \in (a, \infty)$. Let $K(x) = Y(x)\hat{Y}^{-1}(x)$, then

$$K' = E + KFK \quad \text{on} \quad [a, b); \qquad K(a) = 0. \tag{4.21}$$

Because of the fact that $K(x)$ is symmetric at one point, $x = a$, it is easy to show (*9, 82*) that $K(x)$ is symmetric on $[a, b)$.

By Lemma 4.2, det $Y(x) > 0$ $(a, a + \delta)$ for some $\delta > 0$. Suppose that det $Y(x) > 0$ on (a, ∞), i.e., that

$$z_{22}(a) = \infty. \tag{4.22}$$

Then $H(x) = -\hat{Y}(x)\, Y^{-1}(x)$ satisfies

$$H' = F + HEH \quad \text{on} \quad (a, \infty), \tag{4.23}$$

and is symmetric, since $H(x) = -K^{-1}(x)$ near $x = a$.

Assume Howard's property D (*55*), i.e., that

$$\text{min e.v.} \left(\int^x F\right) = \inf_{|\xi|=1} \xi^\dagger \left(\int^x F\right) \xi \to \infty, \quad \text{as} \quad x \to \infty. \tag{4.24}$$

Let $b \in (a, \infty)$; there exists a number $c \in (b, \infty)$ such that

$$H(x) > H(b) + \int_b^x F > 1_2 \qquad \text{for} \quad a < b \leqslant c \leqslant x < \infty, \tag{4.25}$$

where the inequality is in the "positive-definite" sense, i.e.,

$$A > B \Leftrightarrow \xi^\dagger(A - B)\,\xi > 0$$

for every nonzero vector ξ and where 1_n is the $n \times n$ identity. We now parallel Coles' proof of Theorem 1.4.

Let $G(x) = 1_2 + \int_b^x (HEH)$; then $H(x) > G(x) > 0$ and $G'(x) = HEH$. However, it does not follow that $HEH \geqslant GEG$. In fact, it does not follow that $H^2 \geqslant G^2$, unless H and G commute (*15*). In spite of this difficulty, we have $-(G^{-1})' = G^{-1}G'G^{-1} = G^{-1}HEHG^{-1}$ and

$$(4.26) \qquad 0 \leqslant \int_b^\infty (G^{-1}HEHG^{-1}) \leqslant G^{-1}(b) = 1_2 .$$

Taking the trace of (4.25), since $E = \mathcal{E}^\dagger\mathcal{E}$ where

$$\mathcal{E} = \begin{pmatrix} x & -1 \\ 0 & 0 \end{pmatrix},$$

we have

$$(4.27) \qquad 2 \geqslant \int_b^\infty \operatorname{tr}(G^{-1}HEHG^{-1}) \geqslant \int_b^\infty \| \mathcal{E}HG^{-1} \|^2 \geqslant \int_b^\infty \| G^{-1}HEHG^{-1} \|.$$

Also, $\| G^{-1}HEHG^{-1} \| \geqslant \| E \| / \| H^{-1}G \|^2$, where the Euclidean norm

$$\| A \| = \Big(\sum_{i,j} a_{ij}^2\Big)^{1/2} \qquad \text{for} \quad A = (a_{ij}).$$

Dr. A. S. Householder suggested the following argument. Since $H^{-1} < G^{-1}$ on $[c, \infty)$, there is a square root $G^{1/2}$, i.e., $G = (G^{1/2})^2$, then $G^{1/2}H^{-1}G^{1/2} < 1_2$. Since $G^{1/2}H^{-1}G^{1/2}$ and $H^{-1}G$ have the same eigenvalues then $\| H^{-1}G \| < 2$. Consequently, (4.27) implies that

$$(4.28) \qquad \left\| \int_c^x E \right\| \leqslant \int_c^\infty \| E \| < \infty.$$

If we violate (4.28) by assuming (4.24) then $z_{22}(a) < \infty$.

Theorem 4.3. *If u and v are solutions of* (4.6) *such that*

$$\text{(i)} \qquad \text{min e.v.} \int_a^x \begin{pmatrix} -q_4u^2 & -q_4uv \\ -q_4uv & q_2 - q_4v^2 \end{pmatrix} \to \infty \quad \textit{as} \quad x \to \infty$$

and

$$\text{(ii)} \qquad \int_a^\infty \frac{u^2 + v^2}{r^2} = \infty,$$

then $z_{22}(a) < \infty$ for equation (E_4).*

Instead of violating (4.28) by condition (ii) we might allow the determinant of $\int_a^x E$ to become unbounded.

Let $D(x) = \det(\int_a^x E)$; then $D' = \theta/r$ where

$$\theta(x) = \int_a^x \frac{[u(x)\,v(t) - v(x)\,u(t)]^2}{r(t)}\,dt = \int_a^x \frac{v_1{}^2(x, t)}{r(t)}\,dt$$

and $v_1(x, t)$ is the unique solution of

$$(4.30) \qquad (r_1 y')' + q_1 y = 0, \qquad y(t) = 0, \qquad (r_1 y')(t) = 1.$$

Theorem 4.4. *The condition* (ii) *in Theorem 4.3 may be replaced by*

$$\text{(iii)} \quad \det\left(\int_a^x E\right) = \int_a^x \frac{1}{r_2(t)}\left(\int_a^t \frac{v_1{}^2(s, t)}{r_2(s)}\,ds\right)dt \to \infty, \quad \textit{as} \quad x \to \infty.$$

For the special case (3.3), $v_1(x, t) = x - t$. Furthermore, if

$$p(x) = -P(x) \geqslant 0$$

and $Q(x) \leqslant 0$ then conditions for (i) are easily established.

Lemma 4.6. *If* $\int_a^\infty p = \infty$ *and* $p(x) \geqslant 0$ *on* $[a, \infty)$ *then*

$$\text{min. e.v.} \int_a^x p(t)\begin{pmatrix} 1 & t \\ t & t^2 \end{pmatrix} dt \to \infty, \quad \textit{as} \quad x \to \infty.$$

Corollary 4.4.1. *For Eq.* (3.3) *if* (i) $P \leqslant 0$ *and* $Q \geqslant 0$,

$$\text{(ii)} \qquad \int_a^\infty |P| = \infty$$

and

$$\text{(ii)} \qquad \int_a^\infty \frac{1}{r(t)} \int_a^t \frac{(s - t)^2}{r(s)}\,ds\,dt = \infty,$$

then $z_{22}(a) < \infty$.

This is an improvement over Theorem 3.6 of (*10*), since

$$(4.31) \qquad \left(\int_a^t \frac{(s - t)}{r(s)}\,dt\right)^2 \leqslant \int_a^t \frac{(s - t)^2}{r(s)}\,ds \cdot \int_a^t \frac{ds}{r(s)},$$

where, instead of (iii), it was assumed that $\int^{\infty} s^2/r(s)\, ds = \infty$. Dual theorems are obtained by interchanging the roles of E and F.

Theorem 4.5. *For Eq.* (3.3), *if* (i) $P \leqslant 0$ *and* $Q \geqslant 0$,

$$\text{(ii)} \qquad \int_a^{\infty} (1/R) = \infty$$

and

$$\text{(iii)} \qquad \textit{either} \quad \int_a^{\infty} Q = \infty \quad \textit{or} \quad \int_a^{\infty} P(t) \int_a^t P(s)(s-t)^2\, ds = \infty,$$

then $z_{22}(a) < \infty$.

Because of an inequality like (4.31), the second alternative of (iii) is satisfied if

$$\int_a^{\infty} P(t) \left[\int_a^t (t-s)\, P(s)\, ds \right]^2 dt = \infty \quad \text{or} \quad \int_a^{\infty} t^2 P(t)\, dt = \infty,$$

which were the best conditions given previously (*10*). Hunt's oscillation theorem (*56*) for

$$(4.32) \qquad [R(x)\, y^{(n)}]^{(n)} + (-1)^{n+1} P(x) y = 0, \qquad R(x) > 0, \qquad P(x) > 0$$

as given below can be established by paralleling the proof of Theorem 4.3.

Theorem 4.6. *If* $\int^{\infty} P(x)(\overset{x}{I^n} P)^2\, dx = \infty$, *where* $\overset{x}{I^n} P$ *is an* nth-*iterated integral of* $P(x)$, *and* $\int^{\infty} (1/R) = \infty$, *then there exists a nontrivial solution of* (4.32) *with a pair of n-fold zeros on* $[c, \infty)$ *for each* $c \in [a, \infty)$.

Hunt (*57*) later proved a stronger result.

Theorem 4.7. *Suppose that*

$$\int^{\infty} [\overset{x}{I^n}(t^{n-2}/R(t))]\, P(x)\, dx = \infty \qquad \textit{and} \qquad \int^{\infty} [(\overset{x}{I^n} P)/R(x)]\, dx = \infty,$$

then there exists a set of 2n linearly independent solutions, with infinitely many zeros on $[a, \infty)$, *of*

$$[R(x)\, y^{(n)}]^{(n)} + P(x)\, y = 0, \qquad n > 1,$$

where $R(x) > 0$, $P(x) \neq 0$ *and* R & $P \in C[a, \infty)$.

It is not known whether the conclusion of Theorem 4.7 can be extended to have all solutions oscillatory.

4.4. *Quadratic Functionals and Wirtinger Inequalities*

For the self-adjoint equation

$$(3.3) \qquad \mathfrak{L}_4[y] = [(Ry'')' + Qy']' + Py = 0; \qquad R > 0, \quad R, Q \,\&\, P \in C[a, \infty),$$

it is a routine matter to calculate that

$$(4.32) \qquad I_4[y; a, b] \equiv \int_a^b [R(y'')^2 - Q(y')^2 + Py^2] - \int_a^b y\mathfrak{L}_4[y] + [y\mathfrak{D}_3 y - Ry'y'']_a^b,$$

where $\mathfrak{D}_3 y = (Ry'')' + Qy'$.

On the other hand if we emply a method of Coles (*19*) and integrate-by-parts several times the right-hand side of

$$\int_a^b Pw^2 = -\int_a^b w^2[\mathfrak{D}_3 y]'/y,$$

we obtain on $[c, b] \subset [a, \infty)$,

$$(4.33) \qquad I_4[w; c, b] = -[[\mathfrak{D}_3 y]\, w^2/y - Ry''(w')^2/y']_c^b - \int_c^b \frac{y^2}{y'}\left[w' - \frac{y'}{y}w\right]^2 + \int_c^b r\left[w'' - \frac{y''}{y'}w'\right]^2,$$

where $\mathfrak{L}_4[y] = 0$, $y \neq 0$ and $y' \neq 0$ on $[c, b]$ and $w(x)$ is any $\mathfrak{L}_4$-admissible function on $[c, b]$.

By making use of the derivatives and differences of the quotients

(4.34)

$$\lambda_0 = u_3(x, a)/u_2(x, a), \qquad \lambda_1 = u_3'/u_2', \qquad \lambda_2 = u_3''/u_2'', \qquad \lambda_3 = \frac{(Ru_3'')' + Qu_3'}{(Ru_2'')' + Qu_2'}$$

the existence of an appropriate solution $y(x)$ for (4.33) can be established (*11*).

Lemma 4.7. *If $P(x) \leqslant 0$ on $[a, \infty)$, $a < b < \infty$ and $\mu_1(a) = \infty$ then there exists a solution $y(x)$ of (3.3) such that*

$$y(a) = y'(a) = 0, \qquad (Ry'')(b) = 0, \qquad \mathfrak{D}_3 y(b) = [(Ry'')' + Qy'](b) < 0,$$

$$y > 0, \qquad y' > 0, \qquad (Ry'')' + Qy' < 0 \quad on \quad (a, b).$$

Furthermore, $Ry'' > 0$ if $Q \geqslant 0$ on (a, b).

Using the solution guaranteed in Lemma 4.7 in (4.33) we have an analog to the second-order focal point criterion in Theorem 1.9 (*11*).

Theorem 4.8. *If $P(x) \leqslant 0$ on $[a, \infty)$ then in order for $\mu_1(a) = \infty$ it is necessary and sufficient that for each $b \in (a, \infty)$ and each nontrivial admissible function $w(x)$ on $[a, b]$ for which $w(a) = w'(a) = 0$, the zero of w' being of order $> \frac{1}{2}$, it is true that*

$$I_4[w; a, b] = \int_a^b [r(w'')^2 - Q(w')^2] > 0.$$

Recall that Theorem 4.2(d) shows that, under further conditions, $\mu_1(a) = \infty$ if, and only if, $z_{22}(a) = \infty$. Various forms of $I_4[w; a, b] > 0$ are also known as Wirtinger inequalities (*19*).

Corollary 4.8.1. *If $Q(x) \geqslant 0$, $P(x) \leqslant 0$ and $\int_a^\infty (1/R) = \infty$ then Eq. (3.3) has a nontrivial solution with a pair of double zeros, i.e., $z_{22}(a) < \infty$, if, and only if,*

$$I_4[w; a, b] > 0, \qquad a < b < \infty,$$

for each admissible function $w(x)$ of Theorem 4.6.

W. T. Reid (*83–89*) proved Theorem 1.11 for vector-matrix systems, which include (4.1), and his results will be stated here only in terms of the special case (4.1).

Lemma 4.8. *For Eq. (3.3), $z_{22}(a) = \infty$ if, and only if, for each $[c, d] \subset [a, \infty)$,*

$$I[\eta, \xi; c, d] = \int_c^d [\xi^\dagger B \xi + \eta^\dagger C \eta] \geqslant 0$$

for all vector functions η, ξ with $\eta(x)$ absolutely continuous on $[c, d]$ and $\xi(x)$ Lebesgue-measurable and essentially bounded on $[c, d]$ with

$\eta(c) = 0 = \eta(d)$, *and equality only if* $B\xi = 0$ *almost everywhere and* $\eta(x) \equiv 0$ *on* $[c, d]$.

Hinton (*52*) recently made use of the following corollary of Reid's criterion.

Theorem 4.9. *Equation* (3.3) *has no nontrivial solution with a pair of double zeros on* $[a, b]$, $a < b < \infty$ *if, and only if,*

$$I_4[w; a, b] > 0$$

for every nontrivial function $w \in C''[a, b]$, *where* $(Rw'')'$ *is piecewise continuous on* $[a, b]$ *satisfying*

$$w(a) = w'(a) = 0 = w(b) = w'(b).$$

Following Leighton's second-order results (*69*), Hinton established a comparison theorem for fourth-order equations (E_4).

Theorem 4.10. *If* (i) *there is a nontrivial solution of* (E_4) *with two double zeros on* $[a, b]$, $a < b < \infty$,

(ii) $\hat{L}_4$ *is the same operator as* L_4, *except that* r_2, q_2 *and* q_3 *are replaced by* $\hat{r}_2$, $\hat{q}_2$ *and* $\hat{q}_3 \in C[a, b]$, *respectively and*

(iii) *there is a function* $w(x)$ *satisfying the conditions for* w *of Theorem* 4.9 *such that*

$$\int_a^b \{(r_2 - \hat{r}_2)[D_2 y]^2 - (q_2 - \hat{q}_2)(D_1 y)^2 + (q_3 - \hat{q}_3)\, y^2\}$$

$$\geqslant \int_a^b \{r_2[(D_2 y)]^2 - q_2(D_1 y)^2 + q_3 y^2\},$$

then $\hat{L}_4[y] = 0$ *has a nontrivial solution with a pair of double zeros on* $[a, b]$.

Using the preceding comparison theorem, Hinton proved several "oscillation" theorems for Eq. (3.3).

Theorem 4.11. *If* $h(x) \geqslant 0$ *and* $\in C[a, \infty)$ *such that*

$$\int_a^\infty xh(x)\, dx = \infty$$

and $\liminf J(t) = -\infty$ *as* $t \to \infty$, *where*

$$J(t) = \int_a^t \Big\{ R(x)\, h^2(x) - Q(x) \Big[\int_x^t h\Big]^2 + P(x) \Big[\int_x^t (s-x)\, h(s)\, ds\Big]^2 \Big\}$$

$$\times\, ds \Big/ \Big[\int_a^t x h(x)\, dx\Big]^2,$$

then there is a number $b \in (a, \infty)$ *such that* (3.3), $\mathfrak{L}_4[y] = 0$, *has a non-trivial solution with two double zeros on* $[a, b]$.

Corollary 4.9.1. *If, as* $t \to \infty$,

$$\liminf t^{-4} \int_a^t \{R(x) - (t-x)^2 Q(x) + P(x)(t-x)^4/4\}\, dx = -\infty,$$

then the conclusion of Theorem 4.11 holds.

Corollary 4.9.2. *If* $\int_a^\infty [x/R(x)]\, dx = \infty$ *and, as* $t \to \infty$,

$$\liminf \int_a^t \Big\{ P(x) \Big[\int_x^t \frac{s-x}{R(s)}\, ds\Big]^2 - Q(x) \Big[\int_x^t \frac{as}{R(s)}\Big]^2 \Big\}\, dx \Big/ \Big[\int_a^t \frac{s}{R(s)}\, ds\Big]^2 = -\infty,$$

then the conclusion of Theorem 4.11 *holds.*

In (*53*), Hinton has extended his results to

$$(Ry^{(n)})^{(n)} + (-1)^{n-1} Py = 0.$$

Reid (*88*) has developed the quadratic functional criterion for complex self-adjoint quasi-differential equations of order $2n$,

$$D^{\langle 2n\rangle} y = 0, \tag{1.20}$$

where

$$D^{\langle k\rangle} = D^k,\ k = 0, 1, \ldots, n-1;$$

$$D^{\langle n\rangle} = p_{2n} D^n + i p_{2n-1} D^{n-1};$$

$$D^{\langle n+j\rangle} = DD^{\langle n+j-1\rangle} - (-1)^j [i p_{2n-2j+1} - p_{2n-2j} D^{n-j} - i p_{2n-2i-1} D^{n-j-1}],$$

$$j = 1, \ldots, n-1;$$

$$D^{\langle 2n\rangle} = DD^{\langle 2n-1\rangle} - (-1)^n [i p_1 - p_0 D^0];$$

and

$$p_j \in C^{[j/2]}[a, b] \qquad \text{for} \quad j = 0, 1, \ldots, 2n.$$

An example of Reid's results follows.

Theorem 4.12. *If* $p_{2n-1}(x) \equiv 0$, $\int_c^\infty (1/p_{2n}) = \infty$,

$$J(x; \pi) \equiv \sum_{\alpha=1}^{n} p_{2\alpha-2}(x)|\,\pi_\alpha\,|^2 - i \sum_{\alpha=1}^{n-1} p_{2\alpha-1}(x)[\bar{\pi}_\alpha \pi_{\alpha+1} - \bar{\pi}_{\alpha+1}\pi_\alpha] \leqslant 0$$

for all complex n-vectors $\pi = (\pi_i)$, *and* $D^{\langle 2n \rangle} y = 0$ *has no nontrivial solution with a pair of n-fold zeros on* $[c, \infty)$ *for c sufficiently large, then each of the integrals*

$$\int^\infty p_{2\alpha}(t)\, t^{2n-2\alpha-2}\, dt, \qquad \alpha = 0, 1, \ldots, n-1;$$

is convergent.

Since Reid (*90*) has given an elegant summary of variational results for self-adjoint systems, we will not repeat more of his many results here. Also, we have concentrated on nonvariational methods in hopes that our methods may also apply to nonself-adjoint equations.

V. Second-Order Matrix Equations and Related First-Order Systems

5.1. *Examples*

In Section 4.3 we studied a particular example (4.19) of the matrix differential system

$$Y' = E(x)\hat{Y}, \qquad \hat{Y}' = -F(x)Y, \tag{5.1}$$

where E and F are symmetric square matrices of continuous functions on an interval I. Another example is the matrix equation having the coefficients of (4.15), i.e.,

$$E = \begin{pmatrix} 0 & 1/r_1 \\ 1/r_1 & q_1 \end{pmatrix}, \qquad F = \begin{pmatrix} -q_4 & -q_1 \\ -q_1 & 1/r_2 \end{pmatrix}, \tag{5.2}$$

which was utilized by Atkinson (*2*) and Hartman and Wintner (*48*), as a formulation of the special fourth-order scalar self-adjoint equation (3.3).

The *second-order* scalar *complex* differential equation of Subsection 1.7,

$$(5.3) \qquad (ry')' + qy = 0; \qquad r = r_1 + ir_2 \neq 0, \quad q = q_1 + iq_2, \quad r_i \,\&\, q_i \in C(I),$$

is equivalent to the system (5.1) where

$$(5.3') \qquad E = (1/|r|^2)\begin{pmatrix} -r_2 & r_1 \\ r_1 & r_2 \end{pmatrix}, \qquad F = \begin{pmatrix} -q_2 & q_1 \\ q_1 & q_2 \end{pmatrix}.$$

For oscillation theory of (5.3) see (*7*).

The second-order matrix differential equation

$$(5.4) \qquad (R(x)Y')' + Q(x)Y = 0,$$

where R and Q are symmetric matrices of continuous functions on an interval I and $R > 0$ (positive-definite) on I, has been the subject of numerous investigations (*5*, *6*, *55*, *87*). Of course, if $\hat{Y} = RY'$ then (5.4) is equivalent to (5.1) with

$$(5.4') \qquad E = R^{-1}, \qquad F = Q.$$

Note, also, that (5.3) is equivalent to (5.4) with

$$R = \begin{pmatrix} -r_2 & r_1 \\ r_1 & r_2 \end{pmatrix}, \qquad Q = \begin{pmatrix} -q_2 & q_1 \\ q_1 & q_2 \end{pmatrix}.$$

5.2. *Basic Properties of Solutions*

A matrix function $Y(x)$ is said to be a *solution* of (5.1) if there exists a corresponding matrix function $\hat{Y}(x)$ such that $(Y, \hat{Y})$ is a solution pair of the system (5.1).

Let Y_1 and Y_2 be two such matrix solutions of (5.1) on any interval I and define

$$(5.5) \qquad W[Y_1, Y_2] = Y_1^{\dagger}\hat{Y}_2 - \hat{Y}_1^{\dagger}Y_2.$$

This matrix functional reduces to the usual Wronskian for the scalar ($n = 1$) case and, in general, has analogous properties. One property which *does not hold* in general is the identical vanishing of $W[Y, Y]$. In fact the most that can be said is that $W[Y, Y]$ is *skew-symmetric*. In case $W[Y, Y] = 0$ then the column vectors of the matrix solution are said to be *conjoined* (*85*). This terminology was introduced and the following analogs of scalar properties were first proved by Reid and others who were working in the Calculus of variations in the 1930's (*17*, *74*). The author listed these properties later (*5*).

Theorem 5.1. *Let* $U(x)$ *be a matrix solution of* (5.1).

(a) *If* A *is a constant matrix then* UA *is a solution of* (5.1) *such that*

$$W[UA, UA] = 0.$$

(b) *If* $\det U(x) \neq 0$ *on an interval* I *and* $a \in I$ *then*

$$V(x) = U(x) \int_a^x [U^{-1}E(U^{-1})^\dagger], \qquad x \in I,$$

is also a solution of (5.1) *such that*

$$W[V, V] = 0 \qquad \textit{and} \qquad W[U, V] = 1_n .$$

(c) *If* $V(x)$ *is a solution of* (5.1) *such that* $W[V, V] = 0$ *and* $\det(W[U, V]) \neq 0$ *then every solution of* (5.1) *is given by*

$$Y = UA + VB,$$

where A *and* B *are arbitrary constant matrices.*

Such a pair of solutions U, V can always be obtained by specifying the initial conditions

$$U(a) = 1_n , \qquad \hat{U}(a) = 0 \qquad \text{and} \qquad V(a) = 0, \qquad \hat{V}(a) = I_n . \tag{5.6}$$

Recall the fact that if $Y(x)$ is a matrix solution of (5.1) and γ is a constant vector then $\beta = Y\gamma$ is a vector solution of the related vector-matrix equation

$$\beta' = E\hat{\beta}, \qquad \hat{\beta}' = -F\beta. \tag{5.1$'$}$$

Conversely any vector solution β of (5.1′) forms a column of some matrix solution $Y(x)$ of (5.1).

Corollary 5.1.1. *Under the hypotheses of Theorem* 5.1(c), *every vector solution of* (5.1′) *is given by*

$$\beta = U\gamma + V\delta,$$

where γ *and* δ *are arbitrary constant vectors.*

Corollary 5.1.2. *Let $V(x)$ be a (matrix) solution of* (5.1) *on an interval I such that $a \in I$, $V(a) = 0$ and det$(\hat{V}(a)) \neq 0$, then in order for there to exist a number $b \in I$, $b \neq a$, and a nontrivial vector solution β of* (5.1′) *such that*

$$\beta(a) = \beta(b) = 0 \tag{5.7}$$

it is necessary and sufficient that $\det V(b) = 0$.

Therefore, singularities of matrix solutions of (5.1) determine conjugate points [i.e., (5.7) holds] of vector solutions of (5.1′).

5.3. *Matrix Trigonometric Functions*

An interesting special case of (5.1) is that when $E = F$. For each symmetric $n \times n$ matrix $Q(x)$ of continuous functions on $[a, \infty)$ define

$$S = S[x, a; Q], \qquad C = C[x, a; Q] \tag{5.8}$$

to be the solution pair of

$$\begin{cases} S' = QC, & S(a) = 0, \\ C' = -QS, & C(a) = I. \end{cases} \tag{5.9}$$

For the scalar ($n = 1$) case $S = \sin(\int_a^x Q)$ and $C = \cos(\int_a^x Q)$, hence, the matrices (5.8) are said to be *matrix sines and cosines*. Note that if $\det Q \neq 0$ on an interval I then S and C are solutions of the second-order matrix equation

$$(Q^{-1}Y')' + QY = 0 \quad \text{on} \quad I. \tag{5.9′}$$

Furthermore, $W[S, S] = W[C, C] = 0$ and $W[C, S] = 1_n$. The matrix functions (5.5) were introduced by the author (*6*) and were studied further by Reid (*87*) and Etgen (*25–27*). These and other easily-proved results imply certain expected "trigonometric" identities.

Theorem 5.2.

$$\text{(a)} \begin{cases} C^\dagger C + S^\dagger S = 1_n, \\ C^\dagger S = S^\dagger C, \end{cases} \qquad \text{(b)} \begin{cases} CC^\dagger + SS^\dagger = 1_n, \\ CS^\dagger = SC^\dagger. \end{cases}$$

Note that Theorem 5.2 is equivalent to saying that the $2n \times 2n$ matrix

$$\begin{pmatrix} C & -S \\ S & C \end{pmatrix}$$

is orthogonal.

Etgen (*25*) has listed a number of other properties of S and C.

Theorem 5.3. (a) *If* $b \in (a, \infty)$ *such that* $\det(C[b, a; Q]) = 0$ *and* $\det(C[x, a; Q]) > 0$ *on* $[a, b)$ *then* $\det(S[x, a; Q] \neq 0$ *on* (a, b).

(b) *If* $\det(C[x, a; Q]) \neq 0$ *on* $[b, c] \subset [a, \infty)$ *then* $\det(S[x, a; Q]$ *has at most n zeros in* $[b, c]$.

(c) *If* $\det(S[x, a; Q]) \neq 0$ *on* $[b, c] \subset [a, \infty)$ *then* $\det(C[x, a; Q])$ *has at most n zeros in* $[b, c]$.

He also gives an interesting simple example for which

$$\det(C[x, a; 0]) = 0 = \det(S[x, a; Q]$$

for infinitely many values of $x \in [a, \infty)$; i.e.,

$$Q = \begin{pmatrix} \pi & 0 \\ 0 & \pi/2 \end{pmatrix}, \qquad a = 0, \qquad S = \begin{pmatrix} \sin \pi x & 0 \\ 0 & \sin(\pi/2)x \end{pmatrix},$$

$$C = \begin{pmatrix} \cos \pi x & 0 \\ 0 & \cos(\pi/2)x \end{pmatrix}.$$

However, it should be noted that if γ is a constant vector such that, for some x, $S\gamma = C\gamma = 0$ then $\gamma = 0$. By means of a discussion similar to that of Section 4.5, the author (*6*) first proved the following.

Theorem 5.4. (a) *If* $Q(x) > 0$ *on* $[a, \infty)$ *and* $\int^{\infty} \operatorname{tr} Q = \infty$ *then* $\det(S[x, a, Q])$ *has at least one zero on* (a, ∞).

(b) *If* $Q(x) > 0$ on $[a, \infty)$ *and*

$$\int_a^{\infty} \operatorname{tr} Q < \pi/2 \sqrt{n}$$

then $\det(S[x, a; Q]) \neq 0$ *and* $\det(C[x, a; Q]) \neq 0$ *on* (a, ∞).

Reid (*87*) improved (b) by replacing $\pi/2 \sqrt{n}$ by $\pi/2$ and showed the following.

Theorem 5.5. *If $Q(x) > 0$ on $[a, \infty)$ then there exists a number $c \in (a, \infty)$ such that no nontrivial vector solution β of the system*

$$\beta' = Q\hat{\beta}, \qquad \hat{\beta}' = -Q\beta \tag{5.9'}$$

has two zeros on $[c, \infty)$ if, and only if, $\int_a^\infty \operatorname{tr} Q = \infty$.

Later, Etgen [*25*] proved a slightly different version of Theorem 5.5 and, subsequently [*26*], established a "double angle" formula.

Theorem 5.6. *If Y, Z is the solution pair of matrices of*

$$Y' = Q'Z + ZQ, \qquad Z' = -QY - YQ$$

and $Y(a) = 0$, $Z(a) = I$ then Y and Z are symmetric and

$$Y = 2SC^{\dagger}, \qquad Z = CC^{\dagger} - SS^{\dagger}.$$

Finally, let us note [*6*] that S and C satisfy a Lipschitz condition with respect to Q.

Theorem 5.7. *If for $i = 1$ and 2, $Q_i(x)$ is a symmetric matrix of continuous functions on $[a, \infty)$ with $S_i = S[a, x; Q_i]$ and $C_i = C[a, x; Q_i]$ then there exists a positive number M such that*

$$\left.\begin{matrix} \| S_2 - S_1 \| \\ \| C_2 - C_1 \| \end{matrix}\right\} \leqslant M \int_a^x \| Q_2 - Q_1 \|.$$

5.4. *A Matrix Prüfer Transformation*

In 1957 the first Prüfer transformation for second-order matrix differential equations (5.4) was published (*6*) and shortly thereafter Reid (*87*) gave a more general development, which will be given here for the real case.

Let $Y(x)$ be a matrix solution of (5.1); i.e., there is a companion matrix $\hat{Y}(x)$ such that

$$Y' = E(x)\,\hat{Y}, \qquad \hat{Y}' = -F(x)Y, \tag{5.1}$$

where E and F are symmetric (square) matrices of continuous functions on an interval I, and $Y(a) = 0$, $\hat{Y}(a) = 1_n$.
Suppose that there exist continuous matrix functions $P(x)$ and $Q(x)$ such that

$$Y(x) = S^{\dagger}[x; a, Q]\, P(x) \quad \text{and} \quad \hat{Y}(x) = C^{\dagger}[x; a, Q]\, P(x). \tag{5.10}$$

Because of the identities in Theorem 5.2,

$$P^{\dagger}P = Y^{\dagger}Y + \hat{Y}^{\dagger}\hat{Y} > 0 \tag{5.11}$$

and $P(x)$ is nonsingular on I.

Differentiation of (5.10) yields

$$\begin{cases} S^{\dagger}P' + S^{\dagger}QP = EC^{\dagger}P, \\ C^{\dagger}P' - S^{\dagger}QP = -FS^{\dagger}P \end{cases} \tag{5.10'}$$

and Theorem 5.2 allows us to solve for P' and Q, so that

$$\begin{cases} P' = [SEC^{\dagger} - CFS^{\dagger}]P, \\ Q = CEC^{\dagger} + SFS^{\dagger}. \end{cases} \tag{5.12}$$

Conversely, suppose that $P(x)$ satisfies (5.11) and

$$P^{\dagger}P' = Y^{\dagger}E\hat{Y} - \hat{Y}^{\dagger}FY \tag{5.13}$$

and $Q(x)$ satisfies

$$P^{\dagger}QP = \hat{Y}^{\dagger}E\hat{Y} + Y^{\dagger}FY; \tag{5.14}$$

then it is a routine matter to check that (5.10) defines the solution of (5.1) which also satisfies the initial conditions (5.1).

Thus Reid reduced the problem to finding a solution $P(x)$ of (5.11) and (5.13). Let

$$M(x) = Y^{\dagger}(x)\, Y(x) + \hat{Y}^{\dagger}(x)\, \hat{Y}(x) \quad \{>0 \text{ on } [a, \infty)\},$$

$b \in (a, \infty)$ and k be a positive number such that

$$0 < k^2 M(x) < 1_n \quad \text{on} \quad [a, b].$$

Then $U(x) = 1_n - k^2 M(x)$ satisfies

$$0 < U(x) < 1_n \quad \text{on} \quad [a, b].$$

Following Reid, let

$$V(x) = 1_n - \sum_{k=1}^{\infty} c_k U^k(x), \qquad \text{where} \quad (1 - x)^{1/2} = 1 - \sum_{k=1}^{\infty} c_k x^k,$$

and it follows that

(i) $V^2(x) = 1_n - U(x)$,

(ii) $V(x) > 0$,

and

(iii) $V(x) \in C'[a, b]$.

If $N(x) = (1/k)\, V(x)$ then $N^2(x) = M(x)$ on $[a, \infty)$. Furthermore, if $N_1^2(x) = M(x)$ and $N_1(x) > 0$ then $N_1(x) \equiv N(x) > 0 \;\&\in C'[a, b]$, and $N_1(x) \in C'[a, \infty)$.

Lemma 5.1. *There is a unique positive-definite matrix function $P_0(x)$ such that*

$$P_0^2(x) = Y^\dagger Y + \hat{Y}^\dagger \hat{Y}, \qquad P_0(x) > 0 \quad \text{and} \quad P_0 \in C'[a, \infty). \tag{5.15}$$

It is easy to see that all solutions of (5.11) are given by

$$P = WP_0, \tag{5.16}$$

where W is an orthogonal matrix, i.e., $W^T W = 1_n$, and condition (5.13) is equivalent to

$$W' = W\{P_0^{-1}[Y^\dagger E\hat{Y} - \hat{Y}^\dagger FY - P_0 P_0']\, P_0^{-1}\} \quad \text{on} \quad [a, \infty). \tag{5.17}$$

Now (5.17) is a linear equation and the solution W_0 satisfying $W(a) = 1_n$ is orthogonal on $[a, \infty)$; and for

$$P = P_1 = W_0 P_0,$$

$$Q = Q_1 = P_1^{\dagger -1}[\hat{Y}^\dagger E\hat{Y} + Y^\dagger FY]\, P_1^{-1},$$

the solutions Y and $\hat{Y}$ are given by the Prüfer Transformation (5.10).

An alternate approach is to solve the functional equation

$$Q = CEC^\dagger + SFS^\dagger; \qquad C = C[x; a, Q], \qquad S = S[x; a, Q]$$

for $Q(x)$, which is possible because of the Lipschitz conditions of Theorem 5.7, and then solve the other equation of (5.12) for $P(x)$. The extension to Reid's more general case of complex Hermitian coefficients in (5.1) may be obtained easily.

5.5 *Atkinson's Formulation*

Another transformation of (5.1), which is similar to the Prüfer Transformation, has been introduced by Atkinson (*2*) and was later utilized by Etgen (*27*) and Coppel (*21*).

Consider a matrix solution $Y(x)$ of (5.1) such that

$$W[Y, Y] = 0 \quad \text{and} \quad \det Y(x) \not\equiv 0 \quad \text{on} \quad I$$

and, following Atkinson, let

$$Z(x) = (\hat{Y} + iY)(\hat{Y} - iY)^{-1}, \qquad i^2 = -1. \tag{5.18}$$

Differentiation of (5.18) yields that $Z(x)$ is a solution of the complex matrix equation

$$Z' = 2iZG(x), \tag{5.19}$$

where G is the Hermitian matrix

$$G = (\hat{Y}^\dagger + iY^\dagger)^{-1}(\hat{Y}^\dagger E\hat{Y} + Y^\dagger FY)(\hat{Y} - iY)^{-1}. \tag{5.20}$$

Furthermore,

$$Z(x) \quad \text{is a unitary matrix}$$

and if $w_j(x)$, $j = 1, 2, \ldots, n$, denote its eigenvalues then, for at least one j and $c \in I$,

$$w_j(c) = +1 \quad \text{if and only if} \quad \det Y(c) = 0$$

or

$$w_j(c) = -1 \quad \text{if and only if} \quad \det \hat{Y}(c) = 0.$$

Using these ideas Atkinson established some separation theorems for such numbers c. A detailed discussion is given in Chapter 10 of his book (*2*) where the coefficient matrices E and F are Hermitian.

Etgen (*27*) pointed out that for the case where

$$Y(a) = 0 \quad \text{and} \quad \hat{Y}(a) = I$$

Atkinson's transformation (5.18) is related to Etgen's "double angle" formulas of Theorem 5.6, i.e.,

$$Z = (CC^\dagger - SS^\dagger) + 2iSC. \tag{5.21}$$

Coppel (*21*) utilized Atkinson's formulation to establish comparison theorem for more general conjugate points, which correspond to general two-point homogeneous boundary conditions.

5.6. *Oscillation Theorems*

Most of the discussion of the 2×2 matrix equations of Section 4 holds also for the matrix equation (5.1) with E and F $n \times n$ symmetric (or Hermitian) matrices. For example, the proof of Theorem 4.3 also establishes a more general result.

Theorem 5.8. *If in Eq.* (5.1) *over* $I = [a, \infty)$

(i) min e.v. $(\int_a^x F) \to \infty$

and

(ii) max e.v. $(\int_a^x E) \to \infty$ *as* $x \to \infty$, *and* $E(x) \geqslant 0$, *then any matrix solution* $Y(x)$ *for which* $W[Y, Y] = 0$ *has an oscillatory determinant, i.e.,*

$$\det Y(x) \quad \textit{has infinitely many zeros as} \quad x \to \infty.$$

Such a matrix system is said to be oscillatory. Howard (*55*) proved a weaker form of Theorem 5.8 with (ii) replaced by

$$E(x) \geqslant e(x)\, 1_n\,, \qquad \int^{\infty} e = \infty.$$

Tomastik (*104*) recently proved a more general theorem which holds under the assumptions that

(iii) $E \geqslant 0$ *and* $F \geqslant 0$.

Using the integrated form of the Riccati equations (4.21) and (4.23),

$$H(x) = H(b) + \int_b^x F + \int_b^x HEH$$

$$K(x) = K(b) + \int_b^x E + \int_b^x KFK$$

and applying the Courant–Fischer min–max theorem to the eigenvalues of $H(x)$ and $K(x) = -H^{-1}(x)$, he established the following for $E > 0$ and $F > 0$.

Theorem 5.9. *If, as $x \to \infty$, the limits of r eigenvalues of $\int_a^x E$ and s eigenvalues of $\int_a^x F$ are positively infinite and $r + s > n$, then the system* (5.1) *with nonnegative-definite coefficients* (iii) *is oscillatory on* $[a, \infty)$.

Finally he gave an example to show that the inequality $r + s > n$ cannot be relaxed.

It would be interesting to know if the nonnegative conditions (iii) could be relaxed, as they were in the special case, $r = 1$ and $s = n$, of Theorem 5.8.

Acknowledgments

The author is indebted to the Associated Western Universities and, in particular, to the director of the Differential Equations Symposium, Professor R. W. McKelvey, and his committee for the excellent arrangements for the lectures, the very able audience and for extending the invitation to give lectures to one who welcomed the opportunity. Thanks are also due to the author's former students and colleagues at the Universities of Utah and Tennessee for their patience and numerous suggestions for improvement during earlier versions of these lectures.

Bibliography

1. P. Appel, Sur la transformation des equations differentielles lineares, *Compt. Rend.* (*Paris*) **91** (1880), 211–214.
2. F. V. Atkinson, "Discrete and Continuous Boundary Problems." Academic Press, New York, 1964.
3. N. V. Azbelev and Z. B. Caljuk, On the question of distribution of zeros of solutions of linear differential equations of the third order, *Mat. Sborn.* **51**, 475–486 (1960). [English Translation: *AMS Transl.* **42**, 233–245 (1964)].
4. J. H. Barrett, Behavior of solutions of second-order self-adjoint differential equations, *Proc. AMS* **6**, 247–251 (1955).
5. J. H. Barrett, Matrix systems of second order differential equations, *Portugal. Math.* **14**, 79–89 (1955).
6. J. H. Barrett, A Prufer transformation for matrix differential equations, *Proc. Am. Math. Soc.* **8**, 510–518 (1957).
7. J. H. Barrett, Second order complex differential equations with a real independent variable, *Pacific J. Math.* **8**, 187–200 (1958).
8. J. H. Barrett, Disconjugacy of second order linear differential equations with nonnegative coefficients, *Proc. Am. Math. Soc.* **10**, 552–561 (1959).
9. J. H. Barrett, Systems-disconjugacy of a fourth-order differential equation, *Proc. Am. Math. Soc.* **12**, 205–213 (1961).
10. J. H. Barrett, Fourth-order boundary value problems and comparison theorems, *Canadian J. Math.* **13**, 625–638 (1961).

11. J. H. BARRETT, Two-point boundary problems for linear self-adjoint differential equations of the fourth order with middle term, *Duke Math. J.* **29**, 543–554 (1962).
12. J. H. BARRETT, Disconjugacy of generalized linear differential equations of third and fourth order, (Abstract No. 594-36), *Notices AMS* **9**, 469 (1962).
13. J. H. BARRETT, Canonical forms for third-order linear differential equations, *Ann. Mat. Pura Appl.* **65**, 253–274 (1964).
14. J. H. BARRETT, Third-order equations with nonnegative coefficients, *J. Math. Anal. Appl.* **24**, 212–224 (1968).
15. R. BELLMAN, "Introduction to Matrix Analysis." McGraw-Hill, New York, 1960.
16. G. D. BIRKHOFF, On the solutions of ordinary linear homogeneous differential equations of the third order, *Ann. Math.* **12**, 103–127 (1910–11).
17. G. A. BLISS, Lectures on the Calculus of Variations. The University of Chicago Press, Chicago, 1946.
18. E. A. CODDINGTON AND N. LEVINSON, "Theory of Ordinary Differential Equations." McGraw-Hill, New York, 1955.
19. W. J. COLES, A general Wirtinger-type inequality, *Duke Math. J.* **27**, 133–138 (1960).
20. W. J. COLES, An oscillation criterion for second-order linear differential equations, *Proc. AMS* (to appear).
21. W. A. COPPEL, Comparison theorems for canonical systems of differential equations, *J. Math. Anal. Appl.* **12**, 306–315 (1965).
22. R. COURANT AND D. HILBERT, "Methods of Mathematical Physics," Vol. I. Interscience, 1953, New York.
23. M. A. DOBROHOTOVA, On boundedness of solutions of linear differential equations of third order (in Russian), *Yaroslavl. Gosudarstvennyi Pedagogichoskii Institut Uchenye Zapiski* **34**, 19–34 (1960).
24. J. M. DOLAN, Oscillatory Behavior of Solutions of Linear Ordinary Differential Equations of Third Order. Unpublished doctoral dissertation, University of Tennessee, 1967.
25. G. J. ETGEN, Oscillatory properties of certain nonlinear matrix differential systems of second order, *Trans. Am. Math. Soc.* **122**, 289–310 (1966).
26. G. J. ETGEN, "A note on trigonometric matrices," *Proc. Am. Math. Soc.* **17**, 1226–1232 (1966).
27. G. J. ETGEN, "On the determinants of solutions of second order matrix differential systems," *J. Math. Anal. Appl.* **18**, 585–598 (1967).
28. G. FROBENIUS, Über adjugierte lineare Differentialausdrücke, *Journal für Mathematik* **85**, 185–213 (1878).
29. LANDOLINO GIULIANO, "On linear self-adjoint equations of third order," *Boll. Univ. Mat. Ital.* **12**, 16–18 (1957).
30. M. GREGUŠ, On certain new properties of solutions of the differential equation $y''' + Qy' + Q'y = 0$ (in Czech.) *Publ. Fac. Sci. Université Masaryk, Bruo, Czech.* **362**, 237–251 (1955).
31. M. GREGUŠ, On a new boundary-value problem for differential equations of third order (in Russian), *Czech. Math. J.* **7**, 41–47 (1957).
32. M. GREGUŠ, On the linear differential equations of third order with constant coefficients (in Slovak), *Acta Fac. Rerum. Natur. Univ. Comenian. Math.* **2**, 61–65 (1957).
33. M. GREGUŠ, A remark about the dispersions and transformations of a differential

equation of third-order (in Czech.), *Acta Fac. Rerum Natur. Univ. Comenian. Math.* **4**, 205–211 (1959).

34. M. GREGUŠ, Oxzillatorische Eigenschatten der Lösungen der linearen Differentialgleichung dritter Ordnung $y''' + 2Ay' + (A' + b)y = 0$, wo $A = A(x) \leqslant 0$ ist, *Czech. Mat. J.* **9**, 416–427 (1959).

35. M. GREGUŠ, Über einige Eigenschaften der Lösungen der Differentialgleichung $y''' + 2Ay' + (A' + b)y = 0$, $A \leqslant 0$, *Czech. Mat. J.* (86) **11**, 106–115 (1961).

36. M. GREGUŠ, "Oscillatory properties of the solutions of the third-order differential equation of the type $y''' + 2A(x)y' + [A'(x) + b(x)]y = 0$, *Acta Fac. Rerum natur. Univ. Comenian. Math.* **6**, 275–300 (1961).

37. M. GREGUŠ, "Bemerkungen zu der unlösharen Randwert problemen dritter Ordnung," *Acta Fac. Rerum Natur. Univ. Comenian. Math.* **7**, 639–647 (1963).

38. M. GREGUŠ, Über einige Eigenschatten der Lösungen der Differentialgleichung dritter Ordnung, *Acta Fac. Rerum Natur. Univ. Comenian. Math.* **7**, 585–595 (1963).

39. M. GREGUŠ, Über einige Radwertprobleme Dritter Ordnung, *Czech. Math. J.* **13**, 551–560 (1963).

40. M. GREGUŠ, Über die lineare homogene Differentialgleichung dritter Ordnung, *Wiss. Zeit. Martin-Luther Univ. Halle-Wittenberg* **XII/3**, 265–286 (1963).

41. M. GREGUŠ, "Über die asymptotischen Eigenschatten der Lösungen der linearen Differentialgleichung dritter Ordnung," *Ann. Mat. Pura Appl.* **63**, 1–10 (1963).

42. M. GREGUŠ, Über die Eigenschaften der Lösungen einiger quasilinearer Gleichungen 3. Ordnung, *Acta Fac. Rerum Natur. Univ. Comenian. Math.* **10**, 11–22 (1965).

43. W. HAHN, Über Orthogonalpolynome mit drei Parametern, *Deutsche Math.* **5**, 273–278 (1940).

44. T. G. HALLAM, Asymptotic behavior of the solutions of an *n*th-order nonhomogeneous ordinary differential equation, *Trans. Am. Math. Soc.* **122**, 177–194 (1966).

45. M. HANAN, Oscillation criteria for third-order linear differential equations, *Pacific J. Math.* **11**, 919–944 (1961).

46. G. HANDLEMAN AND YIH-O TU, Lateral vibrations of a beam under initial linear axial stress, *J. SIAM* **9**, 455–473 (1961).

47. P. HARTMAN, "Ordinary Differential Equations." Wiley, New York, 1964.

48. P. HARTMAN AND A. WINTNER, On disconjugate differential systems, *Canadian J. Math.* **8**, 72–81 (1956).

49. J. HEIDEL, Qualitative behavior of solutions of a third order nonlinear differential equation, *Pacific J. Math.* **27**, 507–526 (1968).

50. E. HILLE, Nonoscillation theorems, *Trans. Am. Math. Soc.* **64**, 234–252 (1948).

51 D. B. HINTON, Disconjugate properties of a system of differential equations, *J. Differential Eqs.* **2**, 420–437 (1966).

52. D. B. HINTON, Clamped end boundary conditions for fourth-order self-adjoint differential equations, *Duke Math. J.* **34**, 131–138 (1967).

53. D. B. HINTON, A Criterion for *n–n* oscillations in differential equations of order $2n$, *Proc. Am. Math. Soc.* **19**, 511–518 (1968).

54. H. C. HOWARD, Oscillation criteria for fourth-order linear differential equations, *Trans. Am. Math. Soc.* **96**, 296–311 (1960).

55. H. C. HOWARD, Oscillation criteria for matrix differential equations, *Canadian J. Math.* **19**, 184–199 (1967).

56. R. W. HUNT, The behavior of solutions of ordinary self-adjoint differential equations of arbitrary even order, *Pacific J. Math.* **12**, 945–961 (1962).

57. R. W. Hunt, Oscillation properties of even-order linear differential equations, *Trans. Am. Math. Soc.* **115**, 54–61 (1965).
58. E. L. Ince, "Ordinary Differential Equations," Dover Publications, New York, 1944.
59. V. A. Jakubovič, Oscillation properties of solutions of canonical equations, *Mat. Sb.* **56**, 3–42 (1962); [English translation: *Am. Math. Soc. AMS Transl.* **42**, 247–288 1964].
60. E. Kamke, "Differentialgleichungen Lösungsmethoden und Lösungen, *I*." Becker and Erler, Leipzig, 1943 and J. W. Edwards, Ann Arbor, 1945.
61. A. Kneser, Untersuchen über die reelen Nullstellen der Integrale linearer Differentialgleichungen, *Math. Ann.* **42**, 409–435 (1893).
62. V. A. Kondrat'ev, On the oscillation of solutions of linear differential equations of third and fourth order (in Russian), *Trudy Mosk. Mat. Obšč.* **8**, 259–282 (1959); *Dokl. Akad. Nauk. USSR* **118**, 22–24 (1958).
63. V. A. Kondrat'ev, Oscillatory properties of solutions of the equation $y^{(n)} + p(x)y = 0$ (in Russian), *Trudy Mosk. Mat. Obšč.* **10**, 419–436 (1961); *Dolk. Akad. Nauk USSR* **120**, 1180–1182 (1958).
64. A. Lasota, Sur la distance entre les zéros de l'équation differentielle lineaire du troisième ordre, *Ann. Polon. Math.* **13**, 129–132 (1963).
65. A. C. Lazer, The behavior of solutions of the differential equation $y''' + p(x)y' + q(x)y = 0$, *Pacific J. Math.* **17**, 435–466 (1966).
66. W. Leighton, Principal quadratic functionals, *Trans. Am. Math. Soc.* **67**, 253–274 (1949).
67. W. Leighton, The detection of the oscillation of solutions of a second order linear differential equation, *Duke Math. J.* **17**, 57–62 (1950).
68. W. Leighton, On self-adjoint differential equations of second-order, *J. London Math. Soc.* **27**, 37–47 (1952).
69. W. Leighton, Comparison theorems for linear differential equations of second-order, *Proc. Am. Math. Soc.* **13**, 603–610 (1962).
70. W. Leighton and Z. Nehari, On the oscillation of solutions of self-adjoint linear differential equations of the fourth order, *Trans. Am. Math. Soc.* **89**, 325–377 (1958). (Contains an extensive bibliography).
71. A. Ju. Levin, Some questions on the oscillation of solutions of linear differential equations (in Russian), *Dokl. Akad. Nauk SSSR* **148**, 512–515 (1963); *Soviet Math.* **4**, 121–124 (1963).
72. A. Ju. Levin, Distribution of the zeros of solutions of a linear differential equation, (in Russian), *Dokl. Akad. Nauk SSSR 156*, 1281–1284 (1964); *Soviet Math.* **5**, 818–821 (1964).
73. G. Mammana, Decomposizione delle espressioni differenziali lineari omogenee in prodotti di fattori simbolici e applicazione relativa allo studio delle equazioni differenziali lineari, *Math. Z.* **33**, 186–231 (1931).
74. M. Morse, "The Calculus of Variations in the Large." American Mathematical Society Colloquium Publications, New York, 1934.
75. Z. Nehari, Oscillation criteria for second-order linear differential equations, *Trans. Am. Math. Soc.* **85**, 428–445 (1958).
76. Z. Nehari, On the zeros of solutions of *n*th order linear differential equations, *J. London Math. Soc.* **39**, 327–332 (1964).
77. Z. Nehari, Nonoscillation criteria for *n*th order linear differential equations, *Duke Math. J.* **32**, 607–616 (1965).

78. Z. NEHARI, Disconjugate Linear Operators, *Trans. Am. Math. Soc. AMS* **129**, 500–516 (1967).

79. G. POLYA, On the mean-value theorem corresponding to a given linear homogeneous differential equation, *Trans. Am. Math. Soc.* **24**, 312–324 (1922).

80. H. PRÜFER, Neue Herleitung der Sturm-Liouvilleschen Reihenentwicklung stetiger Funktionen, *Math. Ann.* **95**, 499–518 (1926).

81. M. RAB, Asymptotische Eigenschatten der Lösungen Linearer Differentialgleichung dritter Ordnung, *Spisy Publ. Fac. Sci. Masaryk* **374**, 177–184 (1956/4).

82. W. T. REID, A matrix differential equation of the Riccati type, *Am. J. Math.* **68**, 237–246 (1946). [See also *Am. J. Math.* **70**, 460 (1948).]

83. W. T. REID, Oscillation criteria for linear differential equations with complex coefficients, *Pacific J. Math.* **6**, 733–751 (1956).

84. W. T. REID, A comparison theorem for self-adjoint differential equations of second order, *Ann. Math.* **65**, 197–202 (1957).

85. W. T. REID, Oscillation criteria for linear differential systems with complex coefficients, *Pacific J. Math.* **6**, 733–751 (1956).

86. W. T. REID, Principal solutions of nonoscillatory self-adjoint linear differential systems, *Pacific J. Math.* **8**, 147–169 (1958).

87. W. T. REID, A Prüfer transformation for differential systems, *Pacific J. Math.* **8**, 575–584 (1958).

88. W. T. REID, Oscillation criteria for self-adjoint differential systems, *Trans. Am. Math. Soc.* **101**, 91–106 (1961).

89. W. T. REID, Riccati matrix differential equations and nonoscillation criteria for associated linear systems, *Pacific J. Math.* **13**, 664–685 (1963).

90. W. T. REID, Variational methods and boundary problems for ordinary linear differential systems, *in* "The Proceedings of the Japan–United States Seminar on Functional and Differential Equations" (W. A. Harris and Y. Sibuya, Eds.), pp. 267–299. Benjamin, New York, 1967.

91. G. SANSONE, Studi sulle equazioni differenziali lineari omogenee di terzo ordine nel campo real e, *Revista Mat. Fis. Teor.* (*Tucuman*), **6**, 195–253 (1948).

92. G. SANSONE, "Equazioni Differenziali nel Campo Reale," Zanichelli, Bologna, 1956.

93. J. SCHRÖDER, Randwertantgaben vieter Ordnung mit positiver Greenscher Funktion, *Math. Z.* **90**, 429–440 (1965).

94. J. SCHRÖDER, Hinreichende Bedingungen bei Differentialgleichungen vierter Ordnung, *Math. Z.* **92**, 75–94 (1966).

95. B. SCHWARZ, Disconjugacy of Complex Differential Systems, *Trans. Am. Math. Soc.* **125**, 482–496 (1966).

96. G. SEIFERT, A third order boundary value problem arising in aeroelastic wing theory, *Quart. Appl. Math.* **9**, 210–218 (1951).

97. G. SEIFERT, A third order irregular boundary value problem and the associated series, *Pacific J. Math.* **2**, 395–406 (1952).

98. T. L. SHERMAN, Properties of solutions of *n*th order linear differential equations, *Pacific J. Math.* **15**, 1045–1060 (1965).

99. D. SHINN, Existence theorems for the quasi-differential equation of the *n*th order, *Compt. Rend. Acad. Sci. URSS* **18**, 515–518 (1938).

100. R. L. STERNBERG, Variational methods and nonoscillation theorems for systems of differential equations, *Duke Math. J.* **19**, 311–322 (1952).

101. M. Švec, Sur une propriété des integrales de l'equation $y^{(n)} + Q(x)y = 0$, $n = 3, 4$, *Czech. Math. J.* **7**, 450–461 (1957).

102. M. Švec, Some remarks on a third-order linear differential equation, (in Russian), *Czech. Math. J.* **15**, 42–49 (1965).

103. M. Švec, Einige asymptotische und oszillatorische Eigenschaften der Differentialgleichung $y'''p + A(x)y' + B(x)y = 0$, *Czech. Mat. J.* **15**, 378–393 (1965).

104. E. C. Tomastik, Singular quadratic functionals of n dependent variables, *Trans. Am. Math. Soc.* **124**, 60–76 (1966).

105. P. Waltman, Oscillation criteria for third order nonlinear differential equations, *Pacific J. Math.* **18**, 385–389 (1966).

106. W. M. Whyburn, On self-adjoint ordinary differential equations of the fourth order, *Am. J. Math.* **52**, 171–196 (1930).

107. D. Willet, On the Oscillatory Behavior of the Solutions of Second-Order Linear Differential Equations, [AWU Differential Equations Symposium (1967)] unpublished.

108. A. Wintner, A criterion of oscillatory stability, *Quart. Appl. Math.* **7**, 115–117 (1949).

109. M. Zedek, Cayley's decomposition and Polya's W-property of ordinary linear differential equations, *Israel J. Math.* **3**, 81–86 (1965).

110. A. J. Zettl, Adjoint linear differential operators, *Proc. Am. Math. Soc.* **16**, 1239–1241 (1965).

111. A. J. Zettl (with E. A. Coddington), Hermitian and anti-Hermitian properties of Green's matrices, *Pacific J. Math.* **18**, 451–454 (1966).

112. A. J. Zettl, Some identities related to Polya's property W for linear differential equations, *Proc. Am. Math. Soc.* **18**, 992–994 (1967).

Reprinted from *Advances in Mathematics* **2**, Fascicle 3, 307–363, (1968).

Subfunctions and Second-Order Ordinary Differential Inequalities*

LLOYD K. JACKSON

Department of Mathematics, University of Nebraska, Lincoln Nebraska 68508

1. Introduction

Subfunctions and solutions of differential inequalities have been used for many years in dealing with existence theorems and properties of solutions for both ordinary and partial differential equations. In 1915 in one of the earliest such applications Perron [1] used solutions of differential inequalities to establish the existence of a solution of the initial-value problem for the first-order equation $y' = f(x, y)$.

In a remarkable paper published in 1923, Perron [2] used subharmonic functions to study the Dirichlet problem for harmonic functions for bounded plane domains. By using properties of subharmonic functions and the fact that the Dirichlet problem for circles is solvable, a generalized solution of the Dirichlet problem for a bounded domain is obtained. The generalized solution is harmonic in the interior of the domain and the question of whether or not it assumes the specified boundary value at a boundary point can be dealt with separately.

In the next few years a more complete investigation of properties of subharmonic functions was carried out by F. Riesz [3]-[5]. It was

* This paper is based on a series of lectures given during the summer of 1967 at the Rocky Mountain Symposium in Ordinary Differential Equations sponsored by the Associated Rocky Mountain Universities.

observed by a number of authors that results in function theory and potential theory such as the Liouville Theorem and the Phragmen-Lindelof Theorems depend only on properties of subharmonic functions. Subsequently various authors, Tautz [6], Beckenbach and Jackson [7], Inoue [8], Jackson [9], and a number of others, examined the Perron method of attacking the Dirichlet problem to determine the properties of harmonic functions and subharmonic functions which are essential for the success of the method. This led to the successful application of the Perron method to the study of the Dirichlet problem for more general elliptic partial differential equations including certain types of nonlinear equations. These methods also led to Liouville Theorems and Phragmen-Lindelof Theorems for more general elliptic equations (see, for example, [10]-[13]).

If one is concerned with ordinary rather than partial differential equations, the program analogous to generalizing harmonic functions and subharmonic functions is to generalize linear functions and convex functions. There is an extensive literature in this area starting with a paper by Beckenbach [14] published in 1937. The papers [15]-[18] constitute a small sample of work on this theme. In most of these papers the authors deal with second-order ordinary differential equations for which it is assumed that all boundary-value problems are uniquely solvable. It is then shown that certain properties of convex functions carry over to subfunctions with respect to solutions of the differential equations.

There have been a number of papers in which solutions of differential inequalities have been employed in establishing existence theorems for boundary-value problems for ordinary differential equations. In particular, Caplygin and a number of later Soviet mathematicians using Caplygin's methods (for example, Babkin [19]) have obtained solutions of boundary-value problems as uniform limits of sequences of functions satisfying differential inequalities.

None of the papers referred to above appear to be concerned with the Perron method of attacking the boundary-value problem for second-order ordinary equations, that is, with the use of existence theorems in the small and the properties of subfunctions to establish the existence of solutions in the large. The present work is divided into two main divisions. The first sections will deal with differential equations $y'' = f(x, y, y')$ with the property that, when the boundary-value problem $y'' = f(x, y, y')$, $y(a) = A$, $y(b) = B$ has a solution $y(x) \in C^{(2)}[a, b]$, that solution is unique. In this case subfunctions can be defined in a meaningful way and the Perron method can be developed. Also the relationships between subfunctions and solutions of differential inequalities can be developed.

The second part of the work will be concerned with differential equations which are such that boundary-value problems are not necessarily uniquely solvable. In this part, results will be obtained by using solutions of differential inequalities and restrictions on the rate of growth of $f(x, y, y')$ with respect to y'.

2. Preliminary Results

In this section the local existence theorem and some related results will be given. We will always assume that $f(x, y, y')$ is continuous on $[a, b] \times R^2$ and for the moment we will assume that $[a, b]$ is a compact interval. Later we will consider cases where x ranges over an open interval or an infinite interval. In some of our results it would suffice to assume a type of piecewise continuity for $f(x, y, y')$, however, for simplicity we shall always assume f to be continuous on its domain.

Theorem 2.1. *Let $M > 0$ and $N > 0$ be given real numbers and let q be the maximum of $| f(x, y, y') |$ on the compact set $\{(x, y, y') : a \leqslant x \leqslant b, | y | \leqslant 2M, | y' | \leqslant 2N\}$. Then, if $\delta = \mathrm{Min}[(8M/q)^{1/2}, 2N/q]$, any boundary-value problem $y'' = f(x, y, y')$, $y(x_1) = y_1$, $y(x_2) = y_2$ with $[x_1, x_2] \subset [a, b]$, $x_2 - x_1 \leqslant \delta$, $| y_1 | \leqslant M$, $| y_2 | \leqslant M$, $| (y_1 - y_2)/(x_1 - x_2) | \leqslant N$ has a solution $y(x) \in C^{(2)}[x_1, x_2]$. Furthermore, given $\epsilon > 0$ there is a solution $y(x)$ such that $| y(x) - w(x) | < \epsilon$ and $| y'(x) - w'(x) | < \epsilon$ on $[x_1, x_2]$ provided $x_2 - x_1$ is sufficiently small where $w(x)$ is the linear function with $w(x_1) = y_1$, $w(x_2) = y_2$.* ([20], p. 1252).

Proof. The set

$$B[x_1, x_2] = \{z(x) \in C^{(1)}[x_1, x_2] : \| z \| \leqslant 2M, \| z' \| \leqslant 2N\}$$

is a closed convex subset of the Banach space $C^{(1)}[x_1, x_2]$. The mapping $T : C^{(1)}[x_1, x_2] \to C^{(1)}[x_1, x_2]$ defined by

$$(Tz)(x) = \int_{x_1}^{x_2} G(x, t) f(t, z(t), z'(t))\, dt + w(x),$$

where $G(x, t)$ is the Green's function for the boundary-value problem $y'' = 0$, $y(x_1) = y(x_2) = 0$, is completely continuous. For a $z \in B[x_1, x_2]$ we have

$$|(Tz)(x)| \leqslant \tfrac{1}{8}[q(x_2 - x_1)^2] + M,$$

and

$$|(Tz)'(x)| \leqslant \tfrac{1}{2}[q(x_2 - x_1)] + N$$

on $[x_1, x_2]$. Thus $x_2 - x_1 \leqslant \delta$ implies T maps $B[x_1, x_2]$ into itself. It then follows from the Schauder Fixed-Point Theorem that T has a fixed point in $B[x_1, x_2]$. The fixed point is a solution of the stated boundary-value problem. If $y(x)$ is a solution of the boundary-value problem with $y \in B[x_1, x_2]$, then

$$|y(x) - w(x)| \leqslant \tfrac{1}{8}[q(x_2 - x_1)^2]$$

and

$$|y'(x) - w'(x)| \leqslant \tfrac{1}{2}[q(x_2 - x_1)]$$

on $[x_1, x_2]$ and the last assertion of the theorem follows.

Corollary 2.2. *Assume that there exist constants $h > 0$ and $k > 0$ such that $|f(x, y, y')| \leqslant h + k(|y|)^{1/2}$ on $[a, b] \times R^2$. Then every boundary-value problem $y'' = f(x, y, y')$, $y(a) = \alpha$, $y(b) = \beta$ has a solution $y \in C^{(2)}[a, b]$* ([21], Lemma 2.2).

Proof. In this case we can choose $M > 0$ large enough that

$$|\alpha| \leqslant M, \qquad |\beta| \leqslant M, \qquad \left|\frac{\alpha - \beta}{a - b}\right| \leqslant M,$$

$$b - a \leqslant \left[\frac{8M}{h + k(2M)^{1/2}}\right]^{1/2}, \quad \text{and} \quad b - a \leqslant \frac{2M}{h + k(2M)^{1/2}}.$$

The result then follows from Theorem 2.1.

Using Corollary 2.2 and solutions of certain differential inequalities we can obtain solutions of some boundary-value problems for a modified form of the differential equation $y'' = f(x, y, y')$.

Definition 2.3. Let $\alpha(x), \beta(x) \in C^{(1)}[a, b]$ with $\alpha(x) \leqslant \beta(x)$ on $[a, b]$

and let $c > 0$ be such that $|\alpha'(x)| < c$ and $|\beta'(x)| < c$ on $[a, b]$. Then define

$$F^*(x, y, y') = \begin{cases} f(x, y, c) & \text{for} \quad y' \geqslant c, \\ f(x, y, y') & \text{for} \quad |y'| \leqslant c, \\ f(x, y, -c) & \text{for} \quad y' \leqslant -c, \end{cases}$$

and

$$F(x, y, y') = \begin{cases} F^*(x, \beta(x), y') + [y - \beta(x)]^{1/2} & \text{for} \quad y \geqslant \beta(x), \\ F^*(x, y, y') & \text{for} \quad \alpha(x) \leqslant y \leqslant \beta(x), \\ F^*(x, \alpha(x), y') - [\alpha(x) - y]^{1/2} & \text{for} \quad y \leqslant \alpha(x). \end{cases}$$

We will call $F(x, y, y')$ the modification of $f(x, y, y')$ associated with the triple $\alpha(x)$, $\beta(x)$, c. From the definition it follows that $F(x, y, y')$ is continuous on $[a, b] \times R^2$ and $|F(x, y, y')| \leqslant h + (|y|)^{1/2}$ on $[a, b] \times R^2$ where

$$h = \operatorname{Max}\{|f(x, y, y')| : a \leqslant x \leqslant b, \alpha(x) \leqslant y \leqslant \beta(x), |y'| \leqslant c\}$$
$$+ \operatorname*{Max}_{a \leqslant x \leqslant b} |\alpha(x)|^{\frac{1}{2}} + \operatorname*{Max}_{a \leqslant x \leqslant b} |\beta(x)|^{\frac{1}{2}}.$$

Next we define certain types of solutions of differential inequalities which will be used in the later work.

Definition 2.4. A function $\alpha(x)$ is called a $C^{(1)}$-lower solution of the differential equation $y'' = f(x, y, y')$ on an interval I in case $\alpha(x) \in C(I) \cap C^{(1)}(I^0)$, I^0 the interior of I, and

$$\underline{D}\alpha'(x) \equiv \liminf_{\delta \to 0} \frac{\alpha'(x + \delta) - \alpha'(x - \delta)}{2\delta} \geqslant f(x, \alpha(x), \alpha'(x))$$

on I^0. The function $\alpha(x)$ is called an $AC^{(1)}$-lower solution on I in case $\alpha(x) \in C(I) \cap C^{(1)}(I^0)$, $\alpha'(x)$ is absolutely continuous on each compact subinterval of I^0, and $\alpha''(x) \geqslant f(x, \alpha(x), \alpha'(x))$ almost everywhere on I. Similarly, $\beta(x)$ is a $C^{(1)}$-upper solution on I in case $\beta(x) \in C(I) \cap C^{(1)}(I^0)$ and

$$\bar{D}\beta'(x) \equiv \limsup_{\delta \to 0} \frac{\beta'(x + \delta) - \beta'(x - \delta)}{2\delta} \leqslant f(x, \beta(x), \beta'(x))$$

on I^0. $\beta(x)$ is an $AC^{(1)}$-upper solution on I in case $\beta(x) \in C(I) \cap C^{(1)}(I^0)$, $\beta'(x)$ is absolutely continuous on each compact subinterval of I^0, and $\beta''(x) \leqslant f(x, \beta(x), \beta'(x))$ almost everywhere on I.

When we say simply that $\alpha(x)$ is a lower solution or $\beta(x)$ is an upper solution we will mean that they can be of either type.

Theorem 2.5. *Let $\alpha(x)$, $\beta(x) \in C^{(1)}[a, b]$ be, respectively, lower and upper solutions of $y'' = f(x, y, y')$ on $[a, b]$ with $\alpha(x) \leqslant \beta(x)$ on $[a, b]$. Then, if $F(x, y, y')$ is the modification of $f(x, y, y')$ associated with the triple $\alpha(x)$, $\beta(x)$, c and if $\alpha(a) \leqslant \gamma \leqslant \beta(a)$, $\alpha(b) \leqslant \delta \leqslant \beta(b)$, the boundary-value problem*

$$y'' = F(x, y, y'), \qquad y(a) = \gamma, \qquad y(b) = \delta$$

has a solution $y \in C^{(2)}[a, b]$ satisfying $\alpha(x) \leqslant y(x) \leqslant \beta(x)$ on $[a, b]$.

Proof. By Corollary 2.2 the stated boundary-value problem has a solution $y \in C^{(2)}[a, b]$. Thus we only need show that $\alpha(x) \leqslant y(x) \leqslant \beta(x)$ on $[a, b]$. We will show only that $y(x) \leqslant \beta(x)$ since the arguments for $\alpha(x) \leqslant y(x)$ are essentially the same. Assume that $y(x) > \beta(x)$ at some points of $[a, b]$. Then $y(x) - \beta(x)$ has a positive maximum at a point $x_0 \in (a, b)$. It follows that $y'(x_0) = \beta'(x_0)$ and $|y'(x_0)| < c$; hence

$$y''(x_0) = F(x_0, y(x_0), y'(x_0)) = f(x_0, \beta(x_0), \beta'(x_0)) + [y(x_0) - \beta(x_0)]^{1/2}.$$

If $\beta(x)$ is a $C^{(1)}$-upper solution on $[a, b]$,

$$\bar{D}\beta'(x_0) \leqslant f(x_0, \beta(x_0), \beta'(x_0))$$

and

$$\underline{D}[y'(x_0) - \beta'(x_0)] = y''(x_0) - \bar{D}\beta'(x_0) \geqslant [y(x_0) - \beta(x_0)]^{1/2} > 0,$$

which is impossible at a maximum of $y(x) - \beta(x)$. If β is an $AC^{(1)}$-upper solution on $[a, b]$, then, since $F(x, y(x), y'(x))$ and $f(x, \beta(x), \beta'(x))$ are both continuous at $x = x_0$, there is a $\delta > 0$ such that $[x_0 - \delta, x_0 + \delta] \subset (a, b)$ and $[y'(x) - \beta'(x)]' > 0$ almost everywhere on $[x_0 - \delta, x_0 + \delta]$. This again is incompatible with $y(x) - \beta(x)$ having a maximum at x_0. We conclude that $y(x) \leqslant \beta(x)$ on $[a, b]$.

Theorem 2.6. *Assume that in addition to being continuous on $[a, b] \times R^2$ $f(x, y, y')$ is such that solutions of initial value problems for $y'' = f(x, y, y')$ are unique. Let $\alpha(x)$, $\beta(x) \in C^{(1)}[a, b]$ be lower and upper solutions on $[a, b]$ with $\alpha(x) \leqslant \beta(x)$ on $[a, b]$. Then, if $\alpha(x_0) = \beta(x_0)$ and $\alpha'(x_0) = \beta'(x_0)$ at some $x_0 \in [a, b]$, it follows that $\alpha(x) \equiv \beta(x)$ on $[a, b]$* ([21], Lemma 2.4).

Proof. Assume that the hypotheses of the Theorem are satisfied but that $\alpha(x) \not\equiv \beta(x)$ on $[a, b]$. We will consider the case where there is an $[x_0, x_1] \subset [a, b]$ such that $\alpha(x_0) = \beta(x_0)$, $\alpha'(x_0) = \beta'(x_0)$, and $\alpha(x) < \beta(x)$ on $(x_0, x_1]$. Let $F_1(x, y, y')$ be the modification of $f(x, y, y')$ as defined in Definition 2.3 for the interval $[x_0, x_1]$ and the triple $\alpha(x)$, $\beta(x)$, c_1. Then, if $\alpha(x_1) < \delta_1 < \beta(x_1)$, it follows from Theorem 2.5 that the boundary-value problem

$$y'' = F_1(x, y, y'), \qquad y(x_0) = \alpha(x_0), \qquad y(x_1) = \delta_1$$

has a solution $y_1 \in C^{(2)}[x_0, x_1]$ satisfying $\alpha(x) \leqslant y_1(x) \leqslant \beta(x)$ on $[x_0, x_1]$. Therefore, $y_1(x_0) = \alpha(x_0)$, $y_1'(x_0) = \alpha'(x_0)$; hence, by the way $F_1(x, y, y')$ is defined, there is a maximal interval $[x_0, x_2] \subset [x_0, x_1]$ on which $y_1(x)$ is a solution of $y'' = f(x, y, y')$. If $x_2 = x_1$, then

$$\alpha(x_1) < y_1(x_2) = y_1(x_1) = \delta_1 < \beta(x_1).$$

If $x_0 < x_2 < x_1$, it is still true that $\alpha(x_2) < y_1(x_2) < \beta(x_2)$ for, if either inequality were an equality, we would have $| y_1'(x_2) | < c_1$ and the interval $[x_0, x_2]$ would not be maximal.

This being the case, we can construct another modification $F_2(x, y, y')$ of $f(x, y, y')$ on the interval $[x_0, x_2]$ with respect to the triple $\alpha(x)$, $y_1(x)$, c_2. Applying Theorem 2.5 again with $\alpha(x_2) < \delta_2 < y_1(x_2)$, we conclude that there is a solution $y_2 \in C^{(2)}[x_0, x_2]$ of the boundary-value problem

$$y'' = F_2(x, y, y'), \qquad y(x_0) = \alpha(x_0), \qquad y(x_2) = \delta_2$$

satisfying $\alpha(x) \leqslant y_2(x) \leqslant y_1(x)$ on $[x_0, x_2]$. As above, it follows that there is a maximal subinterval $[x_0, x_3] \subset [x_0, x_2]$ on which $y_2(x)$ is a solution of $y'' = f(x, y, y')$ and that $\alpha(x_3) < y_2(x_3) < y_1(x_3)$. This contradicts the assumption that solutions of initial value problems for $y'' = f(x, y, y')$ are unique. We conclude that $\alpha(x) \equiv \beta(x)$ on $[a, b]$.

3. The Relation between Subfunctions and Solutions of Differential Inequalities

In this section we define subfunctions and superfunctions and consider necessary and sufficient conditions for such functions, when sufficiently smooth, to be respectively lower and upper solutions of the differential equation.

Definition 3.1. A function $\varphi(x)$ is said to be a subfunction with respect to solutions of $y'' = f(x, y, y')$ on an interval I in case for any $[x_1, x_2] \subset I$ and any solution $y \in C^{(2)}[x_1, x_2]$ $y(x_i) \geqslant \varphi(x_i)$ for $i = 1, 2$ implies $y(x) \geqslant \varphi(x)$ on $[x_1, x_2]$. The function $\psi(x)$ is said to be a superfunction with respect to solutions of $y'' = f(x, y, y')$ on an interval I in case for any $[x_1, x_2] \subset I$ and any solution $y \in C^{(2)}[x_1, x_2]$ $y(x_i) \leqslant \psi(x_i)$ for $i = 1, 2$ implies $y(x) \leqslant \psi(x)$ on $[x_1, x_2]$.

Theorem 3.2. *Assume that $\varphi \in C(I) \cap C^{(1)}(I^0)$ is a subfunction on I with respect to solutions of $y'' = f(x, y, y')$. Then φ is a $C^{(1)}$-lower solution of the differential equation on I.*

Proof. Let $x_0 \in I^0$. Then, if $h \geqslant 0$, $k \geqslant 0$, $h + k > 0$ are sufficiently small, it follows from Theorem 2.1 that the boundary-value problem

$$y'' = f(x, y, y'), \qquad y(x_0 + h) = \varphi(x_0 + h), \qquad y(x_0 - k) = \varphi(x_0 - k)$$

has a solution $y \in C^{(2)}[x_0 - k, x_0 + h]$. Since φ is a subfunction on I, it follows that

$$\frac{\varphi'(x_0 + h) - \varphi'(x_0 - k)}{h + k} \geqslant \frac{y'(x_0 + h) - y'(x_0 - k)}{h + k} = y''(\xi) = f(\xi, y(\xi), y'(\xi))$$

for some $x_0 - k < \xi < x_0 + h$. Since f is continuous, it also follows from Theorem 2.1 that

$$f(\xi, y(\xi), y'(\xi)) \to f(x_0, \varphi(x_0), \varphi'(x_0)) \qquad \text{as} \qquad h + k \to 0.$$

Thus

$$\liminf_{\substack{h+k\to 0 \\ h\geqslant 0, k\geqslant 0, h+k>0}} \frac{\varphi'(x_0 + h) - \varphi'(x_0 - k)}{h + k} \geqslant f(x_0, \varphi(x_0), \varphi'(x_0))$$

and, in particular, $\underline{D}\varphi'(x_0) \geqslant f(x_0, \varphi(x_0), \varphi'(x_0))$ from which it follows that φ is a $C^{(1)}$-lower solution on I.

We also conclude from the proof of Theorem 3.2 that φ has a finite second derivative almost everywhere and $\varphi''(x) \geqslant f(x, \varphi(x), \varphi'(x))$ almost everywhere on I ([22], p. 128). It does not follow however that φ is necessarily an $AC^{(1)}$-lower solution on I. To see this let $h(x)$ be a continuous strictly increasing function on [0, 1] with $h(0) = 0$, $h(1) = 1$ but with $h(x)$ not absolutely continuous on [0, 1]. Then $\varphi(x) \equiv \int_0^x h(t)\, dt$ is convex on [0, 1], hence $\varphi \in C^{(1)}[0, 1]$ is a subfunction on [0, 1] with respect to solutions of $y'' = 0$. However $\varphi' = h$ is not absolutely continuous on [0, 1] and φ is not an $AC^{(1)}$-lower solution of $y'' = 0$ on [0, 1].

Assuming only the continuity of $f(x, y, y')$ we conclude from Theorem 3.2 that a subfunction of class $C^{(1)}$ is necessarily a $C^{(1)}$-lower solution of the differential equation. Now we turn the question around and look at the possibility of a lower solution of the differential equation being a subfunction. For this to be the case it is obvious that more is required of $f(x, y, y')$ than continuity. This is apparent from the fact that a solution is also a lower solution, hence, if lower solutions are subfunctions, then solutions are subfunctions. From the definition of subfunctions it would then follow that, if a boundary-value problem on an interval $[x_1, x_2]$ has a solution $y \in C^{(2)}[x_1, x_2]$, that solution is unique. This will not be true in general and, therefore, we cannot expect lower solutions to be subfunctions unless stronger conditions are placed on $f(x, y, y')$. Also it is clear that a theorem which gives sufficient conditions for a lower solution to be a subfunction is automatically a theorem giving sufficient conditions for solutions of boundary-value problems to be unique.

Lemma 3.3. *Assume that $f(x, y, y')$ is nondecreasing in y for all fixed x, y'. Let φ be a $C^{(1)}$-lower solution and ψ a $C^{(1)}$-upper solution on a closed interval I and assume that at each point on I^0 at least one of the differential inequalities is a strict inequality. Further assume that $\varphi(x) \leqslant \psi(x) + M$ at the end points of I where $M \geqslant 0$. Then $\varphi(x) < \psi(x) + M$ on I^0.*

Proof. It suffices to consider only the case $M = 0$ since

$$\bar{D}\psi' = \bar{D}[\psi + M]' \leqslant f(x, \psi, \psi') \leqslant f(x, \psi + M, \psi')$$

from which it follows that $\psi^* = \psi + M$ is an upper solution. Hence, assume $\varphi(x) \leqslant \psi(x)$ at the endpoints of I but that also $\varphi(x) \geqslant \psi(x)$ at some points in I^0. Then $\varphi(x) - \psi(x)$ has a nonnegative maximum at some $x_0 \in I^0$ and at x_0

$$\underline{D}[\varphi'(x_0) - \psi'(x_0)] \geqslant \underline{D}\varphi'(x_0) - \bar{D}\psi'(x_0)$$
$$> f(x_0, \varphi(x_0), \varphi'(x_0)) - f(x_0, \psi(x_0), \psi'(x_0)).$$

Since $\varphi'(x_0) = \psi'(x_0)$ and f is nondecreasing in y, we conclude that $\underline{D}[\varphi'(x_0) - \psi'(x_0)] > 0$ which is impossible with $\varphi(x) - \psi(x)$ having a maximum at x_0. We conclude that $\varphi(x) < \psi(x)$ on I^0.

Lemma 3.4. *Assume that on* $[a, b] \times R^2$ $f(x, y, y')$ *is nondecreasing in* y *for each fixed* x, y' *and that* $f(x, y, y')$ *satisfies a Lipschitz condition with respect to* y' *on each compact subset of* $[a, b] \times R^2$. *Let* $[c, d] \subset [a, b]$ *and assume that* $\varphi \in C^{(1)}[c, d]$ *is a* $C^{(1)}$*-lower solution on* $[c, d]$. *Then given* $\epsilon > 0$ *there is a function* $\varphi_1 \in C^{(1)}[c, d]$ *such that* $\varphi(x) - \epsilon \leqslant \varphi_1(x) \leqslant \varphi(x)$ *on* $[c, d]$ *and* $\underline{D}\varphi_1'(x) > f(x, \varphi_1(x), \varphi_1'(x))$ *on* (c, d) ([23], Lemma 2.2).

Proof. By assumption $f(x, y, y')$ satisfies a Lipschitz condition with respect to y' on the compact set

$$K = \{(x, y, y') : c \leqslant x \leqslant d, | \varphi(x) - y | \leqslant 1, | \varphi'(x) - y' | \leqslant 1\}.$$

Let $k > 0$ be an associated Lipschitz coefficient. For a given $\epsilon > 0$ let $\rho(x)$ satisfy the following conditions:

$$\rho''(x) = (k + 1)\, \rho'(x) \qquad \text{on} \qquad [c, d],$$

$$0 < \rho'(x) \leqslant 1 \qquad \text{on} \qquad (c, d),$$

and

$$-\operatorname{Min}[1, \epsilon] \leqslant \rho(x) \leqslant 0 \qquad \text{on} \qquad [c, d].$$

Then setting $\varphi_1 = \varphi + \rho$ we have that on $[c, d]$

$$\varphi(x) - \epsilon \leqslant \varphi_1(x) \leqslant \varphi(x)$$

and on (c, d)

$$\begin{aligned}
\underline{D}\varphi_1'(x) &= \underline{D}[\varphi'(x) + \rho'(x)] \\
&= \underline{D}\varphi'(x) + \rho''(x) \geqslant f(x, \varphi(x), \varphi'(x)) + (k + 1) | \rho'(x) | \\
&\qquad \geqslant f(x, \varphi(x), \varphi'(x)) + f(x, \varphi(x), \varphi'(x) + \rho'(x)) \\
&\qquad\qquad - f(x, \varphi(x), \varphi'(x)) + | \rho'(x) | \\
&\qquad \geqslant f(x, \varphi_1(x), \varphi_1'(x)) + | \rho'(x) | \\
&\qquad > f(x, \varphi_1(x), \varphi_1'(x)).
\end{aligned}$$

Theorem 3.5. *Let* $f(x, y, y')$ *be nondecreasing in* y *for fixed* x, y' *and satisfy a Lipschitz condition with respect to* y' *on each compact subset of* $[a, b] \times R^2$. *Let* $[c, d] \subset [a, b]$ *and assume that* $\varphi, \psi \in C[c, d] \cap C^{(1)}(c, d)$ *are respectively* $C^{(1)}$*-lower and* $C^{(1)}$*-upper solutions on* $[c, d]$. *Then, if* $\varphi(x) \leqslant \psi(x) + M$, $M \geqslant 0$, *at the endpoints of* $[c, d]$, *it follows that* $\varphi(x) \leqslant \psi(x) + M$ *on* $[c, d]$ ([23], Theorem 2.2).

Proof. As in Lemma 3.3 it suffices to consider the case $M = 0$. Hence, assume $\varphi(c) \leqslant \psi(c)$ and $\varphi(d) \leqslant \psi(d)$ but $\varphi(x) > \psi(x)$ at some points in (c, d). Let $\epsilon = \text{Max}[\varphi(x) - \psi(x)]$ on $[c, d]$ and let $[c_1, d_1] \subset [c, d]$ such that $\varphi(c_1) - \psi(c_1) = \varphi(d_1) - \psi(d_1) = \frac{1}{2}\epsilon$, $\varphi(x) - \psi(x) \geqslant \frac{1}{2}\epsilon$ on $[c_1, d_1]$, and the maximum $\varphi(x) - \psi(x) = \epsilon$ is assumed on (c_1, d_1). By Lemma 3.4 there is a function $\varphi_1 \in C^{(1)}[c_1, d_1]$ such that $\varphi - \frac{1}{2}\epsilon \leqslant \varphi_1 \leqslant \varphi$ on $[c_1, d_1]$ and $\underline{D}\varphi_1' > f(x, \varphi_1, \varphi_1')$ on (c_1, d_1). Then $\varphi_1(x) \leqslant \psi(x) + \frac{1}{2}\epsilon$ at $x = c_1$ and $x = d_1$ and $\varphi_1(x) \geqslant \psi(x) + \frac{1}{2}\epsilon$ at some points in (c_1, d_1). This contradicts Lemma 3.3 and we conclude that $\varphi(x) \leqslant \psi(x)$ on $[c, d]$.

Remark. If in Theorem 3.5 we assume that φ is an $AC^{(1)}$-lower solution and ψ is an $AC^{(1)}$-upper solution, the conclusion of the Theorem is still valid. To see this we proceed as in the proof of the Theorem up to the point where the subinterval $[c_1, d_1] \subset [c, d]$ with the stated properties is obtained. Then we note that the proof of Lemma 3.4 leads to the conclusion that there is a function $\varphi_1 \in C^{(1)}[c_1, d_1]$ such that φ_1' is absolutely continuous on $[c_1, d_1]$, $\varphi(x) - \frac{1}{2}\epsilon \leqslant \varphi_1(x) \leqslant \varphi(x)$ on $[c_1, d_1]$, and $\varphi_1''(x) \geqslant f(x, \varphi_1(x), \varphi_1'(x)) + |\rho'(x)|$ almost everywhere on $[c_1, d_1]$. Since $\varphi_1(x) - \psi(x) \leqslant \frac{1}{2}\epsilon$ at $x = c_1$ and $x = d_1$ while $\varphi_1(x) - \psi(x) \geqslant \frac{1}{2}\epsilon$ at some points of (c_1, d_1), $\varphi_1(x) - \psi(x)$ has a positive maximum at some $x_0 \in (c_1, d_1)$. Furthermore $|\rho'(x)| > 0$ on (c_1, d_1), hence there is a $\delta > 0$ such that $[x_0 - \delta, x_0 + \delta] \subset (c_1, d_1)$ and $\varphi_1''(x) - \psi''(x) > 0$ almost everywhere on $[x_0 - \delta, x_0 + \delta]$. This is impossible with $\varphi_1(x) - \psi(x)$ having a maximum at x_0. Thus we conclude again that $\varphi(x) \leqslant \psi(x)$ on $[c, d]$.

Corollary 3.6. *If f satisfies the hypotheses of Theorem 3.5, then a lower solution φ on a subinterval $I \subset [a, b]$ is a subfunction on I.*

We have already observed that a $C^{(1)}$-lower solution is not necessarily an $AC^{(1)}$-lower solution. However, if $f(x, y, y')$ satisfies the hypotheses of Theorem 3.5, then it follows from Corollary 3.6 and the Remark preceding it that an $AC^{(1)}$-lower solution on $I \subset [a, b]$ is a subfunction on I. Then by Theorem 3.2 it will be a $C^{(1)}$-lower solution on I.

Corollary 3.7. *If $f(x, y, y')$ satisfies the conditions of Theorem 3.5 and $y_1, y_2 \in C^{(2)}[x_1, x_2]$ are solutions on $[x_1, x_2] \subset [a, b]$ with $y_1(x_i) = y_2(x_i)$ for $i = 1, 2$, then $y_1(x) \equiv y_2(x)$ on $[x_1, x_2]$.*

Corollary 3.7 is no longer valid if the Lipschitz condition on $f(x, y, y')$ with respect to y' on compact sets is omitted, and, as a matter of fact, it cannot be weakened in any very significant way. For example the boundary-value problem

$$y'' = (\tfrac{3}{2})(\tfrac{5}{2})^{2/3} | y' |^{1/3}, \qquad y(-1) = y(+1) = 1$$

has solutions $y(x) \equiv 1$ and $y(x) = | x |^{5/2}$ of class $C^{(2)}[-1, +1]$.

Corollary 3.8. *If f, φ, ψ satisfy the conditions of Theorem 3.5 with $\varphi(c) = \psi(c)$ and $\varphi(x) \leqslant \psi(x)$ on $[c, d]$, then $\varphi(x) - \psi(x)$ is nondecreasing on $[c, d]$.*

Theorem 3.9. *Let $f(x, y, y')$ be nondecreasing in y for fixed x, y' on $[a, b] \times R^2$ and assume that solutions of initial-value problems for $y'' = f(x, y, y')$ are unique. Then, if φ and ψ are lower and upper solutions on $[c, d] \subset [a, b]$ with $\varphi(x) \leqslant \psi(x) + M$, $M \geqslant 0$, at $x = c$ and $x = d$, it follows that $\varphi(x) \leqslant \psi(x) + M$ on $[c, d]$.*

Proof. Assume that $\varphi(x) > \psi(x) + M$ at some points in (c, d) and let $N = \text{Max}[\varphi(x) - \psi(x)]$ on $[c, d]$. Then $\alpha(x) = \varphi(x)$ and $\beta(x) = \psi(x) + N$ satisfy the hypotheses of Theorem 2.6. It follows that $\alpha(x) \equiv \beta(x)$ on $[c, d]$. From this contradiction we conclude that $\varphi(x) \leqslant \psi(x) + M$ on $[c, d]$.

Corollary 3.10. *If $f(x, y, y')$ satisfies the hypotheses of Theorem 3.9, a lower solution on an interval $I \subset [a, b]$ is a subfunction on I.*

It will again be the case that, if φ is an $AC^{(1)}$-lower solution on I, then φ is a $C^{(1)}$-lower solution on I.

Corollary 3.11. *If $f(x, y, y')$ satisfies the hypotheses of Theorem 3.9, solutions of boundary-value problems of class $C^{(2)}[x_1, x_2]$, $[x_1, x_2] \subset [a, b]$, when they exist are unique.*

In the preceding results of this section conditions have been placed on $f(x, y, y')$ which are sufficient to imply that lower solutions are subfunctions and, consequently, that solutions of boundary-value problems, when they exist, are unique. In the next theorem we take the uniqueness of solutions of boundary-value problems as one of the hypotheses.

Theorem 3.12. *Assume that solutions of boundary-value problems for $y'' = f(x, y, y')$, when they exist, are unique in the sense of Corollary 3.7. Assume also that each initial-value problem for $y'' = f(x, y, y')$ has a solu-*

tion which extends throughout $[a, b]$. *Then, if* $I \subset [a, b]$ *and* $\varphi \in C^{(1)}(I)$ *is a lower solution on* I, φ *is a subfunction on* I [24].

Proof. Assume that φ is not a subfunction on I. Then there is an interval $[c, d] \subset I$ and a solution $y_0 \in C^{(2)}[c, d]$ such that $y_0(c) = \varphi(c)$, $y_0(d) = \varphi(d)$, and $y_0(x) < \varphi(x)$ on (c, d). Now define $F(x, y, y')$ on $[c, d] \times R^2$ by

$$F(x, y, y') = \begin{cases} f(x, y, y') & \text{for} \quad y \geqslant \varphi(x), \\ f(x, \varphi(x), y') - (\varphi(x) - y) & \text{for} \quad y \leqslant \varphi(x). \end{cases}$$

Since $F(x, y, y')$ is continuous on $[c, d] \times R^2$ and $\varphi \in C^{(1)}[c, d]$, it follows from Theorem 2.1 that there is a $\delta > 0$ such that $[x_1, x_2] \subset [c, d]$ and $x_2 - x_1 \leqslant \delta$ implies that the boundary-value problem

$$y'' = F(x, y, y'), \qquad y(x_1) = \varphi(x_1), \qquad y(x_2) = \varphi(x_2)$$

has a solution $y(x) \in C^{(2)}[x_1, x_2]$. Using the fact that φ is a lower solution we can show that $y(x) \geqslant \varphi(x)$ on $[x_1, x_2]$ with the same type of argument as used in the proof of Theorem 2.5. Consequently, for $[x_1, x_2] \subset [c, d]$ and $x_2 - x_1 \leqslant \delta$, the boundary value-problem

$$y'' = f(x, y, y'), \qquad y(x_1) = \varphi(x_1), \qquad y(x_2) = \varphi(x_2)$$

has a solution $y(x) \in C^{(2)}[x_1, x_2]$ with $\varphi(x) \leqslant y(x)$ on $[x_1, x_2]$. Thus φ is a subfunction "in the small".

Now $d - c > \delta$ since otherwise there would be a solution $y(x)$ with $y(c) = \varphi(c)$, $y(d) = \varphi(d)$, and $y(x) \geqslant \varphi(x)$ on $[c, d]$. This solution would be distinct from $y_0(x)$ contradicting our assumption concerning the uniqueness of solutions of boundary value problems. Now for each positive integer n let $P(n)$ be the proposition that there exists an interval $[c_n, d_n] \subset [c, d]$ with $0 < d_n - c_n \leqslant d - c - (n - 1)\,\delta$ and a solution $y_n(x) \in C^{(2)}[c_n, d_n]$ with $y_n(c_n) = \varphi(c_n)$, $y_n(d_n) = \varphi(d_n)$, and $y_n(x) < \varphi(x)$ on (c_n, d_n). Now $P(1)$ is true with $[c_1, d_1] = [c, d]$ and $y_1(x) = y_0(x)$. Assume $P(k)$ is true. Then $d_k - c_k > \delta$ otherwise we would obtain a contradiction of $y_k(x)$ being the distinct solution with boundary values $\varphi(c_k)$ and $\varphi(d_k)$. Let $z_1(x)$ be the solution of the boundary-value problem

$$y'' = f(x, y, y'), \qquad y(c_k) = \varphi(c_k), \qquad y(c_k + \delta) = \varphi(c_k + \delta).$$

Since each initial-value problem has a solution extending throughout $[a, b]$, there is a solution $z_2(x)$ on $[c_k, d_k]$ such that $z_2(x) \equiv z_1(x)$ on

$[c_k, c_k + \delta]$. Now, if $P(k + 1)$ is not true, we must have $z_2(x) \geqslant \varphi(x)$ on $[c_k + \delta, d_k]$. Also we must have $z_2(d_k) > \varphi(d_k)$. If $d_k - c_k - \delta \leqslant \delta$, the boundary-value problem

$$y'' = f(x, y, y'), \qquad y(c_k + \delta) = \varphi(c_k + \delta), \qquad y(d_k) = \varphi(d_k)$$

has a solution $z_3 \in C^{(2)}[c_k + \delta, d_k]$ with $\varphi(x) \leqslant z_3(x)$. Then, since $z_3(d_k) < z_2(d_k)$ and solutions of boundary-value problems are unique,

$$\varphi(x) \leqslant z_3(x) \leqslant z_2(x)$$

on $[c_k + \delta, d_k]$. This implies

$$z_3(c_k + \delta) = z_2(c_k + \delta) \qquad \text{and} \qquad z_3'(c_k + \delta) = z_2'(c_k + \delta).$$

Consequently, $u(x)$ defined by

$$u(x) = \begin{cases} z_1(x) & \text{on} \quad [c_k, c_k + \delta] \\ z_3(x) & \text{on} \quad [c_k + \delta, d_k], \end{cases}$$

is of class $C^{(2)}[c_k, d_k]$ and is a solution on $[c_k, d_k]$ with $u(c_k) = y_k(c_k)$, $u(d_k) = y_k(d_k)$. However $u(x) \not\equiv y_k(x)$ on $[c_k, d_k]$ and this contradicts the uniqueness of solutions of boundary-value problems. We conclude that $d_k - c_k - \delta > \delta$. This being the case the boundary-value problem

$$y'' = f(x, y, y'), \qquad y(c_k + \delta) = \varphi(c_k + \delta), \qquad y(c_k + 2\delta) = \varphi(c_k + 2\delta)$$

has a solution $z_4 \in C^{(2)}[c_k + \delta, c_k + 2\delta]$ with

$$\varphi(x) \leqslant z_4(x) \leqslant z_2(x)$$

on $[c_k + \delta, c_k + 2\delta]$. Again this implies $z_4(c_k + \delta) = z_2(c_k + \delta)$ and $z_4'(c_k + \delta) = z_2'(c_k + \delta)$. Hence

$$v(x) = \begin{cases} z_1(x) & \text{on} \quad [c_k, c_k + \delta], \\ z_4(x) & \text{on} \quad [c_k + \delta, c_k + 2\delta], \end{cases}$$

is such that $v \in C^{(2)}[c_k, c_k + 2\delta]$ and is a solution of the boundary-value problem

$$y'' = f(x, y, y'), \qquad y(c_k) = \varphi(c_k), \qquad y(c_k + 2\delta) = \varphi(c_k + 2\delta).$$

This solution $v(x)$ has an extension $z_5(x)$ to all of $[a, b]$. Since $P(k + 1)$ is assumed to be false we must have $z_5(x) \geqslant \varphi(x)$ on $[c_k + 2\delta, d_k]$ and

$z_5(d_k) > \varphi(d_k)$. Now the above arguments can be repeated and the assumption that $P(k + 1)$ is false allows us to work our way across the interval $[c_k, d_k]$ by subintervals of length δ until we obtain a solution $w \in C^{(2)}[c_k, d_k]$ with $w(x) \geqslant \varphi(x)$ on $[c_k, d_k]$, $w(c_k) = \varphi(c_k) = y_k(c_k)$, and $w(d_k) = \varphi(d_k) = y_k(d_k)$. Then $w(x) \not\equiv y_k(x)$ on $[c_k, d_k]$ which again contradicts the uniqueness of solutions of boundary-value problems.

We conclude that, if $P(k)$ is true, it follows that $P(k + 1)$ is true. Hence, $P(n)$ is true for all $n \geqslant 1$. This leads to the contradiction $0 < d - c - (n - 1)\,\delta$ for all $n \geqslant 1$. Thus φ is a subfunction on I.

If one examines the induction argument used in the proof of Theorem 3.12, it can be seen that the assumption that each initial value problem has a solution extending throughout $[a, b]$ can be weakened. It would suffice to assume that for any $a \leqslant x_0 < b$ and any solution $y(x)$ of an initial-value problem at x_0 either $y(x)$ has an extension to $[x_0, b]$ or $y(x)$ is unbounded on $[x_0, c)$ where $[x_0, c)$ is a maximal interval of existence of $y(x)$. This will be the case if $f(x, y, y')$ satisfies a Nagumo condition. Nagumo conditions will be discussed in later sections.

The question of whether or not Theorem 3.12 remains valid if it is assumed only that $f(x, y, y')$ is continuous and that solutions of boundary-value problems when they exist are unique is still not answered.

In this Section most of the results have been stated in terms of lower solutions and subfunctions. There are corresponding results concerning upper solutions and superfunctions. When it becomes necessary to refer to such a result for superfunctions we shall simply refer to the subfunction statement of the result.

4. Properties of Subfunctions and the Study of Boundary-Value Problems by Subfunction Methods

We shall now use the Perron method in attempting to establish existence theorems for solutions of boundary value problems for second-order ordinary differential equations. First it will be necessary to make a more detailed examination of properties of subfunctions and superfunctions. Then these properties and Theorem 2.1, the existence "in the small" Theorem, will be used to establish existence "in the large". Again in this Section most results will be stated in terms of subfunctions and the obvious analogous results for superfunctions will not be stated.

As remarked earlier we will always assume that $f(x, y, y')$ is continuous on its domain $I \times R^2$. When additional hypotheses are required, they will be stated.

Theorem 4.1. *If φ is a subfunction on an interval $J \subset I$, then φ has right- and left-hand limits in the extended reals at each point in J^0 and has appropriate one-sided limits at finite endpoints of J.*

Proof. It will suffice to consider one case. Let $x_0 \in J^0$ and assume that $\varphi(x_0 - 0) = \lim_{x \to x_0 -} \varphi(x)$ does not exist in the extended reals. Then there exist real numbers α and β such that

$$\liminf_{x \to x_0 -} \varphi(x) < \alpha < \beta < \limsup_{x \to x_0 -} \varphi(x).$$

Let $\{t_n\}$, $\{s_n\}$ be strictly increasing sequences in J such that $t_n < s_n < t_{n+1}$ for $n \geqslant 1$, $\lim t_n = \lim s_n = x_0$, $\lim \varphi(t_n) = \limsup_{x \to x_0 -} \varphi(x)$, and $\lim \varphi(s_n) = \liminf_{x \to x_0 -} \varphi(x)$. With $\epsilon = \frac{1}{4}(\beta - \alpha)$ it follows from Theorem 2.1 that there is a $\delta > 0$ such that for any $[x_1, x_2] \subset [t_1, x_0]$ with $x_2 - x_1 \leqslant \delta$ the boundary-value problem

$$y'' = f(x, y, y'), \qquad y(x_1) = y(x_2) = \tfrac{1}{2}(\alpha + \beta)$$

has a solution $y(x) \in C^{(2)}[x_1, x_2]$ with $|y(x) - \frac{1}{2}(\alpha + \beta)| < \epsilon$ on $[x_1, x_2]$. Let n be a fixed positive integer chosen large enough that $s_{n+1} - s_n \leqslant \delta$, $\varphi(s_n) < \alpha$, $\varphi(s_{n+1}) < \alpha$, and $\varphi(t_{n+1}) > \beta$. It follows that there is a solution of the above boundary-value problem with $[x_1, x_2] = [s_n, s_{n+1}]$ and $|y(x) - \frac{1}{2}(\alpha + \beta)| < \epsilon$ on $[s_n, s_{n+1}]$. Since φ is a subfunction on J and $\varphi(s_n) < y(s_n)$, $\varphi(s_{n+1}) < y(s_{n+1})$, it must be the case that $\varphi(t_{n+1}) \leqslant y(t_{n+1})$. However,

$$y(t_{n+1}) < \tfrac{1}{2}(\alpha + \beta) + \epsilon < \beta < \varphi(t_{n+1}).$$

From this contradiction we conclude that $\varphi(x_0 - 0)$ exists in the extended reals.

Corollary 4.2. *If φ is a bounded subfunction on $J \subset I$, then φ has at most a countable number of discontinuities on J. At each $x_0 \in J^0$*

$$\varphi(x_0) \leqslant \mathrm{Max}[\varphi(x_0 + 0), \varphi(x_0 - 0)].$$

Proof. The first assertion is a classical result that follows from the fact that $\varphi(x_0 + 0)$ and $\varphi(x_0 - 0)$ exist at each $x_0 \in J^0$. The second assertion follows readily from Theorem 2.1 and the fact that φ is a subfunction on J.

Next we consider differentiability of subfunctions. For a function $g(x)$ with a finite right-hand limit $g(x_0 + 0)$ at x_0, we define

$$Dg(x_0 +) = \lim_{x \to x_0+} \frac{g(x) - g(x_0 + 0)}{x - x_0}$$

provided the limit exists. Similarly, for a function $g(x)$ with a finite left-hand limit $g(x_0 - 0)$ at x_0, we define

$$Dg(x_0 -) = \lim_{x \to x_0-} \frac{g(x) - g(x_0 - 0)}{x - x_0}$$

provided the limit exists.

Theorem 4.3. *If φ is a bounded subfunction on $J \subset I$, then $D\varphi(x_0 +)$ and $D\varphi(x_0 -)$ exist in the extended reals for each $x_0 \in J^0$. The appropriate one-sided "derivatives" exist at finite end points of J.*

Proof. As in Theorem 4.1 it will suffice to consider just one case. Assume $x_0 \in J^0$ and that

$$\liminf_{x \to x_0+} \frac{\varphi(x) - \varphi(x_0 + 0)}{x - x_0} < \limsup_{x \to x_0+} \frac{\varphi(x) - \varphi(x_0 + 0)}{x - x_0}.$$

Let m be a real number strictly between these two limits. Then the initial-value problem

$$y'' = f(x, y, y'), \qquad y(x_0) = \varphi(x_0 + 0), \qquad y'(x_0) = m$$

has a solution $y(x) \in C^{(2)}[x_0, x_0 + \delta]$ for some $\delta > 0$. Since

$$\lim_{x \to x_0+} \frac{y(x) - y(x_0)}{x - x_0} = m$$

and m is strictly between the above lower and upper derivatives, there exist x_1, x_2, x_3 such that $x_0 < x_1 < x_2 < x_3 < x_0 + \delta$,

$$\frac{\varphi(x) - \varphi(x_0 + 0)}{x - x_0} < \frac{y(x) - y(x_0)}{x - x_0}$$

at $x = x_1$ and $x = x_3$, and

$$\frac{\varphi(x_2) - \varphi(x_0 + 0)}{x_2 - x_0} > \frac{y(x_2) - y(x_0)}{x_2 - x_0}.$$

It follows that $\varphi(x_1) < y(x_1)$, $\varphi(x_3) < y(x_3)$, and $\varphi(x_2) > y(x_2)$. This contradicts the fact that φ is a subfunction on J. We conclude that $D\varphi(x_0 +)$ exists in the extended reals.

Corollary 4.4. *If φ is a bounded subfunction on $J \subset I$, then φ has a finite derivative almost everywhere on J.*

Proof. This is an immediate consequence of Theorem 4.3 and classical results in the theory of functions of a real variable.

Lemma 4.5. *If $\varphi(x)$ is subfunction on the interval $J \subset I$ and is bounded above on each compact subinterval of J, then $\bar{\varphi}(x) = \limsup_{z \to x} \varphi(z)$ is a subfunction on J.*

Proof. Let $[x_1, x_2] \subset J$ and assume $y(x)$ is a solution of class $C^{(2)}[x_1, x_2]$ with $\bar{\varphi}(x_1) \leqslant y(x_1)$ and $\bar{\varphi}(x_2) \leqslant y(x_2)$. Then $\varphi(x_i) \leqslant \bar{\varphi}(x_i) \leqslant y(x_i)$ for $i = 1, 2$ and, since φ is a subfunction, $\varphi(x) \leqslant y(x)$ on $[x_1, x_2]$. It follows that for each $x_1 < x < x_2$ $\bar{\varphi}(x) = \limsup_{z \to x} \varphi(z) \leqslant y(x)$. Hence $\bar{\varphi}(x) \leqslant y(x)$ on $[x_1, x_2]$ and $\bar{\varphi}$ is a subfunction on J.

Next we consider some lattice properties of subfunctions.

Theorem 4.6. *Assume that the collection of subfunctions $\{\varphi_\alpha : \alpha \in A\}$ on the interval $J \subset I$ is bounded above at each point of J. Then $\varphi_0(x) = \sup_{\alpha \in A} \varphi_\alpha(x)$ is a subfunction on J.*

Proof. Assume $[x_1, x_2] \subset J$ and assume that $y(x) \in C^{(2)}[x_1, x_2]$ is a solution on $[x_1, x_2]$ with $\varphi_0(x) \leqslant y(x)$ at $x = x_1, x_2$. Then from the definition of $\varphi_0(x)$ it follows that $\varphi_\alpha(x) \leqslant y(x)$ at $x = x_1, x_2$ for each $\alpha \in A$. Since each φ_α is a subfunction on J, we conclude that $\varphi_\alpha(x) \leqslant y(x)$ on $[x_1, x_2]$ for each $\alpha \in A$. This implies $\varphi_0(x) \leqslant y(x)$ on $[x_1, x_2]$ and φ_0 is a subfunction on J.

Theorem 4.7. *Let φ be a subfunction on an interval $J \subset I$ and φ_1 a subfunction on an interval J_1 with $J_1 = \bar{J}_1 \cap J$. Furthermore, assume that*

$\varphi_1(x) \leqslant \varphi(x)$ at finite end points of J_1 which are contained in J. Then φ_2 defined by

$$\varphi_2(x) = \begin{cases} \text{Max}[\varphi_1(x), \varphi(x)] & \text{for} \quad x \in J_1, \\ \varphi(x) & \text{for} \quad x \in J - J_1 \end{cases}$$

is a subfunction on J.

Proof. By hypothesis $\varphi_2(x) = \varphi(x)$ is a subfunction on $J - J_1$ and by Theorem 4.6 $\varphi_2(x)$ is a subfunction on J_1. Consequently, to complete the proof we need to show that we have the correct behavior on intervals $[x_1, x_2] \subset J$ which are not contained in either J_1 or $J - J_1$. We will consider just one case since the arguments in the other cases proceed in a similar way. Assume $x_1 \in J_1$, $x_2 \in J - J_1$, and x_3, $x_1 < x_3 < x_2$, is the right-hand end point of J_1. Assume $y(x) \in C^{(2)}[x_1, x_2]$ is a solution on $[x_1, x_2]$ with $\varphi_2(x_1) \leqslant y(x_1)$ and $\varphi_2(x_2) \leqslant y(x_2)$. Then $\varphi(x_1) \leqslant \varphi_2(x_1) \leqslant y(x_1)$ and $\varphi(x_2) = \varphi_2(x_2) \leqslant y(x_2)$, and since φ is a subfunction on J it follows that $\varphi(x) \leqslant y(x)$ on $[x_1, x_2]$. In particular, $\varphi_2(x) = \varphi(x) \leqslant y(x)$ on $(x_3, x_3]$. Also $\varphi_1(x_3) \leqslant \varphi(x_3) \leqslant y(x_3)$ and $\varphi_1(x_1) \leqslant \varphi_2(x_1) \leqslant y(x_1)$, hence, since φ_1 is a subfunction on J_1, $\varphi_1(x) \leqslant y(x)$ on $[x_1, x_3]$. It follows that $\varphi_2(x) = \text{Max}[\varphi(x), \varphi_1(x)] \leqslant y(x)$ on $[x_1, x_3]$. Putting these things together we have $\varphi_2(x) \leqslant y(x)$ on $[x_1, x_2]$. The other possibilities are dealt with in a similar way and we conclude that φ_2 is a subfunction on J.

Theorem 4.8. *Assume that $f(x, y, y')$ is nondecreasing in y on $I \times R^2$ for each fixed x, y' and either satisfies a Lipschitz condition with respect to y' on each compact subset of $I \times R^2$ or is such that solutions of initial-value problems are unique. Then, if $\psi(x)$ is an upper solution on $[x_1, x_2] \subset I$ and $\varphi(x)$ is a bounded subfunction on $[x_1, x_2]$ with $\varphi(x_1 + 0) \leqslant \psi(x_1)$ and $\varphi(x_2 - 0) \leqslant \psi(x_2)$, it follows that $\varphi(x) \leqslant \psi(x)$ on (x_1, x_2).*

Proof. Assume that the stated conditions hold but that $\varphi(x) > \psi(x)$ at some points of (x_1, x_2). First we note that it suffices to consider the case where φ is upper semicontinuous on (x_1, x_2). To see this let $\bar{\varphi}(x) = \limsup_{z \to x} \varphi(z)$ on (x_1, x_2). Then by Lemma 4.5, $\bar{\varphi}(x)$ is a subfunction on (x_1, x_2). By Corollary 4.2 $\bar{\varphi}(x_1 + 0) = \varphi(x_1 + 0) \leqslant \psi(x_1)$ and $\bar{\varphi}(x_2 - 0) = \varphi(x_2 - 0) \leqslant \psi(x_2)$. Furthermore $\varphi(x) > \psi(x)$ at some points in (x_1, x_2) implies $\bar{\varphi}(x) > \psi(x)$ at some points in (x_1, x_2). Consequently, we may assume $\varphi(x)$ is upper semicontinuous on (x_1, x_2). This being the case $\varphi(x) > \psi(x)$ at some points in (x_1, x_2) and

$\varphi(x_1 + 0) \leqslant \psi(x_1)$, $\varphi(x_2 - 0) \leqslant \psi(x_2)$ implies $\varphi(x) - \psi(x)$ has a positive maximum M which is assumed on a compact set $E \subset (x_1, x_2)$.

Now assume $f(x, y, y')$ is nondecreasing in y and that solutions of initial-value problemas are unique. Let $x_0 = \text{lub } E$. Then there exists a $\delta > 0$ and $\epsilon > 0$ such that

$$[x_0 - \delta, x_0 + \delta] \subset (x_1, x_2), \qquad 0 < \epsilon < \psi(x_0 + \delta) + M - \varphi(x_0 + \delta),$$

and such that the boundary-value problem

$$y'' = f(x, y, y'), \qquad y(x_0 - \delta) = \psi(x_0 - \delta) + M,$$
$$y(x_0 + \delta) = \psi(x_0 + \delta) + M - \epsilon$$

has a solution $y(x) \in C^{(2)}[x_0 - \delta, x_0 + \delta]$. This follows from Theorem 2.1. Since $\psi(x) + M$ is an upper solution, it follows from Theorem 2.6 that $y(x) < \psi(x) + M$ on $(x_0 - \delta, x_0 + \delta)$. However,

$$\varphi(x_0 - \delta) \leqslant \psi(x_0 - \delta) + M = y(x_0 - \delta)$$

and

$$\varphi(x_0 + \delta) < \psi(x_0 + \delta) + M - \epsilon = y(x_0 + \delta),$$

which implies $\varphi(x_0) \leqslant y(x_0)$ since φ is a subfunction on (x_1, x_2). This contradicts the fact that $x_0 \in E$ and $\varphi(x_0) = \psi(x_0) + M$. We conclude that with these hypotheses $\varphi(x) \leqslant \psi(x)$ on (x_1, x_2).

Now assume that $f(x, y, y')$ is nondecreasing in y and satisfies a Lipschitz condition with respect to y' on each compact subset of $I \times R^2$. Let $[x_3, x_4] \subset (x_1, x_2)$ be such that $E \subset [x_3, x_4]$ and

$$\varphi(x_3) < \psi(x_3) + \frac{M}{2},$$

$$\varphi(x_4) < \psi(x_4) + \frac{M}{2}.$$

Then by Lemma 3.4 there is a $\psi_1 \in C^{(1)}[x_3, x_4]$ such that

$$\psi(x) \leqslant \psi_1(x) \leqslant \psi(x) + \frac{M}{2} \qquad \text{on} \qquad [x_3, x_4]$$

and

$$\bar{D}\psi_1' < f(x, \psi_1, \psi_1') \qquad \text{on} \qquad (x_3, x_4).$$

It follows that $\varphi(x_3) < \psi_1(x_3) + M/2$, $\varphi(x_4) < \psi_1(x_4) + M/2$, and $\varphi(x) = \psi(x) + M \geqslant \psi_1(x) + M/2$ on $E \subset (x_3, x_4)$. Hence, $\varphi(x) - \psi_1(x)$ has a positive maximum N on a compact subset $E_1 \subset (x_3, x_4)$. Let $x_5 = \text{lub } E_1$. Then there is, by Theorem 2.1, a $\delta > 0$ such that $[x_5 - \delta, x_5 + \delta] \subset (x_3, x_4)$ and such that the boundary-value problem

$$y'' = f(x, y, y'), \qquad y(x_5 \pm \delta) = \psi_1(x_5 \pm \delta) + N$$

has a solution $y(x) \in C^{(2)}[x_5 - \delta, x_5 + \delta]$. By Lemma 3.3

$$y(x) < \psi_1(x) + N \qquad \text{on} \qquad (x_5 - \delta, x_5 + \delta).$$

Since

$$\varphi(x_5 \pm \delta) \leqslant \psi_1(x_5 \pm \delta) + N = y(x_5 \pm \delta)$$

and φ is a subfunction, $\varphi(x_5) \leqslant y(x_5) < \psi_1(x_5) + N$, which contradicts $x_5 \in E_1$. We conclude again that $\varphi(x) \leqslant \psi(x)$ on (x_1, x_2).

We consider now properties of bounded functions which are simultaneously subfunctions and superfunctions. We will need a result concerning solutions of initial value problems which is well known. For the sake of completeness we include a proof of this result.

Lemma 4.9. *If $(x_0, y_0, y_0') \in I \times R^2$, there exist $\delta > 0$, $M_1 > 0$, and $M_2 > 0$ such that every solution of the initial-value problem $y'' = f(x, y, y')$, $y(x_0) = y_0$, $y'(x_0) = y_0$ is defined on*

$$I_\delta = [x_0 - \delta, x_0 + \delta] \cap I.$$

Furthermore, $|y(x)| \leqslant M_1$ and $|y'(x)| \leqslant M_2$ on I_δ for all solutions.

Proof. Given $(x_0, y_0, y_0') \in I \times R^2$ first choose $\delta_1 > 0$ such that $I_{\delta_1} = [x_0 - \delta_1, x_0 + \delta_1] \cap I$ is a compact subinterval of I. Then let

$$M = \text{Max}\{|f(x, y, y')| : x \in I_{\delta_1}, |y - y_0| \leqslant 1, |y' - y_0'| \leqslant 1\}.$$

Let $\delta = \text{Min}[\delta_1, 1/M, 1/(|y_0'| + 1)]$, $M_1 = |y_0| + 1$, and $M_2 = |y_0'| + 1$. Then it is easy to see that every solution of the given initial value problem extends to I_δ and on I_δ all solutions satisfy $|y(x)| \leqslant M_1$, $|y'(x)| \leqslant M_2$.

Theorem 4.10. *Assume that $f(x, y, y')$ is such that $C^{(2)}$ solutions of boundary-value problems, when they exist, are unique. That is, assume that,* *if* $[x_1, x_2] \subset I$ *and* $y_1, y_2 \in C^{(2)}[x_1, x_2]$ *are solutions of* $y'' = f(x, y, y')$ *on* $[x_1, x_2]$ *with* $y_1(x_1) = y_2(x_1)$ *and* $y_1(x_2) = y_2(x_2)$, *then* $y_1(x) \equiv y_2(x)$ *on* $[x_1, x_2]$. *Assume that* $z(x)$ *is bounded on each compact subinterval of* $J \subset I$ *and that* $z(x)$ *is simultaneously a subfunction and a superfunction on* J. *Then* $z(x)$ *is a solution of* $y'' = f(x, y, y')$ *on an open subset of* J *the complement of which has measure zero. Furthermore, if* $x_0 \in J^0$ *is a point of continuity of* $z(x)$ *at which* $z(x)$ *does not have a finite derivative, then either* $Dz(x_0 +) = Dz(x_0 -) = +\infty$ *or* $Dz(x_0 +) = Dz(x_0 -) = -\infty$. *If* $z(x_0 + 0) > z(x_0 - 0)$, $Dz(x_0 +) = Dz(x_0 -) = +\infty$, *and if* $z(x_0 + 0) < z(x_0 - 0)$, $Dz(x_0 +) = Dz(x_0 -) = -\infty$.

Proof. By Corollary 4.4 $z(x)$ has a finite derivative almost everywhere on J. If $x_0 \in J^0$ is a point at which $z(x)$ has a finite derivative, there is a $\delta > 0$ such that $[x_0 - \delta, x_0 + \delta] \subset J$, $|z(x)| \leqslant |z(x_0)| + 1$ on $[x_0 - \delta, x_0 + \delta]$, and

$$\left| \frac{z(x_0 + \eta) - z(x_0 - \eta)}{2\eta} \right| \leqslant |z'(x_0)| + 1 \qquad \text{for} \qquad 0 < \eta \leqslant \delta.$$

It then follows from Theorem 2.1 that there is a δ_1, $0 < \delta_1 \leqslant \delta$, such that the boundary-value problem

$$y'' = f(x, y, y'), \qquad y(x_0 \pm \delta_1) = z(x_0 \pm \delta_1)$$

has a solution $y(x) \in C^{(2)}[x_0 - \delta_1, x_0 + \delta_1]$. Since $z(x)$ is simultaneously a subfunction and a superfunction on J, $z(x) \equiv y(x)$ on $[x_0 - \delta_1, x_0 + \delta_1]$. We conclude that $z(x)$ is a solution of $y'' = f(x, y, y')$ on an open subset of J the complement of which has measure zero.

Next let $x_0 \in J^0$ be a point of continuity of $z(x)$ at which $z(x)$ does not have a finite derivative. By Theorem 4.3, $Dz(x_0 +)$ and $Dz(x_0 -)$ both exist in the extended reals. If both are finite, then by the same argument as used above there is an interval around x_0 in which $z(x)$ is a solution. This contradicts the assumption that $z(x)$ does not have a finite derivative at x_0. Consequently, at least one of $Dz(x_0 +)$, $Dz(x_0 -)$ is not finite. To be specific assume $Dz(x_0 +) = +\infty$ and $Dz(x_0 -) \neq +\infty$. Then there exist numbers $\delta > 0$ and N such that

$$z(x) > w(x) = z(x_0) + N(x - x_0) \qquad \text{on} \qquad x_0 - \delta \leqslant x < x_0 .$$

By Theorem 2.1 there is a δ_1, $0 < \delta_1 \leqslant \delta$, such that the boundary-value problem

$$y'' = f(x, y, y'), \qquad y(x_0 - \delta_1) = w(x_0 - \delta_1), \qquad y(x_0) = z(x_0)$$

has a solution $y_1(x) \in C^{(2)}[x_0 - \delta_1, x_0]$. By Lemma 4.9 there is a $\delta_2 > 0$ such that $y_1(x)$ can be extended to be a solution on $[x_0 - \delta_1, x_0 + \delta_2]$ and such that all solutions of the initial-value problem

$$y'' = f(x, y, y'), \qquad y(x_0) = y_1(x_0), \qquad y'(x_0) = y_1'(x_0)$$

exist on $[x_0, x_0 + \delta_2]$ and satisfy $| y(x) - y(x_0) | \leqslant M(x - x_0)$ on $[x_0, x_0 + \delta_2]$ where $M = | y_1'(x_0) | + 1$. Again applying Theorem 2.1 we conclude that there is an η, $0 < \eta \leqslant \text{Min}[\delta_1, \delta_2]$, such that for $0 < \delta_3 \leqslant \eta$ the boundary-value problem

$$y'' = f x, y, y'), \qquad y(x_0 - \delta_3) = y_1(x_0 - \delta_3),$$
$$y(x_0 + \delta_3) = y_1(x_0) + (M + \epsilon)\, \delta_3$$

has a solution $y_2(x) \in C^{(2)}[x_0 - \delta_3, x_0 + \delta_3]$ where $\epsilon > 0$ is fixed. Since $Dz(x_0 +) = + \infty$, we can assume that δ_3 is chosen so that $0 < \delta_3 \leqslant \eta$ and $y_1(x_0) + (M + \epsilon)\, \delta_3 < z(x_0 + \delta_3)$. Then

$$y_1(x_0 + \delta_3) < y_2(x_0 + \delta_3) < z(x_0 + \delta_3)$$

and

$$y_1(x_0 - \delta_3) = y_2(x_0 - \delta_3) \leqslant z(x_0 - \delta_3).$$

The last inequality follows from the fact that $z(x)$ is a superfunction and $y_1(x_0) = z(x_0)$, $y_1(x_0 - \delta_1) < z(x_0 - \delta_1)$. Since $z(x)$ is a superfunction, we conclude from the above inequalities that $y_2(x) \leqslant z(x)$ on $[x_0 - \delta_3, x_0 + \delta_3]$. From the same inequalities and the fact that solutions of boundary-value problems when they exist are unique, we conclude that $y_1(x) \leqslant y_2(x)$ on $[x_0 - \delta_3, x_0 + \delta_3]$. Thus $y_1(z_0) = z(x_0) = y_2(x_0)$ and $y_1'(x_0) = y_2'(x_0)$, hence, $y_2(x)$ is a solution of the initial-value problem with initial conditions $y(x_0) = y_1(x_0)$, $y'(x_0) = y_1'(x_0)$. But

$$| y_2(x_0 + \delta_3) - y_2(x_0) | = (M + \epsilon)\, \delta_3$$

which contradicts the fact that all solutions of this initial value problem satisfy $| y(x) - y_1(x_0) | \leqslant M(x - x_0)$ on $[x_0, x_0 + \delta_2]$. We are forced to conclude that $Dz(x_0 -) = + \infty$. By similar arguments, in some cases using the fact that $z(x)$ is also a subfunction, the other statements concerning the behavior of $z(x)$ at a point of continuity can be established.

We consider now the behavior of $z(x)$ at points of discontinuity. If $x_0 \in J^0$ is a point of discontinuity of $z(x)$, then by Theorem 4.1 $z(x_0 + 0)$ and $z(x_0 - 0)$ both exist and are finite since $z(x)$ is bounded on each compact interval of J. Furthermore, $z(x_0 + 0) \neq z(x_0 - 0)$ since, by Corollary 4.2,

$$\mathrm{Min}[z(x_0 + 0), z(x_0 - 0)] \leqslant z(x_0) \leqslant \mathrm{Max}[z(x_0 + 0), z(x_0 - 0)].$$

Assume that $z(x_0 + 0) > z(x_0 - 0)$ and that $Dz(x_0 +) \neq +\infty$. Then there is a $\delta > 0$ and an N such that $[x_0, x_0 + \delta] \subset J$ and $z(x) < w(x) = z(x_0 + 0) + N(x - x_0)$ on $x_0 < x \leqslant x_0 + \delta$. By Theorem 2.1 there is a δ_1, $0 < \delta_1 \leqslant \delta$, such that the boundary-value problem

$$y'' = f(x, y, y'), \qquad y(x_0) = z(x_0 + 0), \qquad y(x_0 + \delta_1) = w(x_0 + \delta_1)$$

has a solution $y_1(x) \in C^{(2)}[x_0, x_0 + \delta_1]$. Since

$$y_1(x_0) = z(x_0 + 0) \geqslant z(x_0), \qquad y_1(x_0 + \delta_1) = w(x_0 + \delta_1) > z(x_0 + \delta_1)$$

and $z(x)$ is a subfunction, $y_1(x) \geqslant z(x)$ on $[x_0, x_0 + \delta_1]$. Now proceeding as in the paragraph above and using the fact that $z(x)$ is a subfunction, we can obtain a solution of the initial-value problem

$$y'' = f(x, y, y'), \qquad y(x_0) = y_1(x_0), \qquad y'(x_0) = y_1'(x_0),$$

the graph of which is not contained in the sector to the left of x_0 in which such solutions must be. From this contradiction we conclude that $Dz(x_0 +) = +\infty$. The other assertions about derivatives at points of discontinuity of $z(x)$ are dealt with in a similar way.

It should be noted that Theorem 4.10 is a combination of results concerning subfunctions and superfunctions. Let $\varphi(x)$ be a subfunction on $J \subset I$ and assume that $\varphi(x)$ is bounded on each compact subinterval of J. Then, if $\varphi(x)$ is continuous at $x_0 \in J^0$ and $D\varphi(x_0 -) = +\infty$, it follows that $D\varphi(x_0 +) = +\infty$. If $\varphi(x)$ is continuous at $x_0 \in J^0$ and $D\varphi(x_0 +) = -\infty$, then $D\varphi(x_0 -) = -\infty$. If $\varphi(x_0 + 0) > \varphi(x_0 - 0)$, $D\varphi(x_0 +) = +\infty$. If $\varphi(x_0 + 0) < \varphi(x_0 - 0)$, $D\varphi(x_0 -) = -\infty$. If $D\varphi(x_0 +)$ and $D\varphi(x_0 -)$ are both finite, $\varphi(x)$ is continuous at x_0. Similarly, let $\psi(x)$ be a superfunction on $J \subset I$ which is bounded on each compact subinterval of J. Then, if $\psi(x)$ is continuous at $x_0 \in J^0$ and $D\psi(x_0 +) = +\infty$, it follows that $D\psi(x_0 -) = +\infty$. If $\psi(x)$ is continuous at x_0 and $D\psi(x_0 -) = -\infty$, then $D\psi(x_0 +) = -\infty$. If $\psi(x_0 + 0) > \psi(x_0 - 0)$, $D\psi(x_0 -) = +\infty$. If $\psi(x_0 + 0) < \psi(x_0 - 0)$, $D\psi(x_0 +) = -\infty$. If $D\psi(x_0 +)$ and $D\psi(x_0 -)$ are both finite, $\psi(x)$ is continuous at x_0.

Now we begin the consideration of boundary value problems by the Perron method. We assume now that $f(x, y, y')$ is continuous on $[a, b] \times R^2$.

Definition 4.11. A bounded real-valued function φ defined on $[a, b]$ is said to be an underfunction with respect to the boundary-value problem

$$y'' = f(x, y, y'), \qquad y(a) = A, \qquad y(b) = B$$

in case $\varphi(a) \leqslant A$, $\varphi(b) \leqslant B$, and φ is a subfunction on $[a, b]$ with respect to solutions of $y'' = f(x, y, y')$. The bounded function $\psi(x)$ defined on $[a, b]$ is said to be an overfunction with respect to the boundary-value problem in case $\psi(a) \geqslant A$, $\psi(b) \geqslant B$, and ψ is a superfunction on $[a, b]$ with respect to solutions of $y'' = f(x, y, y')$.

Theorem 4.12. *Assume that $C^{(2)}$ solutions of boundary value problems for $y'' = f(x, y, y')$ on subintervals of $[a, b]$ are unique in the sense specified in the statement of Theorem 4.8. Assume that there exist both an overfunction ψ_0 and an underfunction φ_0 with respect to the boundary-value problem*

$$(*) \qquad y'' = f(x, y, y'), \qquad y(a) = A, \qquad y(b) = B$$

and that $\varphi_0(x) \leqslant \psi_0(x)$ on $[a, b]$. Let Φ be the collection of all underfunctions φ such that $\varphi(x) \leqslant \psi_0(x)$ on $[a, b]$. Then $z(x) = \sup_{\varphi\in\Phi} \varphi(x)$ is simultaneously a subfunction and a superfunction on $[a, b]$.

Proof. It follows from Theorem 4.6 that z is a subfunction on $[a, b]$. Now assume that $z(x)$ is not a superfunction on $[a, b]$. Then there is a subinterval $[x_1, x_2] \subset [a, b]$ and a solution $y(x) \in C^{(2)}[x_1, x_2]$ such that $y(x_1) \leqslant z(x_1)$, $y(x_2) \leqslant z(x_2)$, but $y(x) > z(x)$ at some points of (x_1, x_2). Define $z_1(x)$ on $[a, b]$ by

$$z_1(x) = \begin{cases} \mathrm{Max}[y(x), z(x)] & \text{on} \quad [x_1, x_2] \\ z(x) & \text{on} \quad [a, b] - [x_1, x_2]. \end{cases}$$

Then, by Theorem 4.7, $z_1(x)$ is a subfunction on $[a, b]$ also $z_1(a) = z(a) \leqslant A$, $z_1(b) = z(b) \leqslant B$. Furthermore,

$$y(x_1) \leqslant z(x_1) \leqslant \psi_0(x_1), \qquad y(x_2) \leqslant z(x_2) \leqslant \psi_0(x_2),$$

and ψ_0 a superfunction implies $y(x) \leqslant \psi_0(x)$ on $[x_1, x_2]$. Consequently,

$z_1(x) \leqslant \psi_0(x)$ on $[a, b]$. Hence, $z_1 \in \Phi$ and $z_1(x) \leqslant z(x)$ but $z_1(x) = y(x) > z(x)$ at some points in (x_1, x_2). From this contradiction we conclude that $z(x)$ is a superfunction on $[a, b]$.

Definition 4.13. The function $z(x)$ obtained in Theorem 4.12 depends on the boundary-value problem (*) and on the overfunction $\psi_0(x)$. It will be designated by $z(x; \psi_0)$ and will be called a generalized solution of the boundary-value problem (*).

Since $z(x; \psi_0)$ is simultaneously a subfunction and a superfunction on $[a, b]$, the assertions made in Theorem 4.10 apply to $z(x; \psi_0)$. We now consider the behavior of $z(x; \psi_0)$ at the end points on $[a, b]$.

Theorem 4.14. *Assume that the hypotheses of Theorem 4.12 are satisfied and let $z(x; \psi_0) = z(x)$ be the corresponding generalized solution of (*). Then $z(a) = A$. If $Dz(a+) \neq +\infty$, $z(a+0) \leqslant z(a)$. If $z(a+0) < A$, $Dz(a+) = -\infty$. Hence, if $Dz(a+)$ is finite, $z(a+0) = z(a) = A$. Similar statements apply at $x = b$.*

Proof. It is clear that, if $\varphi(x)$ is a subfunction on $[a. b]$, then $\varphi_c(x)$ defined by $\varphi_c(x) = \varphi(x)$ on $(a, b]$ and $\varphi_c(a) = \varphi(a) + c$, $c > 0$, is a subfunction on $[a, b]$. From this observation and the definition of $z(x)$ it is clear that $z(a) = A$.

Now assume that $Dz(a+) \neq +\infty$ and $z(a+0) > z(a)$. Then there is a δ, $0 < \delta < b - a$, and an N such that

$$z(x) < w(x) = z(a+0) + N(x-a) \qquad \text{on} \qquad a < x \leqslant a + \delta.$$

It follows from Theorem 2.1 that for $0 < \epsilon < z(a+0) - z(a)$ and $0 < \delta_1 \leqslant \delta$ sufficiently small, the boundary-value problem

$$y'' = f(x, y, y'), \qquad y(a) = z(a+0) - \epsilon, \qquad y(a+\delta_1) = w(a+\delta_1)$$

has a solution $y(x) \in C^{(2)}[a, a+\delta_1]$. Since $z(x)$ is a subfunction, $z(x) \leqslant y(x)$ on $[a, a+\delta_1]$ which implies

$$z(a+0) \leqslant y(a+0) = y(a) = z(a+0) - \epsilon.$$

From this contradiction it follows that $Dz(a+) \neq +\infty$ implies $z(a+0) \leqslant z(a)$.

Finally, assume $z(a+0) < A$ and $Dz(a+) \neq -\infty$. Then using the same type of argument as above and the fact that $z(x)$ is also a superfunction on $[a, b]$ we obtain a contradiction. Hence, $z(a+0) < A$ implies $Dz(a+) = -\infty$.

If $Dz(a+)$ is finite, then combining the assertions of the Theorem we have $A = z(a) \geqslant z(a+0) \geqslant A$.

From the preceding results we see that the Perron method of studying the boundary value problem (*) can be separated into two parts. The first part of the problem is to establish the existence of an overfunction ψ_0 and an underfunction φ_0 such that $\varphi_0(x) \leqslant \psi_0(x)$ on $[a, b]$. The second part is to establish conditions under which the generalized solution $z(x; \psi_0)$ is of class $C^{(2)}[a, b]$ and is a solution on $[a, b]$. In view of Theorems 4.10, 4.12, and 4.14, to accomplish this it suffices to show that $Dz(x+)$ is finite on $[a, b)$ and $Dz(x-)$ is finite on $(a, b]$.

Lemma 4.15. *Let $f(x, y, y')$ be nondecreasing in y on $[a, b] \times R^2$ for fixed x, y'. Assume that $f(x, y, y')$ is such that lower and upper solutions of the differential equation are subfunctions and superfunctions, respectively. Then, if $u(x) \in C^{(2)}[a, b]$ is a solution of $y'' = f(x, y, y')$ on $[a, b]$, there exist overfunctions and underfunctions with respect to any boundary-value problem on $[a, b]$.*

Proof. For a sufficiently large $M > 0$ $\psi_0(x) = u(x) + M$ is an overfunction, $\varphi_0(x) = u(x) - M$ is an underfunction, and $\varphi_0(x) \leqslant \psi_0(x)$ on $[a, b]$.

Lemma 4.16. *Assume that $f(x, y, y')$ is nondecreasing in y on $[a, b] \times R^2$ for fixed x, y' and is such that lower and upper solutions of the differential equation are subfunctions and superfunctions. Further assume that there is a $k > 0$ such that $|f(x, 0, y') - f(x, 0, 0)| \leqslant k|y'|$ on $a \leqslant x \leqslant b$ for all y'. Then again there exist overfunctions and underfunctions with respect to every boundary-value problem on $[a, b]$* ([23], Theorem 6.1).

Proof. Let $M = \text{Max}\,|f(x, 0, 0)|$ on $[a, b]$ and let $w(x)$ be the solution of the boundary-value problem

$$w'' = -kw' - M, \qquad w(a) = 0, \qquad w'(b) = 0.$$

Then $w(x) \geqslant 0$ and $w'(x) \geqslant 0$ on $[a, b]$; hence,

$$w'' = -kw' - M = -k|w'| - M \leqslant f(x, 0, w') - f(x, 0, 0) - M.$$

Thus $w'' \leqslant f(x, 0, w') \leqslant f(x, w, w')$ on $[a, b]$ and we conclude that, for a given boundary-value problem, $\psi_0(x) = w(x) + M$ will be an overfunction provided $M > 0$ is sufficiently large.

Similarly, let $v(x)$ be the solution of the boundary-value problem

$$v'' = -kv' + M, \qquad v(a) = 0, \qquad v'(b) = 0.$$

Then $v(x) \leqslant 0$ and $v'(x) \leqslant 0$ on $[a, b]$, from which it follows that

$$v'' = k \mid v' \mid + M \geqslant f(x, 0, v') - f(x, 0, 0) + M \geqslant f(x, 0, v')$$

on $[a, b]$. Then $v'' \geqslant f(x, v, v')$ on $[a, b]$ and, for $M > 0$ sufficiently large, $\varphi_0(x) = v(x) - M$ is an underfunction with respect to a given boundary-value problem. Obviously for such functions we will have $\varphi_0(x) \leqslant \psi_0(x)$ on $[a, b]$.

Theorem 4.17. *Assume that $f(x, y, y')$ is nondecreasing in y on $[a, b] \times R^2$ for fixed x, y' and assume either that $f(x, y, y')$ satisfies a Lipschitz condition with respect to y' on each compact subset of $[a, b] \times R^2$ or that solutions of initial-value problems are unique. In addition assume that there is a $k > 0$ such that $\mid f(x, 0, y') - f(x, 0, 0) \mid \leqslant k \mid y' \mid$ on $[a, b]$ for all y'. Then for any boundary-value problem on $[a, b]$ with an associated overfunction $\psi_0(x)$ the generalized solution $z(x) = z(x; \psi_0)$ belongs to $C^{(2)}(a, b)$ and $z'' = f(x, z, z')$ on (a, b).* ([23], Corollary 6.1).

Proof. First it follows from Lemma 4.16 that, with respect to a given boundary-value problem on $[a, b]$, there is an overfunction $\psi_0(x)$ and an underfunction $\varphi_0(x)$ with $\varphi_0(x) \leqslant \psi_0(x)$ on $[a, b]$. Consequently, the generalized solution $z(x) = z(x; \psi_0)$ is defined. Furthermore, the hypotheses imply that solutions of boundary-value problems when they exist are unique, hence, the conclusions of Theorem 4.10 apply to $z(x)$. Thus it suffices to show that $Dz(x_0 +)$ and $Dz(x_0 -)$ are finite at every point of (a, b). Let $x_0 \in (a, b)$ and assume that $z(x_0 + 0) \geqslant z(x_0 - 0)$. The alternative case can be dealt with in a similar way and will not be discussed.

We break the discussion up into two cases. First assume that $z(x_0 + 0) \geqslant 0$. Let $\psi(x)$ be a solution of $\psi'' = -k\psi' - M$, $M = \text{Max} \mid f(x, 0, 0) \mid$ on $[a, b]$, with $\psi(x_0) = 0$, $\psi'(x) \geqslant 0$ on $[x_0, b]$, and $\psi(b) \geqslant z(b - 0)$. Such a solution can be determined by elementary calculations. Then, as in Lemma 4.16, $\psi_1(x) = \psi(x) + z(x_0 + 0)$ is an upper solution on $[x_0, b]$ with $z(x_0 + 0) \leqslant \psi_1(x_0)$ and $z(b - 0) \leqslant \psi_1(b)$. It follows from Theorem 4.8 that $z(x) \leqslant \psi_1(x)$ on (x_0, b), which implies that $Dz(x_0 +) \leqslant \psi_1'(x_0) < +\infty$. Applying Theorem 4.10 we conclude that $z(x)$ is continuous at x_0. Now let $\psi(x)$ be a solution of $\psi'' = k\psi' - M$ on $[a, x_0]$ such that $\psi(x_0) = 0$, $\psi(a) \geqslant z(a + 0)$, and $\psi'(x) \leqslant 0$ on $[a, x_0]$. Then, again by Theorem 4.8, $z(x) \leqslant \psi(x) + z(x_0)$, which implies $Dz(x_0 -) \geqslant \psi'(x_0) > -\infty$. We conclude that in this case $z(x)$ has a finite derivative at x_0.

Finally assume $z(x_0 + 0) < 0$ and let $\varphi_1(x)$ be a solution of $\varphi'' = k\varphi' + M$ on $[a, x_0]$ such that $\varphi(a) \leqslant z(a + 0)$, $\varphi(x_0) = 0$, and $\varphi'(x) \geqslant 0$ on $[a, x_0]$. Then, by Theorem 4.8, $z(x) \geqslant \varphi_1(x) - z(x_0 - 0)$ on (a, x_0), which implies $Dz(x_0 -) \leqslant \varphi_1'(x_0) < +\infty$. It follows from Theorem 4.10 that $z(x)$ is continuous at x_0. In a similar way we can show that $Dz(x_0 +) > -\infty$. Thus we again conclude that $z(x)$ has a finite derivative at x_0.

Since $z(x)$ has a finite derivative at each point of (a, b), it follows from Theorem 4.10 that $z(x) \in C^{(2)}(a, b)$ and is a solution of the differential equation on (a, b).

Theorem 4.18. *Assume that $f(x, y, y')$ satisfies the hypotheses of Theorem 4.17 on $[a, b] \times R^2$. Then the boundary-value problem*

$$y'' = f(x, y, y'), \qquad y(a) = y(b) = 0$$

has a unique solution $y(x) \in C^{(2)}[a, b]$. Furthermore, on $[a, b]$

$$|\, y(x) \,| \leqslant \frac{M}{k^2}\, [e^{k(b-a)} - e^{1/2k(b-a)} - \tfrac{1}{2}\, k(b - a)]$$

and

$$|\, y(x)| \leqslant \frac{M}{k^2}\, [e^{k(b-a)} - e^{\frac{1}{2}k(b-a)} - \tfrac{1}{2}\, k(b - a)]$$

where M and k are as in Theorem 4.17 ([23], Theorem 6.3).

Proof. Let $\psi_1(x)$ be the solution of the boundary-value problem $\psi'' = -k\psi' - M$; $\psi(a) = 0$, $\psi'(b) = 0$. Then $\psi_1'(x) \geqslant 0$ on $[a, b]$ and $\psi_1(x)$ is an overfunction with respect to the boundary-value problem. Similarly, the solution $\psi_2(x)$ of the boundary-value problem $\psi'' = k\psi' - M$, $\psi(b) = 0$, $\psi'(a) = 0$ satisfies $\psi_2'(x) \leqslant 0$ on $[a, b]$ and is also an overfunction. Consequently, since $\psi_1(b) > 0$ and $\psi_2(a) > 0$, $\psi_0(x) = \mathrm{Min}[\psi_1(x), \psi_2(x)]$ is also an overfunction.

The solution $\varphi_1(x)$ of the boundary-value problem $\varphi'' = -k\varphi' + M$, $\varphi(a) = 0$, $\varphi'(b) = 0$ and the solution $\varphi_2(x)$ of the boundary-value problem

$\varphi'' = k\varphi' + M$, $\varphi(b) = 0$, $\varphi'(a) = 0$ satisfy $\varphi_1'(x) \leqslant 0$, $\varphi_2'(x) \geqslant 0$, and are both underfunctions. Therefore, $\varphi_0(x) = \text{Max}[\varphi_1(x), \varphi_2(x)]$ is an underfunction. Also $\varphi_0(x) \leqslant \psi_0(x)$. It follows from Theorem 4.17 that $z(x; \psi_0)$ is a solution on (a, b). Now $\varphi_0(x) \leqslant z(x; \psi_0) \leqslant \psi_0(x)$ on $[a, b]$, $\varphi_0(a) = \psi_0(a) = \varphi_0(b) = \psi_0(b) = 0$, and $\varphi_0'(a)$, $\varphi_0'(b)$, $\psi_0'(a)$, $\psi_0'(b)$ are finite. It follows that $Dz(a+)$ and $Dz(b-)$ are finite which by Theorem 4.14 implies $z(x; \psi_0) \in C^{(2)}[a, b]$ and is a solution of the boundary-value problem. The fact that the solution is unique is a consequence of either Corollary 3.7 or Corollary 3.11.

The functions $\varphi_0(x)$ and $\psi_0(x)$ can be computed and it can be shown that

$$\underset{a \leqslant x \leqslant b}{\text{Max}}\ \psi_0(x) = -\underset{a \leqslant x \leqslant b}{\text{Min}}\ \varphi_0(x) = \frac{M}{k^2}\left[e^{k(b-a)} - e^{\frac{1}{2}k(b-a)} - \tfrac{1}{2}k(b-a)\right].$$

This establishes the desired bound on $|y(x)|$ for the solution $y(x)$. If $x_0 \in (a, b)$ and $y(x_0) \geqslant 0$, let $\psi_1(x; x_0)$ and $\psi_2(x; x_0)$ be the respective solutions of the boundary-value problems

$$\psi'' = -k\psi' - M, \qquad \psi(x_0) = 0, \qquad \psi'(b) = 0,$$

and

$$\psi'' = k\psi' - M, \qquad \psi(x_0) = 0, \qquad \psi'(a) = 0.$$

Then $y(x) \leqslant \psi_1(x; x_0) + y(x_0)$ on $[x_0, b]$ and $y(x) \leqslant \psi_2(x; x_0) + y(x_0)$ on $[a, x_0]$. It follows that $\psi_2'(x_0; x_0) \leqslant y'(x_0) \leqslant \psi_1'(x_0; x_0)$. Similarly, if $y(x_0) < 0$ and $\varphi_1(x; x_0)$, $\varphi_2(x; x_0)$ are the respective solutions of the boundary-value problems

$$\varphi'' = -k\varphi' + M, \qquad \varphi(x_0) = 0, \qquad \varphi'(b) = 0$$

and

$$\varphi'' = k\varphi' + M, \qquad \varphi(x_0) = 0, \qquad \varphi'(a) = 0,$$

then $y(x) \geqslant \varphi_1(x; x_0) + y(x_0)$ on $[x_0, b]$, $y(x) \geqslant \varphi_2(x; x_0) + y(x_0)$ on $[a, x_0]$, and $\varphi_1'(x; x) \leqslant y'(x_0) \leqslant \varphi_2'(x_0; x_0)$. These functions and derivatives can be computed and we obtain

$$|y'(x)| \leqslant \frac{M}{k}\left(e^{k(b-a)} - 1\right) \qquad \text{on} \qquad [a, b].$$

Corollary 4.19. *Assume $f(x, y, y')$ is nondecreasing in y on $[a, b] \times R^2$ for fixed x, y' and satisfies a uniform Lipschitz condition with* *respect to y' on $[a, b] \times R^2$. Then for any A, B the boundary-value problem*

$$(*) \qquad y'' = f(x, y, y'), \qquad y(a) = A, \qquad y(b) = B$$

has a unique solution $y(x) \in C^{(2)}[a, b]$. ([23], Corollary 6.4; [25]).

Proof. Let $w(x)$ be the linear function with $w(a) = A$, $w(b) = B$. Then the function $g(x, y, y') = f(x, y + w(x), y' + w'(x))$ satisfies the hypotheses of Theorem 4.18, hence, the boundary-value problem

$$y'' = g(x, y, y'), \qquad y(a) = y(b) = 0$$

has a solution $\omega(x)$. Then $y(x) = w(x) + \omega(x)$ is a solution of (*). That the solution is unique follows from Corollary 3.7.

Corollary 4.20. *If $f(x, y)$ is continuous on $[a, b] \times R$ and is non-decreasing in y for fixed x, then for any A, B the boundary-value problem*

$$y'' = f(x, y), \qquad y(a) = A, \qquad y(b) = B$$

has a unique solution $y \in C^{(2)}[a, b]$.

We will now consider an application of Theorem 4.18 to obtain a disconjugacy condition for a linear third order differential equation. The equation

$$y''' + p_0(x)\, y'' + p_1(x)\, y' + p_2(x)\, y = 0 \tag{4.1}$$

with $p_j(x) \in C[a, b]$ is said to be disconjugate on $[a, b]$ in case no non-trivial solution has more than two zeros on $[a, b]$ counting multiplicities of zeros.

A number of disconjugacy conditions have been given which relate bounds on the coefficients in Eq. (4.1) to the interval length $b - a$. Nehari [26] has shown that, if $h = b - a$ and

$$\frac{1}{2}\int_a^b |p_0| + \frac{h}{4}\int_a^b |p_1| + \frac{h^2}{8}\int_a^b |p_2| \leqslant 1,$$

then Eq. (4.1) is disconjugate on $[a, b]$. In 1963 Lasota [27] proved that (4.1) is disconjugate on $[a, b]$ if

$$\frac{1}{4} P_0 h + \frac{1}{\pi^2} P_1 h^2 + \frac{1}{2\pi^2} P_2 h^3 \leqslant 1,$$

where $P_j = \text{Max} \mid p_j(x) \mid$ on $[a, b]$. If $p_1(x) \leqslant 0$ on $[a, b]$, Lasota's condition reduces to

$$\frac{1}{4} P_0 h + \frac{1}{2\pi^2} P_2 h^3 \leqslant 1.$$

The result obtained by the use of Theorem 4.18 will be based on the assumption $p_1(x) \leqslant 0$ on $[a, b]$ and will involve P_0, P_2, and h. We first establish an existence theorem for certain types of boundary value problems for nonlinear third-order equations.

Theorem 4.21. *Assume that $f(x, y, y', y'')$ is continuous on $[a, b] \times R^3$ and satisfies the following conditions:* (i) *f is nondecreasing in y' for each fixed x, y, y'',* (ii) *f satisfies a Lipschitz condition with respect to y'' on each compact subset of $[a, b] \times R^3$, and* (iii) *for each $M > 0$ there is a $k_M > 0$ such that $\mid f(x, y, 0, y'') - f(x, y, 0, 0) \mid \leqslant k_M \mid y'' \mid$ for all $a \leqslant x \leqslant b$, $\mid y \mid \leqslant M$, and all y''. Then there is a $\delta > 0$ such that for any $[x_1, x_2] \subset [a, b]$ with $x_2 - x_1 \leqslant \delta$ there is a solution $y_1(x)$ of $y''' = f(x, y, y', y'')$ on $[x_1, x_2]$ with $y_1(x_1) = y_1'(x_1) = y_1'(x_2) = 0$ and a solution $y_2(x)$ on $[x_1, x_2]$ with $y_2'(x_1) = y_2'(x_2) = y_2(x_2) = 0$.* ([28], Theorem 1).

Proof. Let $M > 0$ be fixed and let

$$B[x_1, x_2] \equiv \{z(x) \in C[x_1, x_2] : \| z \| = \text{Max} \mid z(x) \mid \leqslant M\},$$

where $[x_1, x_2] \subset [a, b]$.

It follows from Theorem 4.18 that for each $z \in B[x_1, x_2]$ the boundary-value problem

$$y'' = f(x, z(x), y, y'), \qquad y(x_1) = y(x_2) = 0$$

has a unique solution $u_z(x) \in C^{(2)}[x_1, x_2]$. Let

$$q_M = \text{Max} \mid f(x, z, 0, 0) \mid \quad \text{for} \quad a \leqslant x \leqslant b, \mid z \mid \leqslant M$$

and

$$E(t) = [e^{k_M t} - e^{\frac{1}{2} k_M t} - \tfrac{1}{2} k_M t].$$

Then it also follows from Theorem 4.18 that

$$\mid u_z(x) \mid \leqslant \frac{q_M}{k_M^2} E(x_2 - x_1)$$

on $[x_1, x_2]$ for each $z \in B[x_1, x_2]$. Now define the mapping $T : B[x_1, x_2] \to C[x_1, x_2]$ by $Tz = w$, where $w(x) = \int_{x_1}^{x} u_z(t)\,dt$. Then $\| w \| \leqslant (q_M/k_M^2)(x_2 - x_1) E(x_2 - x_1)$, consequently, $T(B) \subset B$ if $(q_M/k_M^2)(x_2 - x_1) E(x_2 - x_1) \leqslant M$. $B[x_1, x_2]$ is a closed convex subset of the Banach space $C[x_1, x_2]$ and it is not difficult to show that T is continuous and completely continuous on $B[x_1, x_2]$. Since $E(t)$ is increasing in t, it follows from the Schauder Fixed-Point Theorem that T has a fixed point in $B[x_1, x_2]$ if $x_2 - x_1 \leqslant \delta$ where $(q_M/k_M^2)\,\delta E(\delta) = M$. If $y(x)$ is the fixed point, $y''' = f(x, y, y', y'')$ on $[x_1, x_2]$ and $y(x_1) = y'(x_1) = y'(x_2) = 0$.

If the mapping T is defined by $Tz = w$ with $w(x) = \int_{x_2}^{x} u_z(t)\,dt$, the same estimates hold and in this case for $x_2 - x_1 \leqslant \delta$ with $(q_M/k_M^2)\,\delta E(\delta) = M$ there is a solution $y(x)$ on $[x_1, x_2]$ with $y(x_2) = y'(x_2) = y'(x_1) = 0$.

Theorem 4.22. *Let $h = b - a$. Then, if $P_0 \neq 0$ and*

$$hP_2 E(P_0 h) \leqslant P_0^2,$$

Eq. (4.1) is disconjugate on $[a, b]$ ([28], Theorem 2).

Proof. Consider the equation

$$L[y] = y''' + p_0(x)\,y'' + p_1(x)\,y' + p_2(x)\,y = g(x),$$

where $p_j(x)$, $g(x) \in C[a, b]$ and $p_1(x) \leqslant 0$ on $[a, b]$. It is clear that the hypotheses of Theorem 4.21 are satisfied. In this case

$$q_M \leqslant \| g \| + \| p_2 \| M = \| g \| + P_2 M$$

and

$$k_M = \| p_0 \| = P_0 .$$

Thus given $M > 0$ and $[x_1, x_2] \subset [a, b]$, the boundary-value problems

$$L[y] = g(x), \qquad y(x_1) = y'(x_1) = y'(x_2) = 0 \tag{4.2}$$

and

$$L[y] = g(x), \qquad y'(x_1) = y(x_2) = y'(x_2) = 0 \tag{4.3}$$

have solutions if

$$\frac{(x_2 - x_1)\,[\| g \| + P_2 M]\, E[P_0(x_2 - x_1)]}{P_0^2} \leqslant M.$$

This will be the case if

$$\frac{(x_2 - x_1)\, P_2 E[P_0(x_2 - x_1)]}{P_0^2} \leqslant 1 - \frac{(x_2 - x_1)\, \| g \|\, E[P_0(x_2 - x_1)]}{P_0^2 M}. \tag{4.4}$$

Let $M > 0$ be chosen and fixed. It is known ([28], p. 630) that $L[y] = 0$ is disconjugate on $[a, b]$ if and only if for any $x_0 \in [a, b]$ the solution $u(x; x_0)$ of the initial-value problem $L[y] = 0$, $y(x_0) = y'(x_0) = 0$, $y''(x_0) = 1$ satisfies $u(x; x_0) > 0$ on $[a, b]$ for $x \neq x_0$. Assume $L[y] = 0$ is not disconjugate on $[a, b]$. Then there is an $x_1 \in [a, b]$ such that $u(x; x_1)$ has another zero in $[a, b]$. It follows that there is an $x_2 \in (a, b)$ at which $u'(x_2; x_1) = 0$, to be specific assume $x_1 < x_2$. Then $x_2 - x_1 < h$ and it follows that

$$\frac{(x_2 - x_1)\, P_2 E[P_0(x_2 - x_1)]}{P_0^2} < \frac{h P_2 E[P_0 h]}{P_0^2} \leqslant 1.$$

Let

$$g(x) = \epsilon[6 + 6p_0(x)(x - x_1) + 3p_1(x)(x - x_1)^2 + p_2(x)(x - x_1)^3].$$

Then for $\epsilon > 0$ sufficiently small, inequality (4.4) will be satisfied; consequently, problem (4.2) has a solution. It is easy to see that this solution must be of the form $y(x) = cu(x; x_1) + \epsilon(x - x_1)^3$ where c is a constant. Then $y'(x_2) = cu'(x_2; x_1) + 3\epsilon(x_2 - x_1)^2 = 3\epsilon(x_2 - x_1)^2 \neq 0$, which contradicts $y(x)$'s being a solution of (4.2). If $x_2 < x_1$, we contradict (4.3)'s having a solution. Thus the assumption $u'(x_2; x_1) = 0$ for $x_2 \neq x_1$ leads to a contradiction and we conclude that $u(x; x_0) > 0$ for all $x, x_0 \in [a, b]$ with $x \neq x_0$. Thus $L[y] = 0$ is disconjugate on $[a, b]$.

The inequality of Theorem 4.22 is an improvement over that of Lasota for the $p_1(x) \leqslant 0$ case in that P_0 and h can be large provided P_2 is sufficiently small. By examining the proof of Theorem 4.22 and noting that

$$\lim_{P_0 \to 0} \frac{E(P_0 h)}{P_0^2} = \frac{3}{8} h^2,$$

we see that when $P_0 = 0$ $L[y] = 0$ is disconjugate on $[a, b]$ provided $\frac{3}{8} h^3 P_2 \leqslant 1$. In this case the result is inferior to that of Lasota.

5. Boundary-Value Problems on Infinite Intervals

The subfunction technique can also be used to advantage in dealing with certain types of boundary-value problems on infinite intervals.

In this Section we will consider some illustrations of problems of this type.

Theorem 5.1. *Let f be continuous on $[a, \infty) \times R$, be nondecreasing in y for fixed x, and satisfy $f(x, 0) \equiv 0$ on $[a, \infty)$. Then for any real A the boundary-value problem*

$$y'' = f(x, y), \qquad y(a) = A \tag{5.1}$$

has a unique bounded solution $y(x) \in C^{(2)}[a, \infty)$ ([29], Theorem 3.1).

Proof. We consider only the case $A \geqslant 0$ and, in particular, the case $A > 0$ since $y(x) \equiv 0$ is a bounded solution when $A = 0$. The case $A < 0$ can be dealt with in a similar way.

If $A > 0$, it follows from Corollary 3.6 that $\psi_0(x) \equiv A$ is a superfunction on $[a, \infty)$. Let Φ be the collection of all subfunctions $\varphi(x)$ on $[a, \infty)$ such that $\varphi(x) \leqslant \psi_0(x)$ on $[a, \infty)$. Then Φ is not empty since $\varphi(x) \equiv 0$ belongs to Φ. Let $z(x) = \sup_{\varphi \in \Phi} \varphi(x)$. Then $0 \leqslant z(x) \leqslant \psi_0(x)$ on $[a, \infty)$ and it follows from Theorem 4.17 that $z(x) \in C^{(2)}(a, \infty)$ and is a solution of the differential equation on (a, ∞). By Corollary 4.20 the boundary-value problem

$$y'' = f(x, y), \qquad y(a) = A, \qquad y(a + 1) = 0$$

has a solution $y_1(x)$. By Theorem 4.7 $\varphi_0(x)$ defined by $\varphi_0(x) = y_1(x)$ on $[a, a + 1]$, $\varphi_0(x) = 0$ on $(a + 1, \infty)$ is a subfunction on $[a, \infty)$. Also, since $\psi_0(x) \equiv A$ is a superfunction, $y_1(x) \leqslant \psi_0(x)$ on $[a, a + 1]$. It follows that $\varphi_0(x) \in \Phi$ and $y_1(x) \leqslant z(x) \leqslant \psi_0(x)$ on $[a, a + 1]$. Then by Theorem 4.14, $z(x) \in C^{(2)}[a, \infty)$ and is a solution of the boundary-value problem (5.1).

Now assume that problem (5.1) has two distinct bounded solutions $z_1(x)$, $z_2(x) \in C^{(2)}[a, \infty)$. Then, since by Corollary 3.7 solutions of boundary-value problems on finite intervals are unique, there is an $x_0 \geqslant a$ such that $z_1(x) \equiv z_2(x)$ on $[a, x_0]$ and $z_1(x) \not\equiv z_2(x)$ for all $x > x_0$. To be specific, assume $z_1(x) > z_2(x)$ on (x_0, ∞). Then $z_1''(x) - z_2''(x) = f(x, z_1(x)) - f(x, z_2(x)) \geqslant 0$ on $[x_0, \infty)$, $z_1(x_0) - z_2(x_0) = 0$ and $z_1(x) - z_2(x) > 0$ for $x > x_0$. This obviously implies $z_1(x) - z_2(x)$ is unbounded on $[x_0, \infty)$. From this contradiction we conclude the uniqueness of bounded solutions of (5.1).

The bounded solution $z(x)$ of (5.1) with $A > 0$ satisfies $z'(x) \leqslant 0$ on $[a, \infty)$. Other questions of interest arise concerning the solution $z(x)$.

Is $z(x) > 0$ on $[a, \infty)$? This of course will be the case if initial-value problems for $y'' = f(x, y)$ have unique solutions. Another condition on $f(x, y)$ which implies $z(x) > 0$ is given in [29]. If $z(x) > 0$, is $\lim_{x\to\infty} z(x) > 0$? Does (5.1) also have unbounded solutions? These questions have been considered by a number of authors. Subfunction methods can also be applied to the study of these questions.

The following Theorem with slightly different hypotheses has been proven by other methods by Schuur [30].

Theorem 5.2. *Let $f(x, y, y')$ be continuous on $[a, \infty) \times R^2$, be nondecreasing in y for fixed x, y', be nondecreasing in y' for fixed x, y, and satisfy $f(x, ,0, 0) \equiv 0$ on $[a, \infty)$. Assume either that $f(x, y, y')$ satisfies a Lipschitz condition with respect to y' on each compact subset of $[a, \infty) \times R^2$ or that solutions of initial-value problems for $y'' = f(x, y, y')$ are unique. Then for any real A the boundary-value problem*

$$y'' = f(x, y, y'), \qquad y(a) = A \tag{5.2}$$

has a unique bounded solution on $[a, \infty)$.

Proof. As in Theorem 5.1 it suffices to consider only the case $A > 0$. Since $f(x, 0, 0) \equiv 0, f(x, A, 0) \geqslant 0$ and it follows from either Corollary 3.6 or Corollary 3.10 that $\psi_0(x) \equiv A$ is a superfunction on $[a, \infty)$. Again let $z(x) = \sup_{\varphi\in\Phi} \varphi(x)$, where Φ is the collection of all subfunctions $\varphi(x)$ on $[a, \infty)$ such that $\varphi(x) \leqslant \psi_0(x)$ on $[a, \infty)$. Φ is not empty since it contains $\varphi(x) \equiv 0$. Then $0 \leqslant z(x) \leqslant \psi_0(x) \equiv A$ on $[a, \infty)$ and it follows from Theorem 4.12 and either Corollary 3.7 or Corollary 3.11 that $z(x)$ is simultaneously a subfunction and a superfunction on $[a, \infty)$.

It follows from Theorem 4.10 that $z(x)$ has a finite derivative almost everywhere. Let $x_0 > a$ be a point at which $z(x)$ has a finite derivative. Then by Theorem 4.10 there is an open interval around x_0 on which $z(x)$ is a solution. Let (c, d) be a maximal such interval. Now suppose that $z'(x_1) > 0$ for some $x_1 \in (c, d)$ and that $z'(x) < 0$ at some points of (x_1, d). Then let $x_2 \in (x_1, d)$ be such that $z'(x_2) = 0$ and $z'(x) > 0$ on $[x_1, x_2)$. It follows that

$$0 > -z'(x_1) = z'(x_2) - z'(x_1) = \int_{x_1}^{x_2} f(s, z(s), z'(s))\, ds \geqslant 0.$$

From this contradiction it follows that, if $z'(x_1) > 0$ for $x_1 \in (c, d)$, then $z'(x) > 0$ on $[x_1, d)$. Now assume $c > a$; then, since (c, d) is maximal, it follows from Theorem 4.10 that $z(x)$ cannot have a finite derivative at c and that $Dz(c+) = \pm \infty$. Assume first that $Dz(c+) = +\infty$. Then it follows from the mean-value theorem that $z'(x) > 0$ at points arbitrarily close to c, hence, by the above observation $z'(x) > 0$ on (c, d). However, in this case $z''(x) = f(x, z(x), z'(x)) \geqslant 0$ on (c, d), which is incompatible with $Dz(c+) = +\infty$. Now assume $Dz(c+) = -\infty$. Then by applying the mean-value theorem and the above observation again we conclude that there is a $\delta > 0$ such that $z'(x) < 0$ on $(c, c+\delta]$. Then $z''(x) = f(x, z(x), z'(x)) \leqslant f(x, A, 0)$ on $(c, c+\delta]$ but $Dz(c+) = -\infty$ implies that $z''(x)$ is not bounded above on $(c, c+\delta]$. Thus $Dz(c+)$ must be finite and from this contradiction we conclude that $c = a$. It also follows by the same argument that $Dz(a+)$ is finite. Since $x_0 > a$ was an arbitrary point at which $z(x)$ has a finite derivative, we conclude from Theorems 4.10 and 4.14 that $z(x) \in C^{(2)}[a, \infty)$ and is a solution of (5.2). Furthermore, we conclude that $z'(x) \leqslant 0$ on $[a, \infty)$ for, if $z'(x_0) > 0$, then $z'(x) > 0$ and $z''(x) \geqslant 0$ on $[x_0, \infty)$ and $z(x)$ would be unbounded on $[x_0, \infty)$.

Now assume (5.2) has two distinct bounded solutions $z_1(x)$ and $z_2(x)$. Since it is again the case that with the hypotheses of the Theorem solutions of boundary-value problems on finite intervals are unique, there is an $x_0 \geqslant a$ such that $z_1(x) \equiv z_2(x)$ on $[a, x_0]$ and $z_1(x) > z_2(x)$ on (x_0, ∞). Then there is an $x_1 \geqslant x_0$ at which $z_1'(x_1) - z_2'(x_1) > 0$. Arguing as above and using the monotoneity of f, it follows that $z_1'(x) - z_2'(x) > 0$ on $[x_1, \infty)$ which implies $z_1''(x) - z_2''(x) \geqslant 0$ on $[x_1, \infty)$. This implies $z_1(x) - z_2(x)$ is unbounded on $[x_1, \infty)$ and we conclude that bounded solutions of (5.2) are unique.

Theorem 5.3. *Assume that $f(x, y, y')$ satisfies the hypotheses of Theorem 5.2 except that $f(x, y, y')$ is non-increasing in y' for fixed x, y. Then there is a $\delta > 0$ such that for any A with $|A| < \delta$ the boundary-value problem (5.2) has a bounded solution $y(x) \in C^{(2)}[a, \infty)$.*

Proof. As in Theorem 5.2 the generalized solution can be defined for any A and, if $A > 0$, $0 \leqslant z(x) \leqslant A$ on $[a, \infty)$. We will consider only the case $A > 0$. If $(c, d) \subset [a, \infty)$ is an interval on which $z(x)$ is a solution and if $z'(x_0) < 0$ for some $c < x_0 < d$, it follows in this case that $z'(x) < 0$ on $(c, x_0]$. This leads as in the proof of Theorem 5.2 to the conclusion that $Dz(d-)$ is finite. It follows that $z(x)$ is a solution on (a, ∞).

It follows from Theorem 2.1 that for suitable sufficiently small $A > 0$ and $\delta > 0$ the boundary-value problem

$$y'' = f(x, y, y'), \qquad y(a) = A, \qquad y(a + \delta) = 0$$

has a solution $y(x) \in C^{(2)}[a, a + \delta]$. Then $\varphi_0(x)$ defined by $\varphi_0(x) = y(x)$ on $[a, a + \delta]$ and $\varphi_0(x) \equiv 0$ on $(a + \delta, \infty)$ is a subfunction on $[a, \infty)$. Also, since $\psi_0(x) \equiv A$ is a superfunction, $y(x) \leqslant A$ on $[a, a + \delta]$. Therefore $\varphi_0(x) \leqslant z(x) \leqslant A$ on $[a, \infty)$ and it follows from Theorem 4.14 that for such an A $z(x) \in C^{(2)}[a, \infty)$ and is a solution of (5.2).

If $A_0 > 0$ is such that (5.2) has a solution $z_0(x) \in C^{(2)}[a, \infty)$ with $z_0(a) = A_0$, then for any $0 < A < A_0$, $\varphi_0(x) = z_0(x) - (A_0 - A)$ is an underfunction for the boundary-value problem (5.2) with boundary value A at $x = a$. Thus for any $0 \leqslant A \leqslant A_0$ there is a bounded $C^{(2)}[a, \infty)$ solution of (5.2). This completes the proof of the Theorem.

In this case, bounded solutions are not necessarily unique as evidenced by the problem $y'' = -y'$, $y(0) = A > 0$, which has solutions $y(x) \equiv A$ and $y(x) = Ae^{-x}$.

The arguments used in the proofs of Theorems 5.2 and 5.3 can be applied to boundary-value problems on finite intervals. If $f(x, y, y')$ satisfies the hypotheses of Theorem 5.2, a generalized solution $z(x)$ of the boundary-value problem

$$y'' = f(x, y, y'), \qquad y(a) = A, \qquad y(b) = B$$

is of class $C^{(2)}[a, b)$ and $z(a) = A$. If f satisfies the hypotheses of Theorem 5.3, the generalized solution $z(x) \in C^{(2)}(a, b]$ and $z(b) = B$.

6. A Relation between the Global Existence of Solutions of Initial-Value Problems and the Existence of Solutions of Boundary-Value Problems

As is well known in the case of linear differential equations, the uniqueness of solutions of boundary-value problems implies their existence. In this section we shall see that this is also the case for nonlinear equations provided all solutions of initial value problems exist globally.

Theorem 6.1. *Assume that $I \subset R$ is an interval and that $f(x, y, y')$ is continuous on $I \times R^2$. Assume that for every $(x_0, y_0, y_0') \in I \times R^2$ the initial-value problem*

$$y'' = f(x, y, y'), \qquad y(x_0) = y_0, \qquad y'(x_0) = y_0' \tag{6.1}$$

has a unique solution $y(x) \in C^{(2)}(I)$. *Further, assume that, if for any* $[x_1, x_2] \subset I$ *and any* A, B *the boundary-value problem*

$$y'' = f(x, y, y'), \qquad y(x_1) = A, \qquad y(x_2) = B \tag{6.2}$$

has a solution $y(x) \in C^{(2)}[x_1, x_2]$, *then that solution is unique. Then for any proper subinterval* $[x_1, x_2] \subset I$ *and any* A, B *boundary value problem* (6.2) *has a solution* [41].

Proof. Let $[x_1, x_2] \subset I$ be a proper subinterval. Then either x_1 or x_2 is an interior point of I and to be specific we assume x_1 is. Let $I_1 = I \cap (-\infty, x_1]$. Let A, B be given and consider the corresponding boundary-value problem (6.2). Let $y(x; m)$ be the solution of the initial-value problem (6.1) with initial conditions $y(x_1) = A$, $y'(x_1) = m$. By assumption, $y(x; m)$ is unique and is a solution on all of I. It follows that $y(x_2; m)$ is a continuous function of m so that its range is an interval. To complete the proof it suffices to show that the range of $y(x_2; m)$ is neither bounded above nor bounded below. We will prove only that the range is not bounded above since an exactly analogous argument is used to show that it is not bounded below.

Assume that, on the contrary, the range of $y(x_2; m)$ is bounded above and let $\beta = \text{lub}\{y(x_2; m) : -\infty < m < +\infty\}$. Let $z(x)$ be the solution of the initial-value problem (6.1) with initial conditions $z(x_2) = \beta$, $z'(x_2) = 0$. Then $z(x)$ is a solution on I and $z(x_1) \neq A$ since, if $z(x_1) = A$, $z(x) = y(x; m)$ for some m, but $z(x_2) > y(x_2; m)$. This follows from the fact that uniqueness of solutions of boundary-value problems implies $y(x; m_1) < y(x; m_2)$ for $x > x_1$ and $m_2 > m_1$. Assume first that $z(x_1) < A$. Then for all m, $y(x_1; m) > z(x_1)$ and $y(x_2; m) < z(x_2)$, and it follows again from the uniqueness of solutions of boundary-value problems that $y(x; m) > z(x)$ on I_1 for all m. Let $u(x) = \text{glb}\{y(x; m): -\infty < m < +\infty\}$ for all $x \in I_1$. Then $z(x) \leqslant u(x) \leqslant y(x; 1)$ on I_1. By Theorem 3.12 solutions are superfunctions, consequently, by Theorem 4.6 $u(x)$ is a superfunction on I_1. Now suppose that $u(x)$ is not a subfunction on I_1. Then there is a subinterval $[c, d] \subset I_1$ and a solution $v(x)$ on $[c, d]$ such that $u(c) \leqslant v(c)$, $u(d) \leqslant v(d)$, but $u(x) > v(x)$ at some points in (c, d). With the hypotheses of the theorem solutions of initial-value problems are continuous with respect to initial conditions. Consequently, if $\epsilon > 0$ is sufficiently small the solution $v_1(x)$ with initial conditions $v_1(c) = v(c)$, $v_1'(c) = v(c) + \epsilon$ will satisfy $v(x) < v_1(x)$ on $(c, d]$ but $v_1(x) < u(x)$ at some points of (c, d). Then, for $\eta > 0$ sufficiently small, the solution $v_2(x)$ satisfying initial conditions $v_2(d) = v_1(d)$, $v_2'(d) = v_1'(d) - \eta$ will satisfy $v_2(x) > v_1(x)$ on $[c, d)$ but $v_2(x) < u(x)$ at some points of (c, d). Thus $v_2(x)$ is a solution on $[c, d]$ with $u(c) < v_2(c)$, $u(d) < v_2(d)$, but $u(x) > v_2(x)$ at some points in (c, d). By the definition of $u(x)$ there is an m_1 and an m_2 such that $u(c) < y(c; m_1) < v_2(c)$ and $u(d) < y(d; m_2) < v_2(d)$. Then, if $m_0 = \text{Max}[m_1, m_2]$, $y(c; m_0) \leqslant y(c; m_1) < v_2(c)$ and $y(d; m_0) \leqslant y(d; m_2) < v_2(d)$, but $v_2(x) < u(x) \leqslant y(x; m_0)$ at some points in (c, d). This contradicts the uniqueness of the solutions of boundary-value problems since $v_2(x)$ and $y(x; m_0)$ will agree at two distinct points in (c, d) and differ between the two points. We conclude that $u(x)$ is a subfunction on I_1. It follows from Theorem 4.10 that $u(x)$ is a solution on a nonnull open subset of I_1. Let (a, b) be a maximal open subinterval contained in I_1 on which $u(x)$ is a solution. Then it follows from Theorem 4.10 that, if a is not the left end of I_1, $Du(a+) = \pm\infty$, and, if $b \neq x_1$, $Du(b-) = \pm\infty$. However, there is a solution $y(x)$ on I such that $u(x) \equiv y(x)$ on (a, b) since solutions of initial-value problems are unique and exist globally. Thus in the above cases, neither $Du(a+) = \pm\infty$ nor $Du(b-) = \pm\infty$ is possible and we conclude that $u(x)$ is a solution on $I_1 - \{x_1\}$. It follows that, if $y(x)$ is the solution on I with $y(x) \equiv u(x)$ on $I_1 - \{x_1\}$, then $u(x_1 - 0) = y(x_1) < A$. However, it then follows that, for $x_0 \in I_1$, $x_0 \neq x_1$ and for $\epsilon > 0$ sufficiently small, the solution $y_1(x)$ of the initial-value problem with initial conditions

$$y_1(x_0) = y(x_0) = u(x_0), y_1'(x_0) = y'(x_0) + \epsilon = u'(x_0) + \epsilon$$

satisfies $y_1(x) > y(x)$ for $x > x_0$ and $y(x_1) < y_1(x_1) < A$. But then, for m sufficiently large, $y(x; m)$ and $y_1(x)$ will constitute distinct solutions for some boundary-value problem on some subinterval of $[x_0, x_1]$. From this contradiction we conclude that $z(x_1) < A$ is not possible.

From $z(x_1) > A$ and $z(x_2) = \beta > y(x_2; m)$ for each m, it follows that $y(x; m) < z(x)$ on $[x_1, x_2]$ for all m. In this case we define $w(x) = \text{lub}\{y(x; m) : -\infty < m < +\infty\}$ on $[x_1, x_2]$ and we can argue as above that $w(x)$ is simultaneously a subfunction and a superfunction on $[x_1, x_2]$. This leads to the conclusion that $w(x)$ is a solution on (x_1, x_2) and $w(x_1 + 0) > A$, which leads to a contradiction as before. We conclude that the range of $y(x_2; m)$ is not bounded above. In a similar manner it can be proven that the range is not bounded below. It follows that (6.2) has a solution.

The assumption that solutions of initial value problems are unique can be omitted in Theorem 6.1 if it is assumed that all solutions of all initial-value problems exist on I.

The conclusion of Theorem 6.1 cannot be strengthened to assert that boundary-value problems on I itself have solutions when I is a compact interval. The following example which shows that this is the case was communicated to the author by Keith Schrader.

Consider $y'' = -y + \arctan y$, $-\frac{1}{2}\pi < \arctan y < \frac{1}{2}\pi$, with $I = [0, \pi]$. Since f is independent of y' and $|f_y| \leqslant 1$ on $[0, \pi] \times R$, it follows that all initial-value problems have unique solutions defined on $[0, \pi]$. Now assume that on $[x_1, x_2] \subset [0, \pi]$ there are solutions $y_1(x)$ and $y_2(x)$ with $y_1(x_1) = y_2(x_1)$, $y_1(x_2) = y_2(x_2)$, and $y_1(x) > y_2(x)$ on (x_1, x_2). Then

$$\begin{aligned} w''(x) &= y_1''(x) - y_2''(x) \\ &= -w(x) + \arctan y_1(x) - \arctan y_2(x) \geqslant -w(x) \qquad \text{on} \qquad [x_1, x_2]. \end{aligned}$$

The equation $y'' = -y$ is disconjugate on an interval of length less than π, hence, it follows from Theorem 3.12 that on such an interval a lower solution is a subfunction. Thus, if $[x_1, x_2]$ is a proper subinterval of $[0, \pi]$, $w''(x) \geqslant -w(x)$ on $[x_1, x_2]$ and $w(x_1) = w(x_2) = 0$ implies $w(x) \leqslant 0$ on $[x_1, x_2]$. We conclude that solutions of boundary-value problems on proper subintervals of $[0, \pi]$ are unique. Now assume $[x_1, x_2] = [0, \pi]$. Let $u(x)$ be the solution of the initial-value problem $u'' = -u$, $u(\frac{1}{2}\pi) = w(\frac{1}{2}\pi)$, $u'(\frac{1}{2}\pi) = w'(\frac{1}{2}\pi)$. Then, since $w'' > -w$ on $(0, \pi)$, it follows that $w(x) > u(x)$ on intervals on each side of $x = \frac{1}{2}\pi$. Again using Theorem 3.12 and the fact that $y'' = -y$ is disconjugate on intervals of length less than π, we conclude that $w(0) > u(0)$ and $w(\pi) > u(\pi)$. Since $w(0) = w(\pi) = 0$ and $u(\frac{1}{2}\pi) = w(\frac{1}{2}\pi) > 0$, this would imply that $u(x)$ has two zeros on $(0, \pi)$ which is impossible. Thus solutions of boundary-value problems on $[0, \pi]$, when they exist, are unique. Therefore, the hypotheses of Theorem 6.1 are satisfied. Now suppose that the boundary-value problem

$$y'' = -y + \arctan y, \qquad y(0) = 0, \qquad y(\pi) = 3\pi$$

has a solution $y(x)$ with $y'(0) = m$. Let $v(x)$ be the solution of the initial-value problem

$$v'' = -v + \pi, \qquad v(0) = 0, \qquad v'(0) = m + 1.$$

Then $v'' = -v + \pi > -v + \arctan v$ and it follows from Theorem

3.12 that $v(x)$ is a subfunction on $[0, \pi]$ with respect to solutions of $y'' = -y + \arctan y$. Since $v(x) > y(x)$ on an interval to the right of $x = 0$, it follows that $v(x) > y(x)$ on $(0, \pi]$. Computing $v(x)$ we get $v(x) = (m + 1) \sin x - \pi \cos x + \pi$ and $y(\pi) < v(\pi) = 2\pi$. From this contradiction we conclude that the stated boundary value problem has no solution.

Now we apply Theorem 6.1 to obtain results concerning existence of solutions of boundary value problems when $f(x, y, y')$ satisfies a Lipschitz condition with respect to y and y'. These results have been dealt with in References [31]-[35], but our methods will be quite different.

Assume that $f(x, y, y')$ is continuous on $I \times R^2$ and assume that there exist continuous functions $\ell_1(x)$, $\ell_2(x)$, $k_1(x)$, $k_2(x)$ on I such that

$$G_1(x, y_1 - y_2, y_1' - y_2') \leqslant f(x, y_1, y_1') - f(x, y_2, y_2') \leqslant G_2(x, y_1 - y_2, y_1' - y_2') \tag{6.3}$$

on $I \times R^2$ where

$$G_1(x, y, y') = \begin{cases} k_1(x)\, y + \ell_1(x)\, y' & \text{for} \quad y \geqslant 0, \quad y' \geqslant 0, \\ k_1(x)\, y + \ell_2(x)\, y' & \text{for} \quad y \geqslant 0, \quad y' \leqslant 0, \\ k_2(x)\, y + \ell_2(x)\, y' & \text{for} \quad y \leqslant 0, \quad y' \leqslant 0, \\ k_2(x)\, y + \ell_1(x)\, y' & \text{for} \quad y \leqslant 0, \quad y' \geqslant 0 \end{cases} \tag{6.4}$$

and

$$G_2(x, y, y') = \begin{cases} k_2(x)\, y + \ell_2(x)\, y' & \text{for} \quad y \geqslant 0, \quad y' \geqslant 0, \\ k_2(x)\, y + \ell_1(x)\, y' & \text{for} \quad y \geqslant 0, \quad y' \leqslant 0, \\ k_1(x)\, y + \ell_1(x)\, y' & \text{for} \quad y \leqslant 0, \quad y' \leqslant 0, \\ k_1(x)\, y + \ell_2(x)\, y' & \text{for} \quad y \leqslant 0, \quad y' \geqslant 0. \end{cases} \tag{6.5}$$

From (6.3)-(6.5) it follows that $k_1(x) \leqslant k_2(x)$ and $\ell_1(x) \leqslant \ell_2(x)$ on I and that solutions of initial-value problems for $y'' = f(x, y, y')$ are unique and exist on all of I. The functions G_1 and G_2 can be written in the form

$$G_1(x, y, y') = \tfrac{1}{2}(k_1 + k_2)\, y + \tfrac{1}{2}(k_1 - k_2)\, |\, y\,| + \tfrac{1}{2}(\ell_1 + \ell_2)\, y' + \tfrac{1}{2}(\ell_1 - \ell_2)\, |\, y'\,| \tag{6.6}$$

and

$$G_2(x, y, y') = \tfrac{1}{2}(k_1 + k_2)\, y + \tfrac{1}{2}(k_2 - k_1)\, |\, y\,| + \tfrac{1}{2}(\ell_1 + \ell_2)\, y' + \tfrac{1}{2}(\ell_2 - \ell_1)\, |\, y'\,|\,. \tag{6.7}$$

From these expressions for G_1 and G_2 it is clear that solutions of initial-value problems for $y'' = G_1(x, y, y')$ and $y'' = G_2(x, y, y')$ are unique and exist on all of I.

Theorem 6.2. *If solutions of boundary-value problems for $y'' = G_1(x, y, y')$ when they exist are unique, then solutions of boundary-value problems for $y'' = f(x, y, y')$ when they exist are unique.*

Proof. Assume that solutions of boundary-value problems for $y'' = G_1(x, y, y')$ are unique but that those of $y'' = f(x, y, y')$ are not. Then there is an interval $[x_1, x_2] \subset I$ and solutions $y_1(x)$ and $y_2(x)$ of $y'' = f(x, y, y')$ such that $y_1(x_i) = y_2(x_i)$ for $i = 1, 2$ and $y_1(x) > y_2(x)$ on (x_1, x_2). Then, setting $w(x) = y_1(x) - y_2(x)$ and using inequality (6.3), we have $w'' \geqslant G_1(x, w, w')$ on $[x_1, x_2]$, $w(x_1) = w(x_2) = 0$ and $w(x) > 0$ on (x_1, x_2). It follows from Theorem 3.12 that $w(x)$ is a subfunction on $[x_1, x_2]$ with respect to solutions of $y'' = G_1(x, y, y')$. Let $u(x)$ be the solution of the initial-value problem

$$y'' = G_1(x, y, y'), \qquad y(x_1) = w(x_1), \qquad y'(x_1) = w'(x_1).$$

Since solutions of initial-value problems for $y'' = f(x, y, y')$ are unique, $w'(x_1) > 0$. Consequently, since solutions of boundary-value problems for $y'' = G_1(x, y, y')$ are unique, $u(x_1) = 0$, $u'(x_1) > 0$, and $y(x) \equiv 0$ is a solution, it follows that $u(x) > 0$ on $(x_1, x_2]$. Thus $w(x_1) = u(x_1)$ and $w(x_2) < u(x_2)$ which, since $w(x)$ is a subfunction, implies $w(x) \leqslant u(x)$ on $[x_1, x_2]$. It then follows from Theorem 2.6 that $w(x) \equiv u(x)$ on $[x_1, x_2]$. From this contradiction we conclude that solutions of boundary-value problems for $y'' = f(x, y, y')$ when they exist are unique.

Corollary 6.3. *If solutions of boundary-value problems for $y'' = G_1(x, y, y')$ when they exist are unique, then any boundary-value problem on any proper subinterval $[x_1, x_2] \subset I$ for $y'' = f(x, y, y')$ has a solution.*

Proof. This is an immediate consequence of Theorems 6.1 and 6.2.

Let us consider now the case where $I = [a, b]$ is a compact interval. Assume that solutions of boundary-value problems for $y'' = G_1(x, y, y')$ on subintervals of I when they exist are unique. Then the conclusion of Corollary 6.3 follows but in this case the conclusion applies not only to proper subintervals of I but to the interval I itself. If one examines the proof of Theorem 6.1 and also takes Theorems 4.12 and 4.14 into account, it can be seen that it is sufficient to show that, for any boundary-value problem

$$y'' = f(x, y, y'), \qquad y(a) = A, \qquad y(b) = B, \tag{6.8}$$

there is an overfunction $\psi_0(x)$ and an underfunction $\varphi_0(x)$ with $\varphi_0(x) \leqslant \psi_0(x)$ on $[a, b]$. We will show that such functions can be constructed if solutions of boundary-value problems on subintervals of I for $y'' = G_1(x, y, y')$ when they exist are unique.

Let $y_1(x)$ be the solution of the initial-value problem $y'' = G_1(x, y, y')$, $y(a) = 1$, $y'(a) = 0$ and $y_2(x)$ the solution satisfying initial conditions $y_2(a) = 0$, $y_2'(a) = 1$. Then $y_2(x) > 0$ on $(a, b]$, consequently, for $h > 0$ sufficiently large $v(x) = y_1(x) + hy_2(x) > 0$ on $[a, b]$. Furthermore,

$$\begin{aligned} v'' = y_1'' + hy_2'' &= G_1(x, y_1, y_1') + hG_1(x, y_2, y_2') \\ &= G_1(x, y_1, y_1') + G_1(x, hy_2, hy_2') \\ &\leqslant G_1(x, v, v') \end{aligned}$$

on $[a, b]$. Let $z(x)$ be the solution of the initial-value problem

$$y'' = G_1(x, y, y') + f(x, 0, 0), \qquad y(a) = y'(a) = 0.$$

Then, given any A, B, there is an $r > 0$ such that, if $\psi_0(x) = rv(x) + z(x)$, then $\psi_0(a) \geqslant A$ and $\psi_0(b) \geqslant B$. Then

$$\begin{aligned} \psi_0'' = rv'' + z'' &= rG_1(x, v, v') + G_1(x, z, z') + f(x, 0, 0) \\ &\leqslant G_1(x, \psi_0, \psi_0') + f(x, 0, 0) \\ &\leqslant f(x, \psi_0, \psi_0') - f(x, 0, 0) + f(x, 0, 0) \\ &\leqslant f(x, \psi_0, \psi_0'). \end{aligned}$$

Thus $\psi_0(x)$ is an overfunction with respect to the boundary-value problem (6.8).

Now let $u(x)$ be the solution of the initial-value problem

$$y'' = G_2(x, y, y') + f(x, 0, 0), \qquad y(a) = y'(a) = 0.$$

Let $\varphi_0(x) = -qv(x) + u(x)$ where $v(x)$ is as above and $q > 0$ is chosen large enough that $\varphi_0(a) \leqslant A$, $\varphi_0(b) \leqslant B$, and $\varphi_0(x) \leqslant \psi_0(x)$ on $[a, b]$.

Then

$$\begin{aligned}\varphi_0'' = -qv'' + u'' &= -qG_1(x, v, v') + G_2(x, u, u') + f(x, 0, 0)\\ &= G_2(x, -qv, -qv') + G_2(x, u, u') + f(x, 0, 0)\\ &\geqslant G_2(x, \varphi_0, \varphi_0') + f(x, 0, 0) \geqslant f(x, \varphi_0, \varphi_0')\end{aligned}$$

on $[a, b]$ and it follows that $\varphi_0(x)$ is an underfunction with respect to the boundary-value problem (6.8). We have thus proven the following Theorem.

Theorem 6.4. *If $f(x, y, y')$ is continuous on $I \times R^2$, satisfies inequality (6.3) on $I \times R^2$, and solutions of boundary-value problems for $y'' = G_1(x, y, y')$ when they exist are unique, then all boundary-value problems for $y'' = f(x, y, y')$ on all subintervals of I have solutions and the solutions are unique.*

Now we consider the question of uniqueness of solutions of boundary-value problems for $y'' = G_1(x, y, y')$.

Theorem 6.5. *Solutions of boundary-value problems for $y'' = G_1(x, y, y')$ are unique if and only if for any $[x_1, x_2] \subset I$ the solution of the initial-value problem*

$$y'' = G_1(x, y, y'), \qquad y(x_1) = 0, \qquad y'(x_1) = 1 \tag{6.9}$$

satisfies $y(x) > 0$ on $(x_1, x_2]$.

Proof. Since $y(x) \equiv 0$ is a solution, the condition is obviously necessary.

Now assume that the condition is satisfied for each subinterval of I but that solutions of boundary-value problems are not unique. Then there is an interval $[x_1, x_2] \subset I$ and solutions $y_1(x)$ and $y_2(x)$ such that $y_1(x_i) = y_2(x_i)$ for $i = 1, 2$ and $y_1(x) > y_2(x)$ on (x_1, x_2). Let $w(x) = y_1(x) - y_2(x)$. Then $w(x_1) = w(x_2) = 0$ and $w'(x_1) > 0$ since solutions of initial value problems are unique. Also

$$w'' = G_1(x, y_1, y_1') - G_1(x, y_2, y_2') \geqslant G_1(x, w, w')$$

on $[x_1, x_2]$. If $y(x)$ is the solution of the initial-value problem (6.9), $u(x) = w'(x_1)\, y(x)$ is the solution satisfying initial conditions $u(x_1) = 0$, $u'(x_1) = w'(x_1)$; hence, $u(x) > 0$ on $(x_1, x_2]$. If $w(x) \leqslant u(x)$ on $[x_1, x_2]$, Theorem 2.6 can be applied to obtain the contradiction $w(x) \equiv u(x)$

on $[x_1, x_2]$. It follows that $w(x) > u(x)$ at some points in (x_1, x_2). In this case there is a $0 < \beta < 1$ such that $\beta w(x) \leqslant u(x)$ on $[x_1, x_2]$ and $\beta w(x_0) = u(x_0)$, $\beta w'(x_0) = u'(x_0)$ at some $x_1 < x_0 < x_2$. But then again Theorem 2.6 leads to the contradiction $\beta w(x) \equiv u(x)$ on $[x_1, x_2]$ since we still have $(\beta w)'' \geqslant G_1(x, \beta w, \beta w')$ on $[x_1, x_2]$. It follows that solutions of boundary-value problems for $y'' = G_1(x, y, y')$ are unique.

Let $[a, b] \subset I$ and on $[a, b]$ let $K_1 = \text{Min}\, k_1(x)$, $K_2 = \text{Max}\, k_2(x)$, $L_1 = \text{Min}\, \ell_1(x)$, and $L_2 = \text{Max}\, \ell_2(x)$. Let $G_1^*(x, y, y')$ be defined as in (6.4) using the constants L_1, L_2, K_1, and K_2. Then on $[a, b] \times R^2$

$$G_1^*(x, y_1 - y_2, y_1' - y_2') \leqslant G_1(x, y_1, y_1') - G_1(x, y_2, y_2'),$$

and it follows from Theorem 6.5 that solutions of boundary-value problems for $y'' = G_1(x, y, y')$ on subintervals of $[a, b]$ are unique in case for any $[x_1, x_2] \subset [a, b]$ the solution of the initial-value problem $y'' = G_1^*(x, y, y')$, $y(x_1) = 0$, $y'(x_1) = 1$ satisfies $y(x) > 0$ on $(x_1, x_2]$. This will be the case for any$[x_1, x_2] \subset [a, b]$ if

$$b - a < \alpha(L_1, K_1) + \beta(L_2, K_1),$$

where $\alpha(L_1, K_1)$ is the first positive zero of $u'(x)$ with $u(x)$ the solution of the initial-value problem

$$u'' = K_1 u + L_1 u', \qquad u(0) = 0, \qquad u'(0) = 1$$

and $-\beta(L_2, K_1)$ is the first negative zero of $v'(x)$ with $v(x)$ the solution of the initial-value problem

$$v'' = K_1 v + L_2 v', \qquad v(0) = 0, \qquad v'(0) = -1$$

(see [30], p. 312).

7. Further Existence Theorems for Solutions of Boundary-Value Problems

In this section we shall work with solutions of differential inequalities as in the previous sections but we will not impose conditions which imply the uniqueness of solutions of boundary value problems. In place of such conditions we will impose restrictions on the rate of growth of $f(x, y, y')$ with respect to y', in particular we will assume $f(x, y, y')$ satisfies a Nagumo condition. Nagumo [36] used such growth conditions to prove the existence of solutions of boundary-value problems but his methods also required that solutions of initial-value problems be unique. Our methods will not require this assumption. We will as before always assume that $f(x, y, y')$ is continuous.

Definition 7.1. $f(x, y, y')$ is said to satisfy a Nagumo condition on $[a, b]$ with respect to the pair $\alpha(x), \beta(x) \in C[a, b]$ in case $\alpha(x) \leqslant \beta(x)$ on $[a, b]$ and there exists a positive continuous function $h(s)$ on $[0, \infty)$ such that $|f(x, y, y')| \leqslant h(|y'|)$ for all $a \leqslant x \leqslant b$, $a(x) \leqslant y \leqslant \beta(x)$, $|y'| < \infty$ and

$$\int_\lambda^\infty \frac{s\,ds}{h(s)} > \operatorname*{Max}_{a\leqslant x\leqslant b} \beta(x) - \operatorname*{Min}_{a\leqslant x\leqslant b} \alpha(x) \tag{7.1}$$

where

$$\lambda(b - a) = \operatorname{Max}[|\alpha(a) - \beta(b)|, |\alpha(b) - \beta(a)|]. \tag{7.2}$$

Lemma 7.2. *Assume that $f(x, y, y')$ satisfies a Nagumo condition on $[a, b]$ with respect to the pair $\alpha(x), \beta(x) \in C[a, b]$. Then, for any solution $y(x) \in C^{(2)}[a, b]$ with $\alpha(x) \leqslant y(x) \leqslant \beta(x)$ on $[a, b]$, there is an $N > 0$ depending only on $\alpha(x)$, $\beta(x)$ and $h(s)$ such that $|y'(x)| \leqslant N$ on $[a, b]$.*

Proof. Choose $N > 0$ such that

$$\int_\lambda^N \frac{s\,ds}{h(s)} > \operatorname{Max} \beta(x) - \operatorname{Min} \alpha(x).$$

If $x_0 \in (a, b)$ is such that $(b - a)\,y'(x_0) = y(b) - y(a)$, then by (7.2), $|y'(x_0)| \leqslant \lambda$. Assume that $|y'(x)| \geqslant N$ at some points in $[a, b]$ and to deal with a specific case assume $y'(x) \geqslant N$ at some points. Then there is an interval $[c, d] \subset [a, b]$ such that $y'(c) = N$, $y'(d) = \lambda$, and $\lambda < y'(x) < N$ on (c, d) or $y'(c) = \lambda$, $y'(d) = N$, and $\lambda < y'(x) < N$ on (c, d). Let us consider the former case; then on $[c, d]$,

$$|y''(x)|\,y'(x) = |f(x, y(x), y'(x))|\,y'(x) \leqslant h(y'(x))\,y'(x)$$

and

$$\left|\int_c^d \frac{y''(x)\,y'(x)\,dx}{h(y'(x))}\right| \leqslant \int_c^d \frac{|y''(x)|\,y'(x)\,dx}{h(y'(x))} \leqslant \int_c^d y'(x)\,dx.$$

This leads to the contradiction

$$\int_\lambda^N \frac{s\,ds}{h(s)} \leqslant y(d) - y(c) \leqslant \operatorname{Max} \beta(x) - \operatorname{Min} \alpha(x).$$

Other possibilities can be dealt with in a similar way and we conclude that $|y'(x)| \leqslant N$ on $[a, b]$.

Theorem 7.3. *Assume that $f(x, y, y')$ satisfies a Nagumo condition with respect to the pair $\alpha(x)$, $\beta(x) \in C^{(1)}[a, b]$ which are, respectively, lower and upper solutions of the differential equation on $[a, b]$. Then, for any $\alpha(a) \leqslant c \leqslant \beta(a)$ and $\alpha(b) \leqslant d \leqslant \beta(b)$, the boundary-value problem*

$$y'' = f(x, y, y'), \qquad y(a) = c, \qquad y(b) = d \tag{7.3}$$

has a solution $y(x) \in C^{(2)}[a, b]$ with $\alpha(x) \leqslant y(x) \leqslant \beta(x)$ on $[a, b]$.

Proof. By Lemma 7.2 there is an $N > 0$ depending only on $\alpha(x)$, $\beta(x)$, and the Nagumo function $h(s)$ such that $|y'(x)| \leqslant N$ for any such solution. Let $F(x, y, y')$ be the modification of $f(x, y, y')$ of Definition 2.3 associated with the triple $\alpha(x)$, $\beta(x)$, c_1 where $c_1 > 0$ is chosen so that $N < c_1$ and $|\alpha'(x)| < c_1$, $|\beta'(x)| < c_1$ on $[a, b]$. Then by Theorem 2.5 the boundary-value problem

$$y'' = F(x, y, y'), \qquad y(a) = c, \qquad y(b) = d$$

has a solution $y(x) \in C^{(2)}[a, b]$ with $\alpha(x) \leqslant y(x) \leqslant \beta(x)$ on $[a, b]$. By the mean-value theorem there is an $x_0 \in (a, b)$ such that

$$(b - a)\, y'(x_0) = y(b) - y(a)$$

and it follows that $|y'(x_0)| \leqslant \lambda < N < c_1$. It follows that there is an interval around x_0 in which $y(x)$ is a solution of $y'' = f(x, y, y')$. It follows from Lemma 7.2 that $|y'(x)| \leqslant N < c_1$ in this interval, but $y(x)$ is a solution of $y'' = f(x, y, x')$ as long as $|y'(x)| < c_1$. We conclude that $y(x)$ is a solution of (7.3) on $[a, b]$.

As an example to show that Theorem 7.3 yields results when solutions of boundary-value problems are not necessarily unique consider $y'' = |y'|^p$ where $0 < p < 1$. In this case, $h(s) = s^p + 1$ will serve as a Nagumo function and constants are upper and lower solutions.

Schrader [37] has shown that in certain cases combinations of "one-sided" Nagumo conditions can be used in Theorem 7.3.

Lemma 7.2 and Theorem 7.3 can be used to obtain solutions on infinite intervals.

Theorem 7.4. *Assume that for each $b > a$ $f(x, y, y')$ satisfies a Nagumo condition on $[a, b]$ with respect to the pair $\alpha(x)$, $\beta(x) \in C^{(1)}[a, \infty)$*

where $\alpha(x) \leqslant \beta(x)$ *on* $[a, \infty)$, *and* $\alpha(x)$ *and* $\beta(x)$ *are, respectively, lower and upper solutions on* $[a, \infty)$. *Then for any* $\alpha(a) \leqslant c \leqslant \beta(a)$ *the boundary-value problem*

$$y'' = f(x, y, y'), \qquad y(a) = c \tag{7.4}$$

has a solution $y(x) \in C^{(2)}[a, \infty)$ *with* $\alpha(x) \leqslant y(x) \leqslant \beta(x)$ *on* $[a, \infty)$.

Proof. It follows from Lemma 7.2 and Theorem 7.3 that for each $n \geqslant 1$ there is a solution $y_n(x)$ on $[a, a + n]$ with $y_n(a) = c$, $y_n(a + n) = \beta(a + n)$, and $\alpha(x) \leqslant y_n(x) \leqslant \beta(x)$ on $[a, a + n]$ and there is an $N_n > 0$ such that $| y'(x) | \leqslant N_n$ on $[a, a + n]$ for any solution satisfying $\alpha(x) \leqslant y(x) \leqslant \beta(x)$ on $[a, a + n]$. Thus, for any fixed $n \geqslant 1$, $y_m(x)$ is a solution on $[a, a + n]$ satisfying $\alpha(x) \leqslant y_m(x) \leqslant \beta(x)$ and $| y_m'(x) | \leqslant N_n$ on $[a, a + n]$ for all $m \geqslant n$. Hence, for $m \geqslant n$ the sequences $\{y_m(x)\}$ and $\{y_m'(x)\}$ are both uniformly bounded and equicontinuous on $[a, a + n]$. Then, employing standard diagonalization arguments, one obtains a subsequence which converges uniformly on all compact subintervals of $[a, \infty)$ to a solution $y(x)$. $y(x)$ is the desired solution of (7.4).

Theorem 7.5. *Assume that* $f(x, y, y')$ *satisfies a Nagumo condition on* $[-a, a]$ *for each* $a > 0$ *with respect to the pair* $\alpha(x), \beta(x) \in C^{(1)}(-\infty, +\infty)$, *where* $\alpha(x)$ *and* $\beta(x)$ *are lower and upper solutions on* $(-\infty, +\infty)$ *and* $\alpha(x) \leqslant \beta(x)$ *on* $(-\infty, +\infty)$. *Then there is a solution of* $y'' = f(x, y, y')$ *on* $(-\infty, +\infty)$ *with* $\alpha(x) \leqslant y(x) \leqslant \beta(x)$ *on* $(-\infty, +\infty)$.

Proof. The proof is essentially the same as that of Theorem 7.4.

We consider now some applications of these results. The first application is in the establishment of comparison theorems for solutions of nonlinear equations. These results which constitute a different approach to results previously obtained by Knobloch [38] are contained in [21].

Definition 7.6. A solution $y(x)$ of $y'' = f(x, y, y')$ is said to have property (B) on $[a, b]$ in case there is a sequence of solutions $\{y_n(x)\}$ on $[a, b]$ such that

(i) $y_n \to y$ and $y_n' \to y'$ uniformly on $[a, b]$,

(ii) $\Delta_n = y - y_n \neq 0$ and has the same sign for all $n \geqslant 1$ and $a \leqslant x < b$ or for all $n \geqslant 1$ and $a < x \leqslant b$,

(iii) for each $0 < \delta < \frac{1}{2}(b - a)$ there is a constant $c > 0$ depending on δ but not on n and x such that $|\Delta_n'(x)| \leqslant c\,|\Delta_n(x)|$ for all $n \geqslant 1$ and all $a + \delta \leqslant x \leqslant b - \delta$.

Theorem 7.7. *Assume that $f(x, y, y')$ satisfies a Lipschitz condition with respect to y and y' on each compact subset of $[a, b] \times R^2$ and satisfies a Nagumo condition on $[a, b]$ with respect to the pair $\alpha(x), \beta(x) \in C^{(1)}[a, b]$ which are lower and upper solutions on $[a, b]$. Assume further that $\alpha(a) < \beta(a)$ or $\alpha(b) < \beta(b)$. Then for any $\alpha(a) \leqslant c \leqslant \beta(a)$, $\alpha(b) \leqslant d \leqslant \beta(b)$ the boundary value problem (7.3) has a solution $y(x)$ having property* (B) *on $[a, b]$.*

Proof. Consider the case $\alpha(a) < c \leqslant \beta(a)$, $\alpha(b) \leqslant d \leqslant \beta(b)$. Choose $h > 0$ such that $\alpha(a) < c - h$ and consider the sequence of boundary-value problems

$$y'' = f(x, y, y'), \qquad y(a) = c - \frac{h}{n}, \qquad y(b) = d. \tag{7.5$_n$}$$

By Theorem 7.3 the problem $(7.5)_1$ has a solution $y_1(x)$ with $\alpha(x) \leqslant y_1(x) \leqslant \beta(x)$ on $[a, b]$. Using $y_1(x)$ as a lower solution and $\beta(x)$ as an upper solution, we can apply Theorem 7.3 again to obtain a solution $y_2(x)$ of the problem $(7.5)_2$ with $y_1(x) \leqslant y_2(x) \leqslant \beta(x)$ on $[a, b]$. Proceeding in this way we obtain a sequence $\{y_n(x)\} \subset C^{(2)}[a, b]$ such that $y_n(x)$ is a solution of $(7.5)_n$ for each $n \geqslant 1$ and

$$\alpha(x) \leqslant y_n(x) \leqslant y_{n+1}(x) \leqslant \beta(x)$$

on $[a, b]$ for each $n \geqslant 1$. Furthermore, since solutions of initial value problems are unique and $y_n(a) < y_{n+1}(a)$, it follows that $y_n(x) < y_{n+1}(x)$ on $[a, b)$ for each $n \geqslant 1$. By Lemma 7.2 there is an $N > 0$ such that $|y_n'(x)| \leqslant N$ on $[a, b]$ for all $n \geqslant 1$. Then there is a subsequence which we shall renumber as the original sequence such that $y_n \to y$ and $y_n' \to y'$ uniformly on $[a, b]$ where $y(x)$ is a solution of the boundary-value problem $y'' = f(x, y, y')$, $y(a) = c$, $y(b) = d$. Furthermore, this solution satisfies $\alpha(x) \leqslant y(x) \leqslant \beta(x)$ and $|y'(x)| \leqslant N$ on $[a, b]$.

The solution $y(x)$ just obtained as the uniform limit of the subsequence $\{y_n(x)\}$ obviously satisfies parts (i) and (ii) of property (B). We claim that part (iii) of property (B) is also satisfied by $y(x)$ and the subsequence $\{y_n(x)\}$. To see this let $k > 0$ be a Lipschitz coefficient for $f(x, y, y')$ with respect to y and y' on the compact set

$$\{(x, y, y') : a \leqslant x \leqslant b, \alpha(x) \leqslant y \leqslant \beta(x), |y'| \leqslant N\}.$$

Then with $\Delta_n(x) = y(x) - y_n(x) \geqslant 0$ we have

$$| \Delta_n''(x) | \leqslant k(\Delta_n(x) + | \Delta_n'(x) |) \tag{7.6}$$

on $[a, b]$ for all $n \geqslant 1$. Let $0 < \delta < \frac{1}{2}(b - a)$ be given and let $x_0 \in [a + \delta, b - \delta]$. The hypotheses of Theorem 7.3 are satisfied by the equation $y'' = ky + k | y' |$ on each of the subintervals $[a, x_0]$ and $[x_0, b]$ with $\beta(x) = \Delta_n(x)$ as an upper solution and $\alpha(x) = 0$ as a lower solution. Consequently, the boundary-value problems

$$y'' = ky + k | y' |, \qquad y(a) = 0, \qquad y(x_0) = \Delta_n(x_0)$$

and

$$y'' = ky + k | y' |, \qquad y(b) = 0, \qquad y(x_0) = \Delta_n(x_0)$$

have solutions $y_1(x)$ and $y_2(x)$ with $0 \leqslant y_1(x) \leqslant \Delta_n(x)$ on $[a, x_0]$ and $0 \leqslant y_2(x) \leqslant \Delta_n(x)$ on $[x_0, b]$. Furthermore, it is not difficult to see that $y_1'(x) \geqslant 0$, $y_2'(x) \leqslant 0$, and that these solutions are unique. It follows from the above inequalities that

$$y_2'(x_0) \leqslant \Delta_n'(x_0) \leqslant y_1'(x_0).$$

Computing the solutions $y_1(x)$ and $y_2(x)$ we obtain

$$y_1'(x_0) = \tfrac{1}{2} k \, \Delta_n(x_0) \, [1 + \sqrt{5} \coth \tfrac{1}{2} \sqrt{5} \, k(x_0 - a)]$$

and

$$y_2'(x_0) = \tfrac{1}{2} k \, \Delta_n(x_0) \, [-1 + \sqrt{5} \coth \tfrac{1}{2} \sqrt{5} \, k(x_0 - b)].$$

It follows that part (iii) of property (B) holds for the solution $y(x)$ and subsequence $\{y_n(x)\}$ with the constant

$$c = \tfrac{1}{2} k[1 + \sqrt{5} \coth \tfrac{1}{2} \sqrt{5} \, k\delta].$$

A different elementary proof that part (iii) of property (B) is satisfied is given in [21], Lemma 2.5.

Theorem 7.8. *Assume that $f(x, y, y')$ has continuous first partial derivatives f_y and $f_{y'}$ on $[a, b] \times R^2$. Let $y_0(x)$ be a solution of $y'' = f(x, y, y')$ having property* (B) *on $[a, b]$. Then the linear equation*

$$y'' = f_{y'}(x, y_0(x), y_0'(x)) \, y' + f_y(x, y_0(x), y_0'(x)) \, y \tag{7.7}$$

is disconjugate on (a, b).

Proof. Assume that Eq. (7.7) is not disconjugate on (a, b). Then (7.7) has a nontrivial solution with zeros at x_1, x_2 with $a < x_1 < x_2 < b$.

Now to be specific assume that the sequence of solutions referred to in property (B) satisfies $\Delta_n(x) = y_0(x) - y_n(x) > 0$ on $[a, b)$ for all $n \geqslant 1$. Then applying the mean-value theorem we obtain

$$\Delta_n'' = f_{y'}(x, y_0(x), y_0'(x))\, \Delta_n' + f_y(x, y_0(x), y_0'(x))\, \Delta_n + q_n \mid \Delta_n \mid + p_n \mid \Delta_n' \mid \qquad (7.8)$$

where p_n, $q_n \to 0$ as $n \to \infty$. Let $\epsilon > 0$ be given and pick $\delta > 0$ such that $x_1, x_2 \in (a + \delta, b - \delta)$. Then it follows from part (iii) of property (B) that for n sufficiently large

$$\Delta_n'' \leqslant f_{y'}(x, y_0(x), y_0'(x))\, \Delta_n' + [f_y(x, y_0(x), y_0'(x)) + \epsilon]\, \Delta_n \qquad (7.9)$$

on $[a + \delta, b - \delta]$ with $\Delta_n > 0$ on $[a + \delta, b - \delta]$. We conclude from Theorems 7.3 and 2.6 that there is a solution $z(x)$ of the equality form of the differential inequality (7.9) with $z(x) > 0$ on $[a + \delta, b - \delta]$. It follows that the equality form of (7.9) is disconjugate on $[a + \delta, b - \delta]$ for each $\epsilon > 0$. However, since (7.7) has a nontrivial solution with zeros at $a + \delta < x_1 < x_2 < b - \delta$, it follows that for $\epsilon > 0$ sufficiently small the equality form of (7.9) must have a nontrivial solution with two distinct zeros in $[a + \delta, b - \delta]$. From this contradiction we conclude that (7.7) is disconjugate on (a, b).

With slightly different hypotheses Knobloch [38] proves the existence of at least one solution $y_0(x)$ of $y'' = f(x, y, y')$ such that the corresponding equation (7.7) is disconjugate on the closed interval $[a, b]$. One might conjecture that our results could be strengthened to conclude disconjugacy on $[a, b]$ with the same hypotheses. The following illustration to show that this is not the case is also due to Keith Schrader.

Consider the equation $y'' = -y + y^3$ on the interval $[0, \pi]$. On this interval $\beta(x) \equiv 1$ and $\alpha(x) \equiv 0$ are upper and lower solutions and the hypotheses of Theorem 7.7 are satisfied. Consequently, the boundary-value problem

$$y'' = -y + y^3, \qquad y(0) = y(\pi) = 0$$

has a solution $y_0(x)$ which has property (B) and is such that $0 \leqslant y_0(x) \leqslant 1$ on $[0, \pi]$. It follows that the first variational equation (7.7)

$$y'' = [-1 + 3y_0^2(x)]\, y$$

is disconjugate on $(0, \pi)$. However, this equation is not disconjugate on

$[0, \pi]$ since $y_0(x) \equiv 0$ on $[0, \pi]$. To see this assume $y_0(x) \not\equiv 0$ on $[0, \pi]$. Then, since $y_0(x) \geqslant 0$, $y_0(x) > 0$ on $(0, \pi)$ and $y_0''(x) \geqslant - y_0(x)$ on $[0, \pi]$. Let $y_1(x)$ be the solution of the initial value problem $y'' = - y$, $y(\frac{1}{2}\pi) = y_0(\frac{1}{2}\pi)$, $y'(\frac{1}{2}\pi) = y_0(\frac{1}{2}\pi)$. Then, since $y_0''(\frac{1}{2}\pi) > - y_0(\frac{1}{2}\pi)$, $y_0(x) > y_1(x)$ on an interval to the left and on an interval to the right of $x = \frac{1}{2}\pi$. However on $[0, \frac{1}{2}\pi]$ and on $[\frac{1}{2}\pi, \pi]$ a lower solution of $y'' = - y$ is a subfunction. Consequently, $y_1(0) < y_0(0) = 0$ and $y_1(\pi) < y_0(\pi) = 0$. This implies $y_1(x)$ has two distinct zeros on $(0, \pi)$ which is impossible. We conclude that $y_0(x) \equiv 0$ on $[0, \pi]$.

The above results may be useful in dealing with questions of oscillation of solutions of nonlinear equations. For example, suppose that the equation $y'' + a(x)\, y^{2n-1} = 0$, n a positive integer, has a solution which is positive on $[x_0, \infty)$. Then it follows that it also has a positive solution $y_0(x)$ on $[x_0, \infty)$ such that $y'' + (2n - 1)\, a(x)\, y_0^{2n-2}(x)\, y = 0$ is disconjugate on (x_0, ∞).

Next we consider an application of the results of this Section that was suggested by a recent paper of J. D. Schuur [40]. We consider the third-order linear equation

$$y''' + p_0(x)\, y'' + p_1(x)\, y' + p_2(x)\, y = 0, \tag{7.10}$$

where $p_j(x) \in C[x_0, \infty)$ for $j = 0, 1, 2$. The substitution $z = y'/y$ transforms (7.10) into the nonlinear second-order equation

$$\begin{aligned} z'' &= - 3zz' - p_0(x)\, z' - (z^3 + p_0(x)\, z^2 + p_1(x)\, z + p_2(x)) \\ &= f(x, z, z'). \end{aligned} \tag{7.11}$$

Clearly, if $z(x)$ is a solution of (7.11) on $[x_0, \infty)$ then

$$y(x) = y(x_0) \exp \left[\int_{x_0}^{x} z(s)\, ds \right]$$

is a solution of (7.10) on $[x_0, \infty)$.

Theorem 7.9. *Assume that there exist lower and upper solutions $\alpha(x)$, $\beta(x)$ of (7.11) on $[x_0, \infty)$ such that $\alpha(x) < \beta(x)$ on $[x_0, \infty)$. Then (7.10) has two positive linearly independent solutions on $[x_0, \infty)$ and (7.10) is disconjugate on $[x_0, \infty)$.*

Proof. From the form of equation (7.11) it is clear that $f(x, z, z')$ satisfies a Nagumo condition on $[x_0, x_0 + n]$ with respect to the pair $\alpha(x)$, $\beta(x)$ for each positive integer n. By Theorem 7.4 there is a solution

$z_1(x)$ of (7.11) on $[x_0, \infty)$ such that $z_1(x_0) = \beta(x_0)$ and $\alpha(x) \leqslant z_1(x) \leqslant \beta(x)$ on $[x_0, \infty)$. Furthermore, by Theorem 2.6 $\alpha(x) < z_1(x)$ on $[x_0, \infty)$. Consequently, applying Theorem 7.4 again we conclude that there is a solution $z_2(x)$ on $[x_0, \infty)$ such that $z_2(x_0) = \alpha(x_0)$ and $\alpha(x) \leqslant z_2(x) < z_1(x)$ on $[x_0, \infty)$. Now let $y_1(x) = \exp[\int_{x_0}^{x} z_1(s)\, ds]$, $y_2(x) = \exp[\int_{x_0}^{x} z_2(s)\, ds]$. Then $y_1(x_0) = y_2(x_0) = 1$ and $y_1'(x_0) = \beta(x_0) \neq \alpha(x_0) = y_2'(x_0)$ so that $y_1(x)$ and $y_2(x)$ are positive linearly independent solutions of (7.10) on $[x_0, \infty)$.

Using the procedure for reducing the order of a linear equation when a solution is known and the fact that $y_1(x)$ and $y_2(x)$ are solutions of (7.10), we find that $y_3(x) = y_2(x)\, u(x)$ is a solution of (7.10) on $[x_0, \infty)$ where

$$u(x) = \int_{x_1}^{x} v(s)\, w(s)\, ds, \qquad x_0 \leqslant x_1 < \infty,$$

$$v(x) = \left(\frac{y_1(x)}{y_2(x)}\right)' = \frac{(z_1(x) - z_2(x))\, y_1(x)}{y_2(x)},$$

$$w(x) = \int_{x_1}^{x} \frac{A(s)}{v^2(s)}\, ds,$$

and

$$A(x) = \exp\left(-\int_{x_1}^{x} [3z_2(s) + p_0(s)]\, ds\right).$$

From this it follows that

$$y_3(x_1) = y_3'(x_1) = 0 \qquad \text{and} \qquad y_3''(x_1) = \frac{y_2(x_1)}{v(x_1)} > 0.$$

Hence, $y_3(x)$ is a positive multiple of the Cauchy function for (7.10) with zero at $x = x_1$. Since $w(x) > 0$ for $x > x_1$, $w(x) < 0$ for $x < x_1$, and $v(x) > 0$ for $x \geqslant x_0$, it follows that $y_3(x) > 0$ for $x \geqslant x_0$, $x \neq x_1$. From this we conclude that (7.10) is disconjugate on $[x_0, \infty)$ ([28], p. 630).

The results of Theorem 7.9 apply equally well on any finite interval in place of $[x_0, \infty)$.

Theorem 7.10. *If there exist constants $\alpha < \beta$ such that*

$$\beta^3 + p_0(x)\, \beta^2 + p_1(x)\, \beta + p_2(x) \leqslant 0$$

and

$$\alpha^3 + p_0(x)\,\alpha^2 + p_1(x)\,\alpha + p_2(x) \geqslant 0$$

on $[x_0, \infty)$, *then (7.10) has two positive linearly independent solutions on* $[x_0, \infty)$ *and is disconjugate on* $[x_0, \infty)$ [39], [40].

Proof. In this case $\beta(x) \equiv \beta$ and $\alpha(x) \equiv \alpha$ are respectively upper and lower solutions of (7.11) and the result follows from Theorem 7.9.

Theorem 7.11. *If there is a* $\delta > 0$ *such that*

$$p_1(x) + (x + \delta)\,p_2(x) \leqslant 0$$

and

$$(x + \delta)\,p_0(x) + (x + \delta)^3\,p_2(x) \geqslant 3$$

on $[0, \infty)$, *then (7.10) has two positive linearly independent solutions on* $[0, \infty)$ *and is disconjugate on* $[0, \infty)$.

Proof. In this case a computation shows that $\beta(x) = 1/(x + \delta)$ and $\alpha(x) = -1/(x + \delta)$ are upper and lower solutions of (7.11) on $[0, \infty)$.

References

1. O. Perron, Ein Neuer Existenzbeweis fur die Integrale der Differentialgleichung $y' = f(x, y)$, *Math. Ann.* **76** (1915), 471–484.
2. O. Perron, Eine neue Behandlung der ersten Randwert-aufgabe fur $\Delta u = 0$, *Math. Z.* **18** (1923), 42–54.
3. F. Riesz, Uber Subharmonische Funktionen und ihre Rolle in der Funktionentheorie und in der Potential-theorie. *Acta. Litt. Sci. Szeged.* **2** (1925), 87–100.
4. F. Riesz, Sur les fonctions subharmoniques et leur rapport à la théorie du Potential, I, *Acta Math.* **48** (1926), 329–343.
5. F. Riesz, Sur les fonctions subharmoniques et leur rapport à la théorie du Potential, II, *Acta Math.* **54** (1930), 321–360.
6. G. Tautz, Zur Theorie der elliptischen Differentialgleichungen II, *Math. Ann.* **118** (1943), 733–770.
7. E. F. Beckenbach and L. K. Jackson, Subfunctions of several variables, *Pacific J. Math.* **3** (1953), 291–313.
8. M. Inoue, Dirichlet problem relative to a family of functions, *J. Inst. Polytech., Osaka City Univ.* **7** (1956), 1–16.
9. L. K. Jackson, Subfunctions and the Dirichlet problem, *Pacific J. Math.* **8** (1958), 243–255.
10. J. B. Serrin, Jr., On the Harnack inequality for linear elliptic equations, *J. Analyse Math.* **4** (1955–56), 292–308.
11. S. E. Bohn, Equicontinuity of solutions of a quasi-linear equation, *Pacific J. Math.* **12** (1962), 1193–1202.

12. S. E. BOHN AND L. K. JACKSON, The Liouville theorem for a quasi-linear elliptic partial differential equation, *Trans. Am. Math. Soc.* **104** (1962), 392–397.
13. J. O. HERZOG, Phragman-Lindelof theorems for second order quasi-linear elliptic partial differential equations, *Proc. Am. Math. Soc.* **15** (1964), 721–729.
14. E. F. BECKENBACH, Generalized Convex functions, *Bull. Am. Math. Soc.* **43** (1937), 363–371.
15. F. F. BONSALL, The characterization of generalized convex functions, *Quart. J. Math.* [*Oxford Ser.* (2)] **1** (1950), 100–111.
16. J. W. GREEN, Generalized convex functions, *Proc. Am. Math. Soc.* **4** (1953), 391–396.
17. M. M. PEIXOTO, Generalized convex functions and second order differential inequalities, *Bull. Am. Math. Soc.* **55** (1949), 563–572.
18. W. T. REID, Variational aspects of generalized convex functions, *Pacific J. Math.* **9** (1959), 571–581.
19. B. N. BABKIN, Solution of a boundary value problem for an ordinary differential equation of second order by Caplygin's method, *Prikl. Math. Meh. Akad. Nauk SSSR* **18** (1954), 239–242.
20. L. FOUNTAIN AND L. JACKSON, A generalized solution of the boundary value problem for $y'' = f(x, y, y')$, *Pacific J. Math.* **12** (1962), 1251–1272.
21. L. K. JACKSON AND K. W. SCHRADER, Comparison theorems for nonlinear differential equations, *J. Diff. Eqs.* **3** (1967), 248–255.
22. RALPH P. BOAS, A primer of real functions, *in* "Carus Mathematical Monographs." The Mathematical Association of America, New York, 1961.
23. J. W. BEBERNES, A subfunction approach to boundary value problems for ordinary differential equations, *Pacific J. Math.* **13** (1963), 1053–1066.
24. K. W. SCHRADER, On second order differential inequalities, *Proc. Am. Math. Soc.* **19** (1968), 1007–1012.
25. M. LEES, A boundary value problem for nonlinear ordinary differential equations, *J. Math. Mech.* **10** (1961), 423–430.
26. Z. NEHARI, On an inequality of Lyapunov, *in* "Studies in Mathematical Analysis and Related Topics," pp. 256–261. Stanford University Press, Stanford, California, 1962.
27. A. LASOTA, Sur la distance entre les zéros de l'équation différentielle linéaire du troisième ordre, *Ann. Polon. Math.* **13** (1963), 129–132.
28. R. M. MATHSEN, A disconjugacy condition for $y''' + a_2 y'' + a_1 y' + a_0 y = 0$, *Proc. Am. Math. Soc.* **17** (1966), 627–632.
29. J. W. BEBERNES AND L. K. JACKSON, Infinite interval boundary value problems for $y'' = f(x, y)$, *Duke Math. J.* **34** (1967), 39–48.
30. J. D. SCHUUR, The existence of proper solutions of a second order ordinary differential equation, *Proc. Am. Math. Soc.* **17** (1966), 595–597.
31. W. J. COLES AND T. L. SHERMAN, Two-point Problems for Non-linear Second Order Ordinary Differential Equations. Report 513, Mathematics Research Center, Univ. of Wisconsin, 1964.
32. P. BAILEY AND P. WALTMAN, On the distance between consecutive zeros for second order differential equations, *J. Math. Anal. Appl.* **14** (1966), 23–30.
33. P. BAILEY AND P. WALTMAN, Existence and uniqueness of solutions to the first boundary value problem for non-linear second order differential equations, *Arch. Ratl. Mech. Anal.* **12** (1966), 310–320.
34. P. BAILEY, L. SHAMPINE, AND P. WALTMAN, The first and second boundary value

problems for nonlinear second order differential equations, *J. Diff. Eqs.* **2** (1966), 399–411.

35. P. Bailey, L. Shampine, and P. Waltman, Existence and uniqueness of solutions of the second order boundary value problem, *Bull. Am. Math. Soc.*, **72** (1966), 96–98.
36. M. Nagumo, Uber die differentialgleichung $y'' = f(x, y, y')$, *Proc. Phys.-Math. Soc. Japan*, [*Ser.* 3] **19** (1937), 861–866.
37. K. W. Schrader, Solutions of second order ordinary differential equations, *J. Diff. Eqs.* **4** (1968), 510–518.
38. H. Knobloch, Comparison theorems for nonlinear second order differential equations, *J. Diff. Eqs.* **1** (1965), 1–26.
39. L. Jackson, Disconjugacy conditions for linear third order differential equations, *J. Diff. Eqs.* **4** (1968), 369–372.
40. J. D. Schuur, Asymptotic behavior of a solution of the third order linear differential equation, *Proc. Am. Math. Soc.* **18** (1967), 391–393.
41. A. Lasota and Z. Opial, On the existence and uniqueness of solutions of a boundary value problem for an ordinary second order differential equation, *Colloquium Math.* **18** (1967), 1–5.

Reprinted from *Advances in Mathematics* 2, Fascicle 4, 365–470, (1968).

Topics in Singular Perturbations*

ROBERT E. O'MALLEY, Jr.
New York University, Courant Institute of Mathematical Sciences
New York, New York 10012

Introduction

The area of singular perturbations is a field of increasing interest to applied mathematicians and one without adequate exposition. The only authoritative treatment to date is a single chapter in the highly recommended "Asymptotic Solutions for Ordinary Differential Equations" by Professor Wolfgang Wasow, a substantial contributor to the study of singular perturbations for over twenty years. Otherwise, one must search through the extensive literature.

This article is a slightly revised version of lecture notes prepared and delivered at Bell Telephone Laboratories, Inc., Whippany, New Jersey, during the summer of 1966. The content is selected and strongly reflects the interests, readings, and research of the author. These lecture notes were an attempt to emphasize mathematically sound, constructive techniques for obtaining asymptotic solutions for representative problems. It was hoped that they would, simultaneously, be a satisfactory introduction to the rapidly growing research literature and that they would prove

* This research was supported by the Air Force Office of Scientific Research under Contract No. AF-AFOSR-537-67. Reproduction in whole or in part is permitted for any purpose of the U.S. Government.

useful to those who encounter asymptotic phenomena in physical situations. The favorable response of many readers has been encouraging, but also indicative of the need for such a survey.

Section 1 introduces singular perturbation problems through examples and includes a formal application of the method of matched asymptotic expansions. Section 2 proceeds much more carefully to consider the asymptotic solution of certain boundary value problems for ordinary differential equations with a small parameter multiplying the highest derivatives. Methods introduced there are generalized in chapters three and four to include singular perturbations of eigenvalue problems and of boundary value problems involving two parameters which simultaneously approach zero. The eigenvalue problems generalize the classical study of the vibrations of a string, while the two-parameter problems presented can, in turn, be generalized to problems involving many parameters. The asymptotic solutions of these problems all feature the phenomenon of "loss of boundary conditions."

In Section 5, a different type of singular perturbation phenomena known as relaxation oscillations is encountered, while in Section 6 a more general discussion concerning perturbations of discontinuous solutions is presented. In particular, expansions are obtained for the amplitude and period of the relaxation oscillations of van der Pol's equation. Section 6 also contains Vasil'eva's method for obtaining the asymptotic solution of initial-value problems for systems of nonlinear equations, including detailed calculations for illustrative examples. Lastly, several representative problems for partial differential equations are presented in Section 7. They include examples where the reduced equation is of lower order than the full equation and examples where the reduced equation is of different type, as well as a problem for Oseen flow. Throughout these sections reference is frequently made to relevant literature cited in a concluding bibliography which contains more than one hundred items.

The mathematical and editorial assistance of Dr. J. A. Cochran in this presentation has been exceedingly valuable and is gratefully acknowledged. The assistance of many other friends and colleagues and of Bell Telephone Laboratories is greatly appreciated. Specifically, allow me to mention only Professors G. E. Latta, J. B. Keller, and W. R. Wasow who taught me about singular perturbations and asymptotic methods, and Professor R. W. McKelvey who asked me to include this article in this collection.

1. Regular and Singular Perturbation Problems

Consider a boundary value problem P_ϵ, depending on a small parameter ϵ. Assume that, as $\epsilon \to 0$, the differential equation and the boundary conditions defining P_ϵ approach limiting forms and define a "limiting boundary value problem" P_0. The solution y_ϵ of P_ϵ depends on ϵ, and, under certain conditions, y_ϵ and (some of) its derivatives will approach limits, usually uniformly, as $\epsilon \to 0$ and $\lim y_\epsilon$ will solve the limiting problem P_0. Moreover, if P_ϵ depends on ϵ analytically, we expect that y_ϵ will also depend analytically on ϵ so that y_ϵ can be constructed by the familiar "method of perturbation" as a power series in ϵ. Typically, the solution is expressed as a formal power series in ϵ, the series is substituted into the differential equation and the expressions for the boundary conditions, coefficients of corresponding powers of ϵ are equated, and—with luck—the resulting equations are solvable successively yielding an expansion which converges to the solution, or at least to an asymptotic expansion of the solution ("asymptotic solution") of the boundary value problem as $\epsilon \to 0$. As an example, consider the boundary value problem

$$\frac{d^2u}{dx^2} + (a(x) + \epsilon a_1(x))\frac{du}{dx} + (b(x) + \epsilon b_1(x))\,u = 0,$$

$$u(0) = c_1, \qquad u'(0) = c_2$$

on the interval $x \in [0, 1]$ where the coefficients are smooth and c_1 and c_2 do not depend on ϵ. [Here, and throughout this paper, all functions introduced in defining boundary value problems are assumed to be "sufficiently differentiable." Less differentiability, in general, implies that the asymptotic expansions obtained must be terminated after a finite number of terms. Further, such functions will be considered independent of ϵ, unless otherwise noted.] This problem can be solved by setting

$$u = u_0(x) + \epsilon u_1(x) + \epsilon^2 u_2(x) + \cdots,$$

where, e.g., we ask that u_0 satisfy the limiting boundary value problem

$$u_0'' + a(x)\,u_0' + b(x)\,u_0 = 0,$$

$$u_0(0) = c_1,$$

$$u_0'(0) = c_2.$$

Such problems are known as regular (or ordinary) perturbation problems. For further discussion, see Bellman [1].

Throughout the following, we shall be concerned with (irregular or) singular perturbation problems where $\lim y_\epsilon$, if it exists, will not be attained uniformly. This occurs most frequently for boundary value problems in which the order of the differential equation drops, or its type changes, as $\epsilon \to 0$ so that the boundary conditions for P_ϵ are not appropriate when $\epsilon = 0$ and it is not obvious how P_0 should be defined. Irregular perturbation problems need not be of this type, however, e.g.,

$$y'' - \epsilon^2 y = 0,$$
$$y(0) = 1,$$
$$y'(\infty) = 0.$$

For each fixed $\epsilon > 0$, the solution is $y = e^{-\epsilon x}$, but

$$\lim_{\epsilon \to 0^+} e^{-\epsilon x} = \begin{cases} 1 & \text{if} \quad \epsilon x = o(1), \\ e^{-k} & \text{if} \quad \epsilon x = k + o(1), \\ 0 & \text{if} \quad 1/x = o(\epsilon), \end{cases}$$

so that no single limit will be uniformly valid throughout the x-interval $[0, \infty)$. (Note: The Landau order symbols O and o are used throughout the following discussion. The reader should refer to van der Corput [20] for a discussion of these symbols as well as for the definitions of asymptotic expansions and asymptotic equality.)

Generally, singular perturbation problems feature nonanalytic dependence of y_ϵ on ϵ even in cases where P_ϵ depends on ϵ in a very simple manner. This nonanalytic dependence expresses itself frequently in markedly different behavior as ϵ approaches zero through positive and negative values. Since methods and results are completely analogous, however, henceforth we shall consider $\epsilon > 0$.

For a simple singular perturbation problem to analyze, consider

$$(*) \begin{cases} \epsilon y'' + y' + y = 0, \\ y(0) = \alpha, \qquad y(1) = \beta, \end{cases}$$

for $x \in [0, 1]$. Introducing

$$\rho_1 = \frac{-1 + \sqrt{1 - 4\epsilon}}{2\epsilon}$$

and

$$\rho_2 = \frac{-1 - \sqrt{1 - 4\epsilon}}{2\epsilon}$$

(the roots of the characteristic equation $\epsilon\rho^2 + \rho + 1 = 0$), we obtain the exact solution

$$y = \frac{(\beta - \alpha e^{\rho_2})\, e^{\rho_1 x} + (\alpha e^{\rho_1} - \beta)\, e^{\rho_2 x}}{e^{\rho_1} - e^{\rho_2}}$$

for $x \in [0, 1]$. Since

$$\rho_1 = -1 - \epsilon + O(\epsilon^2)$$

and

$$\rho_2 = -1/\epsilon + 1 + \epsilon + O(\epsilon^2),$$

$$y \to \beta e^{1-x} \quad \text{on} \quad 0 < x \leqslant 1$$

as $\epsilon \to 0$ and convergence is uniform in any x-interval $0 < \delta \leqslant x \leqslant 1$. Note that this limit satisfies the "reduced boundary value problem"

$$u' + u = 0,$$

$$u(1) = \beta.$$

Convergence is, however, nonuniform near $x = 0$ unless $\alpha = \beta e$. Neglecting terms which are asymptotically zero,

$$y \sim \beta e^{-(1-x)\rho_1} + (\alpha - \beta e^{-\rho_1})\, e^{\rho_2 x}$$

and, refining our previous estimate,

$$y = \beta e^{(1-x)} + (\alpha - \beta e)\, e^{-x/\epsilon + x} + O(\epsilon),$$

where the order relation O holds uniformly for all $x \in [0, 1]$.

Following Prandtl's idea, we introduce $t = x/\epsilon$ as a stretching transformation to "blow up" the region of quick transition (physically, the "boundary layer") near $x = 0$ (see Prandtl [91]), and proceed heuristically. Letting a dot represent differentiation with respect to t, we write the "boundary layer equation"

$$\ddot{y} + \dot{y} + \epsilon y = 0$$

and attempt a solution of this equation by the regular perturbation technique. Thus, we ask that

$$y = A + Be^{-t} + O(\epsilon)$$

uniformly for all t in $[0, \infty)$. Asking also that y equal α when $t = 0$, we set $B = \alpha - A$. On the other hand, for x a small, but fixed positive number, $t = x/\epsilon \to \infty$ as $\epsilon \to 0$ so that $y \to A$. Moreover, here (i.e., away from the origin), y should behave like βe^{1-x}, the solution of the reduced boundary value problem, and y should approach βe for small values of x. Matching these two "solutions" then, we set $A = \beta e$ and obtain

$$y = \beta e + (\alpha - \beta e)\, e^{-x/\epsilon} + O(\epsilon)$$

in the "boundary layer" while

$$y = \beta e^{1-x} + O(\epsilon)$$

away from $x = 0$.

Such heuristics are the basis of the method of "matched asymptotic expansions" (or "inner and outer expansions") which was developed and applied to many physically significant problems by Kaplun [54], Kaplun [55], Kaplun and Lagerstrom [57], Proudman and Pearson [92], and others [see Van Dyke [105], Erdélyi [25] and [27], Cole and Kevorkian [18], and especially, Kaplun [56] for further references.] We proceed to give a brief description of the formal application of this method to the preceding problem.

Let $y(x, \epsilon)$ be the solution of

$$(*)\begin{cases} \epsilon y'' + y' + y = 0, \\ y(0) = \alpha, \quad y(1) = \beta, \end{cases} \qquad \text{for} \quad x \in [0, 1].$$

Suppose y possesses an asymptotic expansion y^{o} as $\epsilon \to 0$ which is of the form

$$y^{\text{o}}(x, \epsilon) = \sum_{n=0}^{\infty} \epsilon^n g_n(x),$$

$$y^{\text{o}}(1, \epsilon) = \beta,$$

and, *suppose* this "outer expansion" is valid on the interval $0 < \delta \leqslant x \leqslant 1$ where both $\delta = o(1)$ and $\epsilon = o(\delta)$ as $\epsilon \to 0$. Introducing the stretching transformation $t = x/\epsilon$, consider the "inner expansion" y^{i} of the form

$$y^{\text{i}}(t, \epsilon) = \sum_{n=0}^{\infty} \epsilon^n f_n(t),$$

$$y^{\text{i}}(0, \epsilon) = \alpha,$$

and suppose this expansion is valid for $x = \epsilon t = o(1)$ as $\epsilon \to 0$. Then both the inner and outer expansions hold in an "overlap region" [where both $x = o(1)$ and $\epsilon/x = o(1)$] where they can be matched. Expressing y^i in terms of the outer variable x and expanding the result for small ϵ and fixed x, we obtain $(y^i)^o$, the outer expansion of the inner expansion. The inner expansion of the outer expansion, $(y^o)^i$, is analogously defined. Having assumed the existence of an overlap region, the expansions $(y^i)^o$ and $(y^o)^i$ are related to each other there by the stretching transformation. However as Kaplun and Lagerstrom [57] admit, "there is no *à priori* reason for their regions of validity to overlap." Thus results obtained by formal manipulation should not be *à priori* assumed asymptotically correct. Nevertheless, we proceed formally to demonstrate the procedure.

We find that

$$g_0(x) = \beta e^{(1-x)}$$

and

$$g_1(x) = (1 - x)\,\beta e^{(1-x)},$$

so that the two-term outer expansion is

$$y_2^o = \beta e^{(1-x)}[1 + \epsilon(1 - x)].$$

Writing this in terms of t, expanding for small ϵ, and terminating the result after two terms,

$$\begin{aligned}(y_2^o)_2^i &= \beta e[1 - \epsilon t + \epsilon] \\ &= \beta e[1 - x + \epsilon].\end{aligned}$$

Likewise, the two-term inner expansion is of the form

$$y_2^i = [a_0 + (\alpha - a_0)\,e^{-t}] + \epsilon[-(a_0 t + a_1) + e^{-t}((\alpha - a_0)\,t + a_1)]$$

for a_0 and a_1 undetermined constants.

Writing this in terms of x, expanding for small ϵ, and terminating after two terms, yields

$$(y_2^i)_2^o = a_0(1 - x) - \epsilon a_1\,.$$

Thus, matching $(y_2^o)_2^i$, and $(y_2^i)_2^o$, we select

$$a_0 = -a_1 = \beta e$$

and the two-term inner and outer solutions are completely determined (formally!).

Knowing the solutions y^{o} and y^{i}, we obtain a composite expansion y^{c} which, hopefully, is uniformly valid for $x \in [0, 1]$. The simplest method is to let

$$y^{\mathrm{c}} \sim \begin{cases} y^{\mathrm{i}} + y^{\mathrm{o}} - (y^{\mathrm{o}})^{\mathrm{i}}, \\ y^{\mathrm{i}} + y^{\mathrm{o}} - (y^{\mathrm{i}})^{\mathrm{o}}. \end{cases}$$

Expressing previous results in terms of the outer variable x, we have

$$y_2{}^{\mathrm{c}} = \beta e^{(1-x)}[1 + \epsilon(1 - x)] + e^{-x/\epsilon}[(\alpha - \beta e)(1 + x) - \epsilon\beta e],$$

which should be compared with the known exact solution.

A careful analysis of the method of matched asymptotic expansions has recently been presented by Fraenkel [31]. This paper distinguishes between the idea of overlapping and the "asymptotic matching principle", and analyzes independently the two corresponding techniques for establishing the relationship between the inner and outer expansions.

2. The Asymptotic Solution of Boundary Value Problems for Ordinary Differential Equations Containing a Parameter

Many boundary value problems P_ϵ depend on a small positive parameter ϵ in such a way that the full differential equation is of higher order than the reduced one obtained by setting the small parameter ϵ equal to zero. (For surveys of such problems which have physical significance, see Friedrichs [36], Carrier [10], and Segel [95].) Obvious questions arise. Does the boundary value problem P_ϵ have a limiting solution as $\epsilon \to 0$? If so, (a) does the limiting solution satisfy the reduced equation, and (b) which, if any, of the boundary conditions will be satisfied by the limiting solution? Indeed, when the limiting solution exists and solves the reduced equation (i.e., in cases of "regular degeneration"), we expect the solution of the original problem to, in general, converge nonuniformly near the boundary due to the loss of boundary conditions required in converging to the limiting solution. Such regions of nonuniform convergence are known as boundary layers, in reference to Prandtl's boundary layer theory for viscous flow past a body at high Reynolds number (see Prandtl [91]). Since the limiting solution fails, in general, to be uniformly valid up to the boundary, the straightforward (or regular) perturbation procedure fails and we are faced with a singular perturbation problem.

Wasow [109], following the earlier results of Birkhoff [2], Noaillon [79], and Turrittin [101], stated in a most elegant fashion sufficient conditions for convergence to a limiting solution for homogeneous linear ordinary differential equations depending linearly on a parameter. Latta [63] went further by giving a complete uniformly valid asymptotic expansion for solutions of such equations as well as for some linear partial differential equations. As might be expected, Latta's success was based upon making a change of variables to "blow up" the boundary region in order to accurately describe the nonuniformly converging terms in the solution of the boundary value problem. Cochran [14] formalized this technique by introducing such variables as new independent variables and, by so doing, was able to solve turning point problems and some nonlinear problems as well as those solved by Latta. An extension of these techniques to certain boundary value problems where the orders of differentiation in the boundary conditions may be smaller at $\epsilon = 0$ then for $\epsilon > 0$ has been obtained by O'Malley and Keller [88]. Among other papers discussing these and similar problems Visik and Lyusternik [107] and Harris [46] are especially recommended.

Specifically, let us consider the limiting behavior, as ϵ tends to zero, of solutions of boundary value problems for homogeneous equations of the form

$$\epsilon M\Phi + N\Phi = 0$$

with suitable boundary conditions where $M\Phi$ and $N\Phi$ are linear differential expressions involving one or more independent variables with m, the order of M, greater than n, that of N. Restricting our consideration, for now, to ordinary differential equations on the x-interval $[0, 1]$ without turning points, let

$$My = y^{(m)} + \alpha_1(x)\, y^{(m-1)} + \cdots + \alpha_k(x)\, y^{(m-k)},$$

and

$$Ny = \beta(x)[y^{(n)} + \beta_1(x)\, y^{(n-1)} + \cdots + \beta_j(x)\, y^{(n-j)}]$$

where all coefficients are real and $\beta(x) \neq 0$ for $x \in [0, 1]$.

Further, let us solve

$$\epsilon My + Ny = 0 \tag{2.1}$$

subject to the m boundary conditions

$$\begin{cases} y^{(\lambda_i)}(0) = l_i\,, & i = 1, 2,\ldots, r \\ y^{(\tau_i)}(1) = l_{r+i}\,, & i = 1, 2,\ldots, m - r \end{cases} \tag{2.2}$$

where

$$m > \lambda_1 > \lambda_2 > \cdots > \lambda_r \geqslant 0$$

and

$$m > \tau_1 > \tau_2 > \cdots > \tau_{m-r} \geqslant 0.$$

Before studying the general problem, however, we determine the asymptotic solution of the two-point boundary value problem

$$\begin{cases} \epsilon y'' + a(x)\, y' + b(x)\, y = 0, \\ y(0), \quad y(1) \qquad \text{prescribed}, \end{cases} \tag{2.3}$$

where $a(x)$ is positive for $x \in [0, 1]$.

In studying singular perturbation problems, we continually examine simple illustrative examples—especially constant coefficient problems which can be explicitly integrated. The detailed examination of such problems and their asymptotic solutions is most valuable in gaining direction for studying more general problems. Hence, recall that in Section 1 the solution of the boundary value problem

$$\epsilon y'' + y' + y = 0,$$

$$y(0) = \alpha, \quad y(1) = \beta$$

was found to be

$$y = \frac{(\beta - \alpha e^{\rho_2})\, e^{\rho_1 x} + (\alpha e^{\rho_1} - \beta)\, e^{\rho_2 x}}{e^{\rho_1} - e^{\rho_2}}$$

for $x \in [0, 1]$ where ρ_1 and ρ_2 are the roots of the characteristic equation

$$\epsilon \rho^2 + \rho + 1 = 0.$$

Here, $y \to \beta e^{1-x}$, the solution of the "reduced boundary value problem"

$$u' + u = 0,$$

$$u(1) = \beta,$$

uniformly on any x interval $0 < \delta \leqslant x \leqslant 1$, but, in general, convergence is nonuniform near $x = 0$. Further, since

$$y \sim \beta e^{-(1-x)\rho_1} + (\alpha - \beta e^{-\rho_1})\, e^{\rho_2 x}$$

we see that this nonuniform convergence is due to the exponential decay of $e^{\rho_2 x}$, ρ_2 being that root of the auxiliary polynomial which becomes singular as $\epsilon \to 0$.

After studying this problem where $a(x) = b(x) = 1$, we expect convergence of the solution $y(x)$ of (2.3) on each interval $0 < \delta \leqslant x \leqslant 1$ to

$$u(x) = y(1) \exp\left[\int_x^1 \frac{b(t)}{a(t)}\, dt\right],$$

the solution of the reduced boundary value problem

$$\begin{cases} a(x)\, u' + b(x)\, u = 0, \\ u(1) = y(1). \end{cases}$$

Moreover, we expect nonuniform convergence near $x = 0$ where we expect the boundary layer behavior to be essentially related to the singular root of the auxiliary equation

$$\epsilon D^2 + a(x)\, D + b(x) = 0,$$

i.e.,

$$D_1 = -\frac{a(x)}{2\epsilon}\left(1 + \sqrt{1 - \frac{4\epsilon b(x)}{a^2(x)}}\right)$$

Expanding this root as a function of the small parameter ϵ,

$$\begin{aligned} D_1 &= -\frac{d_1(x, \epsilon)}{\epsilon} = -\frac{1}{\epsilon}\left[a(x) - \frac{\epsilon b(x)}{a(x)} - \frac{\epsilon^2 b^2(x)}{a^3(x)} - \cdots\right] \\ &= -\frac{\tilde{d}_1(x, \epsilon)}{\epsilon} + O(\epsilon), \end{aligned}$$

where

$$\tilde{d}_1(x, \epsilon) \equiv a(x) - \epsilon b(x)/a(x)$$

is positive for ϵ sufficiently small. Again, generalizing from the case when $a(x) = b(x) \equiv 1$, we assume the solution $y(x)$ of (2.3) has the form

$$y(x) \sim A(x, \epsilon) + B(x, \epsilon) \exp\left[-\frac{1}{\epsilon}\int_0^x \tilde{d}_1(s, \epsilon)\, ds\right] \tag{2.4}$$

where

$$A(x, \epsilon) \sim \sum_{r=0}^{\infty} a_r(x)\, \epsilon^r,$$

and

$$B(x, \epsilon) \sim \sum_{r=0}^{\infty} b_r(x)\, \epsilon^r.$$

Note: (1) The function $1/\epsilon \int_0^x \tilde{d}_1(s, \epsilon)$ is analogous to the stretching transformation used by Latta and the new independent variable introduced by Cochran to investigate the region of nonuniformity.

(2) Since the right member of (2.4) asymptotically equals $a_0 + O(\epsilon)$ in any x-interval $0 < \delta \leqslant x \leqslant 1$, we expect that $a_0(x)$ will equal

$$y(1) \exp\left[\int_x^1 \frac{b(t)}{a(t)}\, dt\right].$$

(3) Here and below the expansions obtained will not necessarily be asymptotic in the sense of Poincaré, but in the more general sense of van der Corput [20] for the asymptotic sequence $\{\epsilon^n\}$. Note that the coefficients in the power series expansion (2.4) of $y(x, \epsilon)$ depend on ϵ through the exponential factor. As in all such expansions, then, the function $y(x, \epsilon)$ does not uniquely determine the coefficients of its expansion.

Substituting (2.4) into (2.3) and, for each integer r, formally equating to zero the coefficients of ϵ^r and

$$\epsilon^r \exp\left[-\frac{1}{\epsilon}\int_0^x \tilde{d}_1(s, \epsilon)\, ds\right],$$

we obtain

$$a(x)\, a_r'(x) + b(x)\, a_r(x) = -a_{r-1}''(x)$$

and

$$a(x)\, b_r'(x) + a'(x)\, b_r(x) = c_{r-1}(x),$$

where

$$c_{r-1}(x) = b_{r-1}''(x) + 2b_{r-1}'(x)\, \frac{b(x)}{a(x)} + b_{r-1}(x)\left(\frac{b^2(x)}{a^2(x)} + \left(\frac{b(x)}{a(x)}\right)'\right)$$

with

$$a_{-1}(x) = c_{-1}(x) \equiv 0.$$

Since $y(1) \sim A(1, \epsilon)$, we set $a_0(1) = y(1)$ and $a_k(1) = 0$ for $k \geqslant 1$,

Thus,

$$a_0(x) = y(1) \exp\left[-\int_1^x \frac{b(s)}{a(s)}\, ds\right]$$

and later coefficients are determined successively by the formula

$$a_r(x) = -\int_1^x \frac{a''_{r-1}(t)}{a(t)} \exp\left[-\int_t^x \frac{b(s)}{a(s)}\, ds\right] dt.$$

Likewise, since $y(0) = A(0, \epsilon) + B(0, \epsilon)$, we set $b_0(0) = y(0) - a_0(0)$ and $b_k(0) = -a_k(0)$ for $k \geqslant 1$. Integrating, the b_r's are determined successively from the formulas

$$b_0(x) = \frac{a(0)}{a(x)} [y(0) - a_0(0)]$$

and

$$b_r(x) = \frac{1}{a(x)} \left[-a_r(0)\, a(0) + \int_0^x c_{r-1}(s)\, ds\right] \qquad \text{for} \quad r \geqslant 1.$$

Thus, to establish the validity of (2.4), we need only prove the following theorem:

Theorem 2.1. *Let $y(x)$ solve the boundary value problem*

$$\epsilon y'' + a(x)\, y' + b(x)\, y = 0,$$

$$y(0), \quad y(1) \qquad \textit{prescribed},$$

where $a(x)$ is positive for $x \in [0, 1]$, and let $a_r(x)$ and $b_r(x)$ be the functions defined successively above.

Further, for each positive integer N, let

$$y(x) = \sum_{r=0}^{N} a_r(x)\, \epsilon^r + \left(\sum_{r=0}^{N} b_r(x)\, \epsilon^r\right) \exp\left[-\frac{1}{\epsilon}\int_0^x \left(a(s) - \epsilon \frac{b(s)}{a(s)}\right) ds\right] + \epsilon^{N+1} R_N(x, \epsilon). \tag{2.5}$$

Then,

$$R_N(x, \epsilon) = O(1) \qquad \textit{for all} \qquad x \in [0, 1].$$

Proof of Asymptotic Convergence

The reader who is interested only in the construction of formal asymptotic solutions may, without much loss, omit this proof. The serious student of asymptotics, however, will realize that formal procedures require justification. Thus, the stated results will be established in a manner which emphasizes methods of proof which are applicable generally. Results from integral equations theory used may be found, e.g., in Tricomi [100], while a general discussion of the use of integral equations to prove asymptotic convergence will be found in Erdélyi [28].

Before establishing asymptotic convergence we first derive some subsidiary lemmas.

Lemma 1. *Consider the initial-value problem:*

$$\begin{cases} \epsilon R'' + a(x) R' + b(x) R = f(x, \epsilon) \\ R(0) = 0, \quad R'(0) = c/\epsilon \end{cases} \tag{2.6}$$

for $x \in [0, 1]$ *where* $0 < \delta \leqslant a(x)$, $\epsilon > 0$, *and* $a, b, f,$ *and* c *are bounded. Then* (2.6) *has a unique bounded solution* R *for all* $x \in [0, 1]$.

Proof. Integrating the differential equation,

$$R'(x) = \frac{c}{\epsilon} \exp\left[-\frac{1}{\epsilon}\int_0^x a(s)\, ds\right] + \int_0^x \left[\frac{f(t) - b(t)\, R(t)}{\epsilon}\right] \exp\left[-\frac{1}{\epsilon}\int_t^x a(s)\, ds\right] dt.$$

Integrating again and changing the order of integration:

$$(*) \qquad R(x) = \frac{c}{\epsilon}\int_0^x \exp\left[-\frac{1}{\epsilon}\int_0^z a(s)\, ds\right] dz + \frac{1}{\epsilon}\int_0^x \left(\int_t^x \exp\left[-\frac{1}{\epsilon}\int_t^z a(s)\, ds\right] dz\right) (f(t) - b(t)\, R(t))\, dt.$$

This integral equation is, however, uniquely solvable by successive approximations. We merely set

$$R_0(x) = \frac{c}{\epsilon}\int_0^x \exp\left[-\frac{1}{\epsilon}\int_0^z a(s)\, ds\right] dz + \frac{1}{\epsilon}\int_0^x \int_t^x \exp\left[-\frac{1}{\epsilon}\int_t^z a(s)\, ds\right] dz\, f(t)\, dt$$

and

$$R_j(x) = R_0(x) - \frac{1}{\epsilon}\int_0^x \int_t^x b(t)\, R_{j-1}(t) \exp\left[-\frac{1}{\epsilon}\int_t^z a(s)\, ds\right] dz\, dt \qquad j \geqslant 1$$

and let $R(x) \equiv \lim_{j\to\infty} R_j(x)$. Convergence follows from the estimates

$$0 \leqslant \int_0^x \exp\left[-\frac{1}{\epsilon}\int_0^z a(s)\, ds\right] dz \leqslant \int_0^x e^{-\delta z/\epsilon}\, dz = \frac{\epsilon}{\delta}\left[1 - e^{-\delta x/\epsilon}\right]$$

and

$$\int_0^x \int_t^x \exp\left[-\frac{1}{\epsilon}\int_t^z a(s)\, ds\right] g(t)\, dz\, dt \leqslant \frac{2\epsilon}{\delta}\int_0^x g(t)\, dt \qquad \text{for } g \text{ positive.}$$

Having established the existence and uniqueness of R, its boundedness also follows from the same estimates. Specifically,

$$|\, R(x)| \leqslant \psi(x) \equiv \frac{|\, c\,|}{\delta}\left(1 - e^{-\delta x/\epsilon}\right) + \frac{2M}{\delta}\int_0^x (1 + |\, R(t)|)\, dt$$

where M is a sufficiently large positive constant. (Below, M will represent a generic positive constant, i.e., M will not necessarily be the same constant each time it appears.) This implies that

$$\psi'(x) \leqslant \frac{|\, c\,|}{\epsilon}\, e^{-\delta x/\epsilon} + \frac{2M}{\delta}\,(1 + \psi(x))$$

and

$$\psi(0) = 0$$

which, in turn, implies that

$$|\, R(x)| \leqslant \psi(x) \leqslant \int_0^x \left(\frac{|\, c\,|}{\epsilon}\, e^{-\delta t/\epsilon} + \frac{2M}{\delta}\right) \cdot e^{(2M/\delta)(x-t)}\, dt,$$

so $R(x)$ is bounded throughout $[0, 1]$.

Note, further, that the solution of (2.6) depends continuously on the value of c. Specifically, if $\tilde{R}(x)$ and $\tilde{\tilde{R}}(x)$ represent solutions of (2.6) with c values $\tilde{c}$ and $\tilde{\tilde{c}}$, respectively, the previous estimates imply that

$$|\, \tilde{R}(x) - \tilde{\tilde{R}}(x)| \leqslant \frac{|\, \tilde{c} - \tilde{\tilde{c}}\,|}{\epsilon}\int_0^x e^{-\delta t/\epsilon}\, e^{(2M/\delta)(x-t)}\, dt.$$

Using these estimates, we easily show that

$$R(x) = c\left[-\frac{1}{a(x)}\exp\left[-\frac{1}{\epsilon}\int_0^x a(s)\,ds\right] + \frac{1}{a(0)}\exp\left[-\int_0^x \frac{b(s)}{a(s)}\,ds\right]\right]$$
$$+ \int_0^x \frac{f(t,\epsilon)}{a(t)}\exp\left[-\int_t^x \frac{b(s)}{a(s)}\,ds\right]dt + O(\epsilon)$$

Since R is unique and depends continuously on c, there is a unique value for c such that the solution R of the initial-value problem (2.6) solves the boundary value problem

$$\epsilon R'' + a(x)\,R' + b(x)\,R = f(x,\epsilon),$$
$$R(0) = R(1) = 0.$$

Clearly, for any bounded solution R, c must satisfy the equation

$$c\int_0^1 \exp\left[-\frac{1}{\epsilon}\int_0^z a(s)\,ds\right]dz = -\int_0^1\int_t^1 (f(t) - b(t)\,R(t))$$
$$\cdot \exp\left[-\frac{1}{\epsilon}\int_t^z a(s)\,ds\right]dz\,dt.$$

Thus, we have proved the existence half of the following statement.

Lemma 2. *For all sufficiently small values of* ϵ, *there exists a unique bounded solution of the boundary value problem*

$$\begin{cases}\epsilon R'' + a(x)\,R' + b(x)\,R = f(x,\epsilon), \\ R(0) = R(1) = 0,\end{cases} \tag{2.7}$$

where $0 < \delta \leqslant a(x)$, *and* a, b, *and* f *are bounded for* $x \in [0, 1]$.

Proof. To establish uniqueness, we need only show that the homogeneous problem

$$\epsilon R'' + a(x)\,R' + b(x)\,R = 0,$$
$$R(0) = R(1) = 0$$

has only the trivial solution. Integrating this equation and applying the boundary conditions

$$R(x) = \int_0^1 \psi(z,\epsilon)\,R(z)\,dz \int_0^x \exp\left[-\frac{1}{\epsilon}\int_0^t a(s)\,ds\right]dt$$
$$-\int_0^x\int_t^x \frac{b(t)\,R(t)}{\epsilon}\exp\left[-\frac{1}{\epsilon}\int_t^s a(r)\,dr\right]ds\,dt$$

where

$$\psi(z, \epsilon) \equiv \frac{b(z) \int_z^1 \exp\left[-\frac{1}{\epsilon}\int_z^t a(s)\, ds\right] dt}{\epsilon \int_0^1 \exp\left[-\frac{1}{\epsilon}\int_0^t a(s)\, ds\right] dt},$$

or

$$R(x) = \int_0^1 C(x, z, \epsilon)\, R(z)\, dz + \int_0^x B(x, z, \epsilon)\, R(z)\, dz$$

where C and B are defined in the obvious way.

Introducing $H(x, y, \epsilon)$, the resolvent kernel corresponding to the Volterra kernel $B(x, y, \epsilon)$ with eigenparameter $\lambda = 1$, multiply the above equation by $H(t, x, \epsilon)$ and integrate from $x = 0$ to $x = t$. Then, using the Fredholm identity

$$B(t, \tau, \epsilon) + H(t, \tau, \epsilon) = \int_\tau^t H(t, x, \epsilon)\, B(x, \tau, \epsilon)\, dx$$

and the integral equation, we obtain

$$\begin{aligned} R(t) &= \int_0^1 \left[C(t, \tau, \epsilon) + \int_0^t C(x, \tau, \epsilon)\, H(t, x, \epsilon)\, dx\right] R(\tau)\, d\tau \\ &\equiv \varphi(t, \epsilon) \int_0^1 \psi(\tau, \epsilon)\, R(\tau)\, d\tau \end{aligned}$$

where

$$\begin{aligned} \varphi(t, \epsilon) \equiv \int_0^t \exp\left[-\frac{1}{\epsilon}\int_0^z a(s)\, ds\right] dz \\ + \int_0^t \left(\int_0^x \exp\left[-\frac{1}{\epsilon}\int_0^z a(s)\, ds\right] dz\right) H(t, x, \epsilon)\, dx. \end{aligned}$$

This, however, is a Fredholm integral equation (with degenerate kernel) having only the trivial solution $R \equiv 0$ as $\epsilon \to 0$. Q.E.D.

Note. (1) The convenient use of the resolvent kernel in this proof follows an analogous argument in Cochran [14] while the technique is applied to more general two-point boundary value problems in Cochran [15]. A related theorem for analogous nonlinear singular perturbation problems is found in Willett [118].

(2) The preceding theorem follows immediately since R_N satisfies a boundary value problem of the form (2.7).

(3) If $a(x) < 0$, the transformation $z = 1 - x$ will allow the reader to solve the boundary value problem

$$\epsilon y'' + a(x)\, y' + b(x)\, y = 0,$$
$$y(0), \quad y(1) \qquad \text{prescribed}$$

as a boundary value problem of type (2.3) for $z \in [0, 1]$. Clearly, the boundary layer behavior will then occur in the vicinity of $x = 1$.

Higher-Order Problems

Now consider the equation

$$\epsilon My + Ny = 0 \tag{2.1}$$

for $m > n$ and $\beta(x) \neq 0$ for $x \in [0, 1]$ subject to the m boundary conditions

$$\begin{cases} y^{(\lambda_i)}(0) = l_i\,, & i = 1, 2,..., r, \\ y^{(\tau_i)}(1) = l_{r+i}\,, & i = 1, 2,..., m - r. \end{cases} \tag{2.2}$$

Letting $y_l(x, \epsilon)$, $l = 1, 2,..., m$, be a fundamental system of linearly independent solutions of (2.1), any solution $y(x, \epsilon)$ will be a linear combination of the y_l's with coefficients which are independent of x. The linear combination will be a solution of the problem (2.1)–(2.2), if the coefficients are chosen to satisfy the boundary conditions. Moreover, the set of m linear equations resulting will be uniquely solvable provided the determinant of the coefficient matrix is nonvanishing.

We seek a particular fundamental set of solutions of the form

$$y(x, \epsilon) = G(x, \epsilon) \exp\left[\frac{1}{\kappa}\int_t^x \tilde{d}(s, \kappa)\, ds\right]$$

where

$$\kappa = \epsilon^{1/(m-n)}$$

and t is a constant. Differentiating j times with respect to x, then,

$$y^{(j)}(x, \epsilon) = \left\{G^{(j)} + \frac{j}{\kappa} G^{(j-1)}\tilde{d} + \cdots + \frac{1}{\kappa^{j-1}}\left[jG'\tilde{d}^{j-1} + \frac{j(j-1)}{2} G\tilde{d}^{j-2}\tilde{d}'\right]\right.$$
$$\left. + \frac{1}{\kappa^j} G\tilde{d}^j\right\} \exp\left[\frac{1}{\kappa}\int_t^x \tilde{d}(s, \kappa)\, ds\right].$$

Substituting into (2.1) and analyzing the result, we obtain the desired fundamental system. Note that the auxiliary equation

$$\epsilon[D^m + \alpha_1(x) D^{m-1} + \cdots + \alpha_k(x) D^{m-k}] \\ + \beta(x)[D^n + \beta_1(x) D^{n-1} + \cdots + \beta_j(x) D^{n-j}] = 0 \tag{2.8}$$

associated with (2.1) has $m - n$ solutions of the form

$$\frac{d(x, \kappa)}{\kappa} = \frac{\tilde{d}(x, \kappa)}{\kappa} + O(\kappa)$$

where

$$\tilde{d}(x, \kappa) = \omega^r h(x) + \kappa \left[\frac{\beta_1(x) - \alpha_1(x)}{m - n}\right]$$

for $h(x)$ positive, r an integer, and ω a complex number of modulus one such that

$$(\omega^r h(x))^{m-n} = -\beta(x).$$

To each of these $m - n$ complex solutions of (2.8), we let correspond a formal complex solution of (2.1), namely,

$$y = \frac{A(x, \kappa)}{(h(x))^{(m+n-1)/2}} \exp\left[\frac{1}{\kappa}\int_t^x \tilde{d}(s, \kappa)\, ds\right], \tag{2.9}$$

where A has a power series expansion in κ whose first term is independent of x and whose higher-order terms may each be successively determined to within an additive constant, and t is either 0 or 1.

Note. Since the roots of (2.8) occur in conjugate complex pairs, it is perhaps more natural to associate two formal real solutions of (2.1) with each conjugate pair of roots of (2.8).

Let p be the number of determinations $\tilde{d}$ for which

$$\operatorname{Re} \tilde{d}(x, \epsilon) \tag{2.10}$$

is negative for ϵ sufficiently small and let q be the number for which it is positive. When

$$p + q = m - n$$

we call (2.10) nonexceptional, while we call it exceptional if

$$p + q < m - n.$$

Note. (1) Here, p is equal to the number of roots of the characteristic equation (2.8) which approach $-\infty$ (throughout $[0, 1]$) as $\epsilon \to 0$, while q is the number which approach $+\infty$. Moreover, $p + q = m - n - 2$ in the exceptional case.

(2) For more general equations, (2.10) may be replaced by the integral

$$\operatorname{Re} \int_0^1 \tilde{d}(s, \epsilon)\, ds$$

(cf., O'Malley and Keller [88]).

Corresponding to the q distinct roots with (2.10) positive, define q linearly independent solutions $y_1, y_2, \ldots, y_q$ of the above form, all with $t = 1$, and corresponding to the p distinct roots with (2.10) negative, p linearly independent solutions, $y_{m-n-p+1}, \ldots, y_{m-n}$, all with $t = 0$. Clearly, $y_1, y_2, \ldots, y_q$, and their derivatives are all exponentially small away from $x = 1$, while $y_{m-n-p+1}, \ldots, y_{m-n}$ exhibit this boundary layer behavior at $x = 0$. In the exceptional case, we also define solutions $y_{q+1}, y_{q+2}, \ldots, y_{m-n-p}$ by the formula (2.9) but with t, as yet, undetermined. In addition to these $m - n$ "boundary layer type" solutions, n additional linearly independent asymptotic solutions of (2.1) can be obtained as regular perturbations of any complete set of fundamental solutions of the reduced equation $Nz = 0$. We denote them by $y_{m-n+k}(x, \epsilon)$, $k = 1, 2, \ldots, n$, with each solution a series in powers of ϵ for which each variable coefficient is determined up to an additive constant by the differential equation (2.1) and the fundamental solution z_k of $Nz = 0$ which it perturbs. That $y_1, \ldots, y_m$ form a fundamental system for (2.1) follows from the results of Turrittin [102], as outlined in Wasow [116].

We define the reduced boundary value problem to consist of the reduced equation $Nz = 0$ plus n of the boundary conditions (2.2), and state which $m - n$ boundary conditions are omitted in the following

Cancellation Law. (1) *Cancel p boundary conditions at $x = 0$ and q boundary conditions at $x = 1$, starting from those containing the highest derivative.*

(2) *In the exceptional case, also cancel from the boundary conditions those $m - n - p - q$ of highest order of differentiation. These boundary conditions must all belong to the same end point (that end point being denoted by $\tilde{t}$), and their selection must be possible without ambiguity.*

All together, then, let P be the number of boundary conditions canceled at $x = 0$ and let Q be the number canceled at $x = 1$. Having obtained a complete set of formal asymptotic linearly independent solutions of (2.1), we expect to write the limiting solution of the boundary value problem (if it exists) as a linear combination of these solutions. Thus, we are led to setting

$$y(x) = \kappa^{\alpha_1} \sum_{k=1}^{Q} y_k(x, \kappa) + \kappa^{\alpha_2} \sum_{k=Q+1}^{m-n} y_k(x, \kappa) + \sum_{k=1}^{n} y_{m-n+k}(x, \kappa),$$

where the y_k, $k \leqslant m - n$, are the linearly independent solutions defined above, and the $y_{m-n+k}(x, \kappa)$ are multiplies (with coefficients power series in κ with constant coefficients) of the solutions $y_{m-n+k}(x, \epsilon)$ defined above. In the exceptional case, we set $t = \tilde{t}$ in the expressions for $y_{q+1}, y_{q+2}, \ldots, y_{m-n-p}$. α_1 and α_2 are defined in the following manner:

Nonexceptional case:

$$\alpha_1 = \tau_Q, \qquad \alpha_2 = \lambda_P.$$

Exceptional case:

$$\tilde{t} = 0: \quad \alpha_1 = L, \qquad \alpha_2 = \lambda_P,$$

$$\tilde{t} = 1: \quad \alpha_1 = \tau_Q, \qquad \alpha_2 = L,$$

with $L = \min(\tau_Q, \lambda_p)$. Proceeding in the manner of O'Malley and Keller we obtain the following.

Theorem 2.2. *Consider the differential equation*

$$\epsilon My + Ny = 0$$

(*with My and Ny as defined above*) *subject to the boundary conditions*

$$y^{(\lambda_i)}(0) = l_i, \qquad i = 1, 2, \ldots, r,$$

$$y^{(\tau_i)}(1) = l_{r+i}, \qquad i = 1, 2, \ldots, m - r,$$

where

$$m > \lambda_1 > \lambda_2 > \cdots > \lambda_r \geqslant 0,$$

and

$$m > \tau_1 > \tau_2 > \cdots > \tau_{m-r} \geqslant 0.$$

Further, let $z(x)$ *satisfy the reduced problem*

$$Nz = 0,$$
$$z^{(\lambda_i)}(0) = l_i\,, \qquad i = P+1, P+2,..., r,$$
$$z^{(\tau_j)}(1) = l_{r+j}\,, \qquad j = Q+1, Q+2,..., m-r,$$

where $P \leqslant r$ *and* $Q \leqslant m - r$.

Suppose

1. *The cancellation law is meaningful,*
2. *The reduced problem has a unique solution* $z(x)$,
3. (a) $\lambda_1, \lambda_2, ..., \lambda_P$ *are distinct modulo* $(m - n)$,
   (b) $\tau_1, \tau_2, ..., \tau_Q$ *are distinct modulo* $(m - n)$.

Then, a unique solution $y(x)$ *exists for* ϵ *sufficiently small and has the form*

$$y(x) = \kappa^{\alpha_1} \sum_{k=1}^{0} y_k(x, \kappa) + \kappa^{\alpha_2} \sum_{k=Q+1}^{m-n} y_k(x, \kappa) + u(x, \kappa)$$

where $u(x, \kappa)$ *is a regular perturbation of* $z(x)$.

4. *In the exceptional case, also suppose*
   (a) $\alpha_1 > 0$ *if* $\tilde{t} = 1$,
   (b) $\alpha_2 > 0$ *if* $\tilde{t} = 0$.

Then

$$y(x) \sim z(x) \qquad on \quad 0 < x < 1.$$

Note. (1) These results extend the conclusions of Wasow [109] (cf., e.g., the exceptional case) and are, in turn, extended in O'Malley and Keller [88]. These papers all deal with scalar equations with separated (or uncoupled) boundary conditions. An example with coupled boundary conditions is analyzed in Friedman [34].

(2) That hypotheses 1, 2, 3, and 4 are essentially necessary can be seen by considering a series of problems for constant coefficient equations whose solutions are, except for very special boundary values (e.g., all zero)

divergent as $\epsilon \to 0$. See, e.g., Wasow [109] and O'Malley [81] for lists of examples.

(3) This procedure of finding an asymptotic fundamental system and taking a suitable linear combination of these asymptotic solutions to obtain the limiting solution (if it exists) is the method used by Wasow [109] to obtain the first term in each such product series of the linear combination. This method, without modification (as Latta [63]) points out), is extremely laborious for obtaining further terms in the expansions needed. Furthermore, one does not always need a complete system of asymptotic fundamental solutions to obtain the uniform expansion needed. For example, for $O(\kappa^L)$ accuracy, only the first L terms of $u(x, \kappa)$ are of interest (and the solution y—to this order of approximation—is a regular perturbation of $z(x)$ uniformly on $0 \leqslant x \leqslant 1$). Likewise, if $L = \lambda_p$ and $\tilde{t} = 1$, for $O(\kappa^{\tau_Q})$ accuracy, no terms of the expansions for $y_1, y_2, \ldots, y_Q$ need be calculated.

A Modified Approach

In dealing with some more complicated singular perturbation problems featuring turning-points and nonlinearities, Cochran [14] introduced new variables (or stretching transformations) in terms of which he obtained the asymptotic solutions desired. To illustrate the technique, which sometimes proceeds heuristically, we reconsider the linear problem

$$\begin{cases} \epsilon y'' + a(x)\, y' + b(x)\, y = 0, \\ y(0), \quad y(1) \quad \text{prescribed}, \\ \text{for} \quad a(x) > 0 \quad \text{on} \quad 0 \leqslant x \leqslant 1. \end{cases} \tag{2.3}$$

Experience (and our previous calculation) would predict that the solution of this boundary value problem will feature nonuniform convergence in the vicinity of $x = 0$ and that this boundary layer behavior may be described in terms of the two variables x and $1/\epsilon \int_0^x a(s)\, ds$.

Thus, we introduce the stretching transformation

$$\eta = \frac{1}{\epsilon} \int_0^x [g_0(s) + \epsilon g_1(s)]\, ds \tag{2.11}$$

with the g_i's undetermined, but with g_0 positive, and we formally set

$$y(x) \equiv y(x, \eta) = \sum_{k=0}^{\infty} y_k(x, \eta)\, \epsilon^k, \tag{2.12}$$

where we ask that the functions y_k be bounded independent of ϵ for all values of x and η and that certain "secular" terms be eliminated. We proceed to demonstrate the procedure.

Formally, substitute (2.12) into (2.3) and equate to zero the coefficient of each power of ϵ. The coefficient of ϵ^{-1}, the lowest power appearing, yields

$$g_0^2(y_{0\eta\eta} + (a/g_0)\, y_{0\eta}) = 0. \tag{2.13a}$$

Integrating twice with respect to η, we take

$$y_0(x, \eta) = A_0(x) + B_0(x)\, e^{-(a/g_0)\eta}.$$

Since g_1 is arbitrary, however, we take $B_0(x) = b_0$, a constant, without loss of generality. The coefficient of ϵ^0 yields

$$\begin{aligned} y_{1\eta\eta} g_0^2 + y_{1\eta} a g_0 + 2y_{0x\eta} g_0 + 2y_{0\eta\eta} g_1 g_0 \\ + y_{0\eta} g_0' + a y_{0x} + a g_1 y_{0\eta} + b y_0 = 0. \end{aligned} \tag{2.13b}$$

Using (2.13a) and integrating with respect to η,

$$\begin{aligned} g_0^2[y_1 e^{(a/g_0)\eta}]_\eta + B_0 \left[2g_0 \left(-\frac{a}{g_0}\right)' \eta - g_1 a + g_0'\right] + \eta e^{(a/g_0)\eta}[aA_0' + bA_0] \\ + e^{(a/g_0)\eta} \left[\int_0^\eta \left[aB_0 \left(-\frac{a}{g_0}\right)' \eta e^{-(a/g_0)\eta} + bB_0 e^{-(a/g_0)\eta}\right] d\eta\right] = \tilde{A}_1(x)\, e^{(a/g_0)\eta}. \end{aligned}$$

For boundedness of y_1, we need to pick

$$aA_0' + bA_0 = 0,$$

so

$$A_0(x) = a_0 \exp\left[-\int_1^x \frac{b(s)}{a(s)}\, ds\right].$$

Likewise, to eliminate secular terms of the form $\eta^2 \exp[-(a/g_0)\eta]$ in y_1, we require that

$$g_0 B_0 (a/g_0)' = 0$$

and, therefore, that $g_0(x) = a(x)$. Thus,

$$a^2 y_1 e^\eta + b_0 \eta[-g_1 a + a' - b] = \tilde{A}_1(x)\, e^\eta + \tilde{B}_1(x).$$

To eliminate a further secular term of the form $\eta e^{-\eta}$, we set

$$g_1(x) = \frac{a'(x)}{a(x)} - \frac{b(x)}{a(x)}$$

and obtain

$$y_1(x, \eta) = A_1(x) + B_1(x)\, e^{-\eta}.$$

Thus

$$\eta = \frac{1}{\epsilon} \int_0^x \left[a(s) + \epsilon \left((\ln a(s))' - \frac{b(s)}{a(s)}\right)\right] ds$$

and

$$y_0(x, \eta) = a_0 \exp\left[- \int_1^x \frac{b(s)}{a(s)}\, ds\right] + \frac{b_0}{a(x)} \exp\left[- \frac{1}{\epsilon} \int_0^x \left(a(s) - \frac{b(s)}{a(s)}\right) ds\right],$$

where we need to select $a_0 = y(1)$ and

$$b_0 = a(0) \left[y(0) - y(1) \exp\left[- \int_1^0 \frac{b(s)}{a(s)}\, ds\right]\right]$$

in order to satisfy the boundary conditions. Appropriate specialization of the form of $y_1(x, \eta)$ has thus yielded a complete determination of $y_0(x, \eta)$. Higher-order terms are successively obtained in a similar manner.

Surely, this example was more easily solved previously. However, being familiar with this procedure we can solve more complicated examples. Specifically, Cochran [14] successfully attacked the nonlinear problem

$$\epsilon y'' + yy' - y = 0,$$
$$y(0), \quad y(1) \quad \text{prescribed}$$

and the turning-point problem

$$\epsilon y'' - (\tfrac{1}{2} - x)\, y' - y = 0,$$
$$y(0), \quad y(1) \quad \text{prescribed},$$

among others, and O'Malley [85] considered nonlinear problems of the form

$$\epsilon y'' + f(x, y)\, y' + g(x, y) = 0,$$
$$y(0), \quad y(1) \quad \text{prescribed},$$

In particular, complete expansions valid for $0 \leqslant x \leqslant 1$ were obtained in the special case where f and g are infinitely differentiable and where $f(x, y) = a(x) \neq 0$. Note that similar analysis is reported in Erdélyi [29] and an engineering application is given in O'Malley [86]. Note also that earlier results on this problem were obtained by Coddington and Levinson [16], Wasow [113], Erdélyi [26], and Willett [118].

3. Singular Perturbation of Eigenvalue Problems

The Vibrating String

Recall the eigenvalue problem which arises in considering the vibrations of a string with clamped end points (refer, e.g., to Rayleigh [93]). If the string has negligible stiffness, we consider the problem

$$\begin{cases} -y'' = \lambda^2 y, \\ y(0) = y(1) = 0 \end{cases} \tag{3.1}$$

on the x-interval $[0, 1]$. If, however, stiffness effects are introduced, we consider the higher-order problem

$$\begin{cases} \epsilon y^{\mathrm{IV}} - y'' = \lambda^2 y, \\ y(0) = y(1) = y'(0) = y'(1) = 0, \end{cases} \tag{3.2}$$

with ϵ positive and proportional to the stiffness. As the string stiffness goes to zero ($\epsilon \to 0$), we expect that the eigenvalues and eigenfunctions of (3.2) should converge (but not necessarily uniformly) to those of (3.1), i.e., $n^2\pi^2$ and $\sin n\pi x$, respectively, for $n = 1, 2, \ldots$.

We associate with (3.2) the auxiliary equation

$$\epsilon\rho^4 - \rho^2 - \lambda^2 = 0$$

which has two singular roots (as $\epsilon \to 0$) of the form $\pm 1/\kappa + O(\kappa)$ for

$$\kappa = \epsilon^{1/2} > 0$$

and λ^2 bounded. This, naturally, leads us to introduce stretching coordinates x/κ and $(1 - x)/\kappa$ and to formally let

$$y(x, \epsilon) = A(x, \kappa) + B(x, \kappa)\, e^{-x/\kappa} + C(x, \kappa)\, e^{-(1-x)/\kappa} \tag{3.3}$$

and

$$\lambda^2 \sim \lambda^2(\kappa) \equiv \sum_{j=0}^{\infty} \lambda_j^2 \kappa^j$$

where A, B, and C have power series expansions in κ with jth coefficients $a_j(x)$, $b_j(x)$, and $c_j(x)$, respectively.

Substituting (3.3) into (3.2) and equating coefficients of the functions 1, $e^{-x/\epsilon}$ and $e^{-(1-x)/\epsilon}$, we ask that

$$A'' + \lambda^2 A = \kappa^2 A^{\mathrm{IV}},$$

$$2B' = \kappa(5B'' - \lambda^2 B) - 4\kappa^2 B''' + \kappa^3 B^{\mathrm{IV}},$$

and

$$2C' = -\kappa(5C'' - \lambda^2 C) - 4\kappa^2 C''' - \kappa^3 C^{\mathrm{IV}}.$$

This, in turn, implies the following differential equations for the coefficients $a_j(x)$, $b_j(x)$, and $c_j(x)$:

$$a_j'' + \lambda_0^2 a_j = a_{j-2}^{\mathrm{IV}} - \sum_{l=1}^{j} \lambda_l^2 a_{j-l} \equiv -\lambda_j^2 a_0 + \alpha_{j-1},$$

$$2b_j' = 5b_{j-1}'' - \sum_{l=0}^{j-1} \lambda_l^2 b_{j-1-l} - 4b_{j-2}''' + b_{j-3}^{\mathrm{IV}} \equiv 2\beta_{j-1},$$

and

$$2c_j' = -5c_{j-1}'' + \sum_{l=0}^{j-1} \lambda_l^2 c_{j-1-l} - 4c_{j-2}''' - c_{j-3}^{\mathrm{IV}} \equiv 2\gamma_{j-1},$$

for all integers $j \geqslant 0$ where a_j, b_j, and c_j with negative coefficients are defined to be identically zero.

Proceeding further, formally

$$0 = y(0, \epsilon) \sim A(0, \kappa) + B(0, \kappa),$$

$$0 = \kappa y'(0, \epsilon) \sim \kappa[A'(0, \kappa) + B'(0, \kappa)] - B(0, \kappa),$$

$$0 = y(1, \epsilon) \sim A(1, \kappa) + C(1, \kappa),$$

and

$$0 = \kappa y'(1, \epsilon) \sim \kappa[A'(1, \kappa) + C'(1, \kappa)] + C(1, \kappa),$$

which leads to the successively determined boundary conditions:

$$a_j(0) = -b_j(0) = -a_{j-1}'(0) - b_{j-1}'(0),$$

and

$$a_j(1) = -c_j(1) = a'_{j-1}(1) + c'_{j-1}(1).$$

From these formal procedures, we can inductively determine the expansions $A(x, \kappa)$, $B(x, \kappa)$, $C(x, \kappa)$ and $\lambda^2(\kappa)$. For $j = 0$,

$$b'_0(x) = 0, \qquad b_0(0) = 0,$$

$$c'_0(x) = 0, \qquad c_0(1) = 0$$

imply that $b_0(x) = c_0(x) \equiv 0$, while

$$\begin{aligned} a''_0 + \lambda_0{}^2 a_0 &= 0, \\ a_0(0) = a_0(1) &= 0 \end{aligned} \tag{3.4a}$$

implies that there is a nontrivial solution for $a_0(x)$ only if $\lambda_0 = n\pi$, $n = 1, 2, 3,...$, and then $a_0(x) = A_0 \sin n\pi x$, A_0 a constant. For a unique, nontrivial determination of $a_0(x)$ we normalize by asking that

$$\int_0^1 a_0{}^2(x)\, dx = 1$$

which fixes $A_0 = \sqrt{2}$.

For $j = 1$:

$$2b'_1 = 5b''_0 - \lambda_0{}^2 b_0\,, \qquad b_1(0) = a'_0(0) + b'_0(0),$$

$$2c'_1 = -5c''_0 + \lambda_0{}^2 c_0\,, \qquad c_1(1) = -a'_0(1) - c'_0(1)$$

imply that $b_1(x) = \sqrt{2}\, n\pi$ and $c_1(x) = (-1)^n \sqrt{2}\, n\pi$, while

$$\begin{cases} a''_1 + \lambda_0{}^2 a_1 = -\lambda_1{}^2 a_0\,, \\ a_1(0) = -b(0), \qquad a_1(1) = -c_1(1), \end{cases} \tag{3.4b}$$

and (3.4a) imply that

$$[a'_0(x)\, a_1(x) - a'_1(x)\, a_0(x)]' = \lambda_1{}^2 a_0{}^2(x).$$

Integrating from 0 to 1, then,

$$\lambda_1{}^2 = 4n^2\pi^2.$$

Integrating (3.4b) yields

$$a_1(x) = A_1 \sin n\pi x + (B_1 + 2\sqrt{2}\, n\pi x) \cos n\pi x$$

and applying either boundary condition fixes $B_1 = -\sqrt{2}\, n\pi$. To obtain $a_1(x)$ uniquely, we ask that it be orthogonal to $a_0(x)$, i.e.,

$$\int_0^1 a_1(x)\, a_0(x)\, dx = 0,$$

which yields $A_1 = \sqrt{2}$. Hence, at the second step, we have

$$\lambda^2(\kappa) = n^2\pi^2 + 4\kappa n^2\pi^2 + \cdots,$$
$$A(x, \kappa) = \sqrt{2} \sin n\pi x + \sqrt{2}\, \kappa[\sin n\pi x - n\pi \cos n\pi x + 2n\pi x \cos n\pi x] + \cdots,$$
$$B(x, \kappa) = 0 + \sqrt{2}\, \kappa n\pi + \cdots,$$

and

$$C(x, \kappa) = 0 + (-1)^n \sqrt{2}\, \kappa n\pi + \cdots.$$

At jth step: If $a_l(x)$, $b_l(x)$, $c_l(x)$ and λ_l^2 are known for $l < j$, the differential equations and boundary conditions imply that

$$b_j(x) = b_j(0) + \int_0^x \beta_{j-1}(x)\, dx,$$

$$c_j(x) = c_j(1) + \int_1^x \gamma_{j-1}(x)\, dx,$$

and

$$a_j'' + \lambda_0^2 a_j = -\lambda_j^2 a_0 + \alpha_{j-1}(x). \tag{3.4c}$$

From (3.4a) and (3.4c)

$$[a_0'(x)\, a_j(x) - a_j'(x)\, a_0(x)]' = \lambda_j^2 a_0^2(x) - a_0(x)\, \alpha_{j-1}(x)$$

so that

$$\lambda_j^2 = a_0'(1)\, a_j(1) - a_0'(0)\, a_j(0) + \int_0^1 a_0(x)\, \alpha_{j-1}(x)\, dx.$$

Moreover, integrating (3.4c) and applying the boundary conditions, $a_j(x)$ is determined up to an arbitrary term $A_j \sin nx$. To fix A_j uniquely, we merely require

$$\int_0^1 a_j(x)\, a_0(x)\, dx = 0.$$

Thus, all coefficients in the series $A(x, \kappa)$, $B(x, \kappa)$, $C(x, \kappa)$ and $\lambda^2(\kappa)$ may be successively obtained. That the resulting expansions for $y(x, \epsilon)$

and λ^2 are, indeed, asymptotically correct is proved in Miranker [74]. Although the expansion $\lambda^2(\kappa)$ is analytic at $\kappa = 0$, the eigenvalues are actually of the form

$$\lambda^2 = n^2\pi^2 + S_n(\kappa, e^{-1/\kappa}), \qquad n = 1, 2, \ldots$$

where $S_n(\xi, \eta)$ is a power series in ξ and η with $S_n(0, 0) = 0$. Since $e^{-1/\kappa} \sim 0$, λ^2 is asymptotically represented simply by a power series in κ.

Higher-Order Problems

Generalizing, we consider eigenvalue problems for the equation

$$\epsilon My + Ny = \lambda y \tag{3.5}$$

where M and N are differential operators on the x-interval $[0, 1]$ of even orders $2m$ and $2n$, respectively, with $m > n$. Such problems have been dealt with in Moser [77]. Specifically, let

$$My = \sum_{j=1}^{2m} \alpha_j(x)\, y^{(j)}(x)$$

and

$$Ny = \sum_{j=1}^{2n} \beta_j(x)\, y^{(j)}(x),$$

where

$$(-1)^m\, \alpha_{2m}(x) > 0 \qquad \text{and} \qquad (-1)^n\, \beta_{2n}(x) > 0 \tag{3.6}$$

throughout $[0, 1]$. Further, let m boundary conditions be prescribed at each end point of the form

$$\begin{cases} S_i y = y^{(\sigma_i)}(0) + \sum\limits_{j<\sigma_i} s_{ij} y^{(j)}(0) = 0, \\ T_i y = y^{(\tau_i)}(1) + \sum\limits_{j<\tau_i} t_{ij} y^{(j)}(1) = 0, \qquad i = 1, 2, \ldots, m, \end{cases} \tag{3.7}$$

where

$$2m > \sigma_1 > \sigma_2 > \cdots > \sigma_m \geqslant 0$$

and

$$2m > \tau_1 > \tau_2 > \cdots > \tau_m \geqslant 0.$$

We associate with (3.5) the auxiliary equation

$$\epsilon \sum_{j=1}^{2m} \alpha_j(x) D^j + \sum_{j=1}^{2n} \beta_j(x) D^j = \lambda, \tag{3.8}$$

which has $2(m-n)$ singular roots (as $\epsilon \to 0$) of the form

$$\frac{d(x, \kappa)}{\kappa} = \frac{1}{\kappa} [g(x) + O(\kappa^{2m})],$$

where

$$\kappa = \epsilon^{1/2(m-n)} > 0$$

and

$$g(x) = \left(- \frac{\beta_{2n}(x)}{\alpha_{2m}(x)}\right)^{1/2(m-n)}.$$

Let $g_1, g_2, \ldots, g_{m-n}$ enumerate those roots g with negative real parts, and let $g_{m-n+1}, g_{m-n+2}, \ldots, g_{2(m-n)}$ enumerate the roots with positive real parts. [Hypothesis (3.6) implies that no determination of g has real part zero.] Since, then, $m-n$ singular roots of the characteristic equation approach $-\infty$ as $\epsilon \to 0$ while $m-n$ other roots approach $+\infty$, we are naturally led to define the "reduced eigenvalue problem," i.e., the eigenvalue problem for $\epsilon = 0$, by

$$\begin{cases} Ny = \lambda y, \\ S_i y = 0, \qquad i = m-n+1, \ldots, m, \\ T_i y = 0 \end{cases} \tag{3.9}$$

where $m-n$ boundary conditions (involving the highest-order derivatives) have been omitted at each end point. Clearly, for (3.9) to be reasonable, we must require that

$$\sigma_{m-n+1} < 2n \qquad \text{and} \qquad \tau_{m-n+1} < 2n. \tag{3.10}$$

To determine the eigenvalues of the reduced problem (3.9), we introduce a fundamental system $z_1, z_2, \ldots, z_{2n}$ of the reduced equation $Nz = \lambda z$, noting that each z_i is an entire function of λ. The eigenvalues λ_0 of the reduced problem (3.9) are, then, the zeros of the entire function

$$F(\lambda) = \det \begin{bmatrix} S_i z_k \\ T_i z_k \end{bmatrix} \qquad (i = m-n+1, \ldots, m, \quad k = 1, 2, \ldots, 2n)$$

with corresponding (normalized) eigenfunctions $u_0(x)$.

A complete set of fundamental asymptotic solutions of the differential equation (3.5) may be formally constructed of the form

$$A_j(x, \kappa), \qquad j = 1, 2, \ldots, 2n,$$

$$B_j(x, \kappa) \exp\left[\frac{1}{\kappa}\int_0^x g_j(s)\,ds\right], \qquad j = 1, 2, \ldots, m-n,$$

and

$$C_j(x, \kappa) \exp\left[\frac{1}{\kappa}\int_1^x g_{m-n+j}(s)\,ds\right], \qquad j = 1, 2, \ldots, m-n$$

where A_j, B_j, and C_j represent power series in κ with variable coefficients $a_{jl}(x)$, $b_{jl}(x)$, and $c_{jl}(x)$, respectively. Without loss of generality, the series $A_j(x, \kappa)$ may be considered as real with $a_{j0}(x) = z_j$; and for λ real and $\bar{g}_p(x) = g_q(x)$, $\overline{B_p} = B_q$ or $\overline{C_{p-m+n}} = C_{q-m+n}$. Continuing, we obtain the following.

Theorem 3.1. *If λ_0 is a simple eigenvalue of the reduced eigenvalue problem* (3.9) (*i.e., if $F(\lambda_0) = 0, F'(\lambda_0) \neq 0$*) *and if*

$$\sigma_1, \sigma_2, \ldots, \sigma_{m-n}$$

are distinct modulo $2(m-n)$ and the same is true for

$$\tau_1, \tau_2, \ldots, \tau_{m-n},$$

then, for sufficiently small $\epsilon > 0$, there exists a uniquely determined eigenvalue $\lambda = \lambda(\epsilon)$ in a neighborhood of $\lambda = \lambda_0$ for the original eigenvalue problem (3.5), (3.7). *Further, this eigenvalue has an asymptotic expansion of the form*

$$\lambda \sim \sum_{j=0}^{\infty} \Lambda_j \kappa^j$$

where

$$\kappa = \epsilon^{1/2(m-n)} \quad \text{and} \quad \Lambda_0 = \lambda_0.$$

The corresponding eigenfunction y belonging to $\lambda = \lambda(\epsilon)$ is asymptotically of the form

$$y \sim A(x, \kappa) + \kappa^{\sigma_{m-n}} \sum_{j=1}^{m-n} B_j(x, \kappa) \exp\left[\frac{1}{\kappa}\int_0^x g_j(s)\,ds\right]$$

$$+ \kappa^{\tau_{m-n}} \sum_{j=1}^{m-n} C_j(x, \kappa) \exp\left[\frac{1}{\kappa}\int_1^x g_{m-n+j}(s)\,ds\right]$$

throughout the closed interval [0, 1]. *Here,* $A(x, \kappa)$ *is determined uniquely up to a factor depending on* κ. *Making an appropriate choice of this factor,*

$$A(x, \kappa) \equiv \sum_{j=0}^{\infty} a_j(x) \kappa^j$$

with $a_0(x) = u_0(x)$, *the normalized eigenfunction corresponding to* λ_0.

Moser [77] considers the case where the operators M and N are formally self-adjoint and the boundary conditions (3.7) are self-adjoint with respect to both M and N. He first shows that these requirements are sufficient to guarantee that (3.10) holds. General results for self-adjoint operators (see Rellich [94]), then, imply the existence of infinitely many discrete eigenvalues $\lambda_n(\epsilon)$, $n = 1, 2, \ldots$, which are analytic functions of ϵ for $\epsilon > 0$. Then, by investigating the behavior of $\lambda_n(\epsilon)$ in the neighborhood of $\epsilon = 0$, Moser was able to show the following.

Theorem 3.2. *Under the general assumptions of Theorem* 3.1 *and in the self-adjoint case, where the reduced eigenvalue problem* (3.9) *has only simple eigenvalues,*

$$\lim_{\epsilon \to 0} \lambda_n(\epsilon) = \lambda_n(0)$$

exists for each $n = 1, 2, \ldots$. *Moreover, the values* $\{\lambda_n(0)\}$ *form the full set of eigenvalues of* (3.9). *Thus, the eigenvalues* $\lambda_n(\epsilon)$ *have asymptotic expansions for small values of* ϵ *given by Theorem* 3.1.

Handelman, Keller, and O'Malley [42] study singular perturbations of eigenvalue problems for linear differential equations where the boundary conditions involve the small parameter ϵ and the eigenvalue parameter λ in such a way that the highest order of differentiation involved may be smaller at $\epsilon = 0$ than for $\epsilon > 0$. This reduction in order of the boundary conditions has a pronounced effect on the asymptotic behavior of the eigenvalues and eigenfunctions. In particular, it is shown that eigenvalues of the full problem can, under appropriate conditions, have limits as ϵ approaches zero which are not eigenvalues of the reduced problem. Among other papers concerned with eigenvalue problems, the reader should note Harris [44], which analyzes systems of first-order equations, and Boyce and Handelman [8] and Handelman and Keller [41], which discuss applications.

4. Two-Parameter Singular Perturbation Problems for Ordinary Differential Equations

Thus far we have considered the limiting behavior (as $\epsilon \to 0$) of solutions y_ϵ of boundary value problems for equations of the form

$$\epsilon My + Ny = 0,$$

where M and N are ordinary differential operators with m, the order of M, greater than n, that of N. Physical problems are frequently of a more complicated nature, however, and often involve several small interrelated parameters, e.g., ϵ, ϵ^3, $\epsilon \log^2 \epsilon$, $\epsilon^{1/2}$, or even $f(\epsilon)$. Thus, we are naturally led to consider two-parameter singular perturbation problems for equations of the form

$$\epsilon My + \mu Ny + Ly = 0,$$

where ϵ and μ are small, positive, interrelated parameters simultaneously approaching zero and $m > n > l$ where m, n, and l are the orders of the differential operators M, N, and L, respectively. Such problems, as well as analogous problems for partial differential equations, are considered in O'Malley [81] and the subsequent papers O'Malley [82], which considers second order problems, and O'Malley [83], which considers higher-order problems.

Second-Order Problems

A. *An Initial-Value Problem*

Consider, first, the constant coefficient initial-value problem.

$$\epsilon y'' + \mu a y' + by = 0,$$

$$y(0), \quad y'(0) \qquad \text{prescribed}$$

on the x-interval $[0, 1]$ for a and b positive constants.

Case 1: $\epsilon/\mu^2 \to 0$ *as* $\mu \to 0$.

The general solution is

$$y(x) = \frac{1}{2\sqrt{1 - \dfrac{4\epsilon b}{\mu^2 a^2}}} \left\{ \left[y(0) \left(1 + \sqrt{1 - \frac{4\epsilon b}{\mu^2 a^2}} + \frac{2\epsilon y'(0)}{\mu a} \right) \right] \right.$$

$$\cdot \exp\left[-\frac{\mu a}{2\epsilon} \left(1 - \sqrt{1 - \frac{4\epsilon b}{\mu^2 a^2}} \right) x \right]$$

$$+ \left[y(0) \left(-1 + \sqrt{1 - \frac{4\epsilon b}{\mu^2 a^2}} \right) - \frac{2\epsilon y'(0)}{\mu a} \right]$$

$$\left. \cdot \exp\left[-\frac{\mu a}{2\epsilon} \left(1 + \sqrt{1 - \frac{4\epsilon b}{\mu^2 a^2}} \right) x \right] \right\},$$

which converges to $u(x) \equiv 0$ on every interval $[\delta, 1]$, $\delta > 0$, but convergence is nonuniform at $x = 0$ for $y(0) \neq 0$. Further, the limiting behavior of $y(x)$ is determined primarily by the factors in the exponents, which are, simply, the roots of the auxiliary polynomial

$$\epsilon D^2 + \mu a D + b = 0.$$

Note, that these roots both approach $-\infty$ as $\mu \to 0$, that they can be expanded in powers of the small parameter ϵ/μ^2, and that the exponential decay of $y(x)$ is determined by the singular portion of these roots, e.g., if $\epsilon = \mu^4$, by $-b/\mu a$ and $-1/\mu^3(a - \mu^2 b/a)$.

Case 2: $\mu^2/\epsilon \to 0$ *as* $\epsilon \to 0$.

The solution for $\epsilon > 0$ is

$$y(x) = e^{-\frac{\mu a x}{2\epsilon}} \left\{ \left[\frac{y'(0)\sqrt{\epsilon} + \dfrac{\mu a}{2\sqrt{\epsilon}} y(0)}{\sqrt{b - \dfrac{\mu^2 a^2}{4\epsilon}}} \right] \sin\left(\frac{x}{\sqrt{\epsilon}} \sqrt{b - \frac{\mu^2 a^2}{4\epsilon}} \right) \right.$$

$$\left. + y(0) \cos\left(\frac{x}{\sqrt{\epsilon}} \sqrt{b - \frac{\mu^2 a^2}{4\epsilon}} \right) \right\}$$

and is therefore oscillatory as $\epsilon \to 0$ unless $\epsilon/\mu = o(1)$. Then, however, y converges nonuniformly to zero in $(0, 1]$ as $\epsilon \to 0$, as might be expected since the auxiliary polynomial has conjugate complex solutions

$$-\frac{\mu a}{2\epsilon} \pm \frac{i}{\sqrt{\epsilon}} \sqrt{b - \frac{\mu^2 a^2}{4\epsilon}}\,.$$

whose real part, in this case, approaches $-\infty$ as $\epsilon \to 0$.

With this background, we can predict the limiting behavior for the solution of the variable coefficient initial-value problem

$$\begin{cases} \epsilon y'' + \mu a(x)\, y' + b(x)\, y = 0, \\ y(0), \quad y'(0) \qquad \text{prescribed} \\ \text{for} \quad a(x) > 0, \quad b(x) > 0 \qquad \text{and} \qquad x \in [0, 1] \end{cases} \tag{4.1}$$

by considering the roots of the auxiliary equation

$$\epsilon D^2 + \mu a(x)\, D + b(x) = 0. \tag{4.2}$$

In particular, regular degeneration [i.e., nonuniform convergence on $(0, 1)$ to $u(x) \equiv 0$, the solution of the reduced boundary value problem, as $\epsilon, \mu \to 0$] occurs in both Case 1: $\epsilon/\mu^2 \to 0$ as $\mu \to 0$ and Case 2: μ^2/ϵ and ϵ/μ both $\to 0$ as $\epsilon \to 0$.

For Case 1: Let

$$-\frac{d_1\left(x, \dfrac{\epsilon}{\mu^2}\right)}{\mu} = -\frac{\mu a(x)}{2\epsilon}\left(1 - \sqrt{1 - \frac{4\epsilon b(x)}{\mu^2 a^2(x)}}\right)$$

$$= -\frac{1}{\mu}\left(\frac{b(x)}{a(x)} + \frac{\epsilon}{\mu^2}\frac{b^2(x)}{a^3(x)} + \frac{2\epsilon^2}{\mu^4}\frac{b^3(x)}{a^5(x)} + \cdots\right)$$

and

$$-\frac{\mu d_2\left(x, \dfrac{\epsilon}{\mu^2}\right)}{\epsilon} = -\frac{\mu a(x)}{2\epsilon}\left(1 + \sqrt{1 - \frac{4\epsilon b(x)}{\mu^2 a^2(x)}}\right)$$

$$= -\frac{\mu}{\epsilon}\left[a(x) - \frac{\epsilon}{\mu^2}\frac{b(x)}{a(x)} - \frac{\epsilon^2}{\mu^4}\frac{b^2(x)}{a^3(x)} - \cdots\right]$$

be the solutions of the auxiliary equation (4.2), where we note that d_1

and d_2 are both positive for ϵ/μ^2 sufficiently small. Generalizing from the constant coefficient problem, we set

$$y(x) = A\left(x, \frac{\epsilon}{\mu}, \frac{\epsilon}{\mu^2}\right) \exp\left[-\frac{1}{\mu}\int_0^x d_1\left(s, \frac{\epsilon}{\mu^2}\right) ds\right]$$
$$+ B\left(x, \frac{\epsilon}{\mu}, \frac{\epsilon}{\mu^2}\right) \exp\left[-\frac{\mu}{\epsilon}\int_0^x d_2\left(s, \frac{\epsilon}{\mu^2}\right) ds\right].$$

Substituting into (4.1) and formally equating to zero the coefficients of the linearly independent exponentials, we have:

$$\begin{cases} \dfrac{\epsilon}{\mu} A'' - \dfrac{2\epsilon}{\mu^2} d_1 A' - \dfrac{\epsilon}{\mu^2} d_1' A + aA' = 0, \\ \dfrac{\epsilon}{\mu} B'' - 2d_2 B' - d_2' B + aB' = 0. \end{cases} \tag{4.3}$$

Thus, we take A and B to have double power-series expansions in ϵ/μ and ϵ/μ^2 with variable coefficients $a_{rs}(x)$ and $b_{rs}(x)$, respectively. Moreover, these coefficients can be successively obtained uniquely by formally equating coefficients of each pair of powers $\{(\epsilon/\mu)^r, (\epsilon/\mu^2)^s\}$ in (4.3) and formally applying the initial conditions. Thus, we obtain

$$A\left(x, \frac{\epsilon}{\mu}, \frac{\epsilon}{\mu^2}\right) = y(0) + \frac{\epsilon}{\mu^2}\left(y(0)\int_0^x \frac{1}{a(s)}\left(\frac{b(s)}{a(s)}\right)' ds + \frac{y(0)\,b(0)}{a^2(0)}\right) + \frac{\epsilon}{\mu}\frac{y'(0)}{a(0)} + \cdots$$

and

$$B\left(x, \frac{\epsilon}{\mu}, \frac{\epsilon}{\mu^2}\right) = \frac{1}{a(x)}\left[0 - \frac{\epsilon}{\mu^2}\frac{y(0)\,b(0)}{a(0)} - \frac{\epsilon}{\mu} y'(0) + \cdots\right].$$

Note. (1) The functions

$$\frac{1}{\mu}\int_0^x d_1\left(s, \frac{\epsilon}{\mu^2}\right) ds \qquad \text{and} \qquad \frac{\mu}{\epsilon}\int_0^x d_2\left(s, \frac{\epsilon}{\mu^2}\right) ds$$

may be interpreted as boundary layer stretching transformations used to investigate the region of nonuniform convergence near $x = 0$. Roughly speaking, since $A(x, 0, 0) = y(0)$ and $B(x, 0, 0) = 0$, the lowest-order boundary condition dropped by the limiting solution at $x = 0$ [here, $y(0)$], is associated with the less singular stretching transformation

$$\frac{1}{\mu}\int_0^x d_1\left(s, \frac{\epsilon}{\mu^2}\right) ds.$$

(2) The complicated double series may collapse considerably when ϵ is a known function of μ, and this knowledge may be used to advantage in more efficiently determining the formal expansion.

To justify the formal procedure, we should prove the following theorem.

Theorem 4.1. *Let $y(x)$ solve the initial-value problem*

$$\epsilon y'' + \mu a(x)\, y' + b(x)\, y = 0,$$
$$y(0), \quad y'(0) \qquad \textit{prescribed}$$

for $\epsilon/\mu^2 \to 0$ as $\mu \to 0$ and for $x \in [0, 1]$ where $a(x)$ and $b(x)$ are positive.

Further, let the formal procedure outlined above define $a_{rs}(x)$ and $b_{rs}(x)$ uniquely.

Then, for each $N \geqslant 1$,

$$y(x) = \left(\sum_{\substack{r,s\geqslant 0\\ r+s\leqslant N}} a_{rs}(x)\left(\frac{\epsilon}{\mu}\right)^r \left(\frac{\epsilon}{\mu^2}\right)^s\right) \exp\left[-\frac{1}{\mu}\int_0^x d_1\left(s, \frac{\epsilon}{\mu^2}\right) ds\right]$$
$$+ \left(\sum_{\substack{r,s\geqslant 0\\ r+s\leqslant N}} b_{rs}(x)\left(\frac{\epsilon}{\mu}\right)^r \left(\frac{\epsilon}{\mu^2}\right)^s\right) \exp\left[-\frac{\mu}{\epsilon}\int_0^x d_2\left(s, \frac{\epsilon}{\mu^2}\right) ds\right]$$
$$+ O\left(\left(\frac{\epsilon}{\mu^2}\right)^{N+1}\right) \qquad \textit{throughout} \quad [0, 1].$$

Proof. The proof follows immediately from the definitions of d_1, d_2, a_{rs}, and b_{rs}, and the following estimate.

Lemma. *Let $R(x)$ satisfy the initial-value problem*

$$\epsilon R'' + \mu a(x)\, R' + b(x)\, R = \frac{\epsilon}{\mu} g(x, \mu),$$
$$R(0, \mu) = 0, \qquad R'(0, \mu) = \frac{f(\mu)}{\mu},$$

for $\epsilon/\mu^2 \to 0$ as $\mu \to 0$, f and g bounded, and $x \in [0, 1]$ where $a(x)$ and $b(x)$ are positive.

Then $R(x) = O(\epsilon/\mu^2)$ as $\mu \to 0$ uniformly for $x \in [0, 1]$.

Proof. The argument (see O'Malley [82]) is much like that for

Lemma 1 of Chapter 2. One shows that R satisfies an integral equation of the form

$$R(x, \mu) = \int_0^x R(s, \mu)\, K(x, s)\, ds + R_0(x, \mu)$$

where

$$R_0(x, \mu) = O\left(\frac{\epsilon}{\mu^2}\right) = \int_0^x K(x, s)\, ds.$$

Case 2: μ^2/ϵ *and* ϵ/μ *both* $\to 0$ *as* $\epsilon \to 0$.

Considering the roots of the auxiliary equation (4.2), we assume the form of solution

$$y(x) = \left[A\left(x, \tau, \frac{\mu}{\tau}\right) \sin\left(\frac{1}{\tau}\int_0^x p\left(s, \left(\frac{\mu}{\tau}\right)^2\right) ds\right) + B\left(x, \tau, \frac{\mu}{\tau}\right) \cos\left(\frac{1}{\tau}\int_0^x p\left(s, \left(\frac{\mu}{\tau}\right)^2\right) ds\right)\right] \exp\left[-\frac{\mu}{2\tau^2}\int_0^x a(s)\, ds\right],$$

where

$$\tau = \sqrt{\epsilon}$$

and

$$p\left(x, \left(\frac{\mu}{\tau}\right)^2\right) = \sqrt{b(x) - \left(\frac{\mu}{\tau}\right)^2 \frac{a^2(x)}{4}} = \sqrt{b(x)}\left[1 - \frac{1}{8}\left(\frac{\mu}{\tau}\right)^2 \frac{a^2(x)}{b(x)} - \cdots\right]$$

and A and B have asymptotic double power series expansions in τ and μ/τ whose variable coefficients a_{rs} and b_{rs} are successively uniquely determined by the differential equation and the initial conditions.

To show asymptotic convergence, we prove the following

Lemma 1. *Let* $R(x)$ *satisfy the initial-value problem*:

$$\tau^2 R'' + \mu a(x)\, R' + b(x)\, R = \tau g(x),$$

$$R(0) = 0,$$

$$R'(0) = \frac{f\left(\tau, \frac{\mu}{\tau}\right)}{\tau},$$

where $a(x)$ and $b(x)$ are positive, g and f are bounded, and both τ^2/μ and $\mu/\tau \to 0$ as $\tau \to 0$.

Then R is bounded for all $x \in [0, 1]$.

Proof. (See O'Malley [82].)

Thus, we obtain

Theorem 4.2. *Let $y(x)$ solve the initial-value problem*

$$\epsilon y'' + \mu a(x)\, y' + b(x)\, y = 0,$$

$$y(0), \quad y'(0) \qquad \textit{prescribed},$$

for $x \in [0, 1]$, $a(x)$ and $b(x)$ positive, when ϵ/μ and μ^2/ϵ both $\to 0$ as $\mu \to 0$.

Further, let the formal procedure outlined above define $a_{rs}(x)$ and $b_{rs}(x)$. Then, for each $N \geqslant 1$,

$$y(x) = \left\{\left(\sum_{\substack{r,s\geqslant 0\\ r+s\leqslant N}} a_{rs}(x)\, \tau^r \left(\frac{\mu}{\tau}\right)^s\right) \sin\left(\frac{1}{\tau}\int_0^x p\left(s, \left(\frac{\mu}{\tau}\right)^2\right) ds\right)\right.$$

$$\left. + \left(\sum_{\substack{r,s\geqslant 0\\ r+s\leqslant N}} b_{rs}(x)\, \tau^r \left(\frac{\mu}{\tau}\right)^s\right) \cos\left(\frac{1}{\tau}\int_0^x p\left(s, \left(\frac{\mu}{\tau}\right)^2\right) ds\right)\right\}$$

$$\cdot \exp\left[-\frac{\mu}{2\tau^2}\int_0^x a(s)\, ds\right] + O(\tau^{N+1}) \qquad \textit{uniformly for} \quad x \in [0, 1],$$

where

$$\tau = \sqrt{\epsilon}$$

and

$$p\left(x, \left(\frac{\mu}{\tau}\right)^2\right) = \sqrt{b(x) - \left(\frac{\mu}{\tau}\right)^2 \frac{a^2(x)}{4}}.$$

Note. If $a(x) < 0 < b(x)$, the transformation $z = 1 - x$ will allow the reader to solve the terminal-value problem

$$\epsilon y'' + \mu a(x)\, y' + b(x)\, y = 0,$$

$$y(1), \quad y'(1) \qquad \text{prescribed}$$

for $x \in [0, 1]$ for both Case 1 and Case 2.

B. *A Two-Point Boundary Value Problem*

Consider the constant coefficient boundary value problem

$$\epsilon y'' + \mu a y' - b y = 0,$$
$$y(0), \quad y(1) \qquad \text{prescribed}$$

for $x \in [0, 1]$ and a and b positive. Note that the sign of the coefficient of y has been reversed from that which occurred in the equation considered above.

For Case 1, where $\epsilon/\mu^2 \to 0$ as $\mu \to 0$, expanding the exact solution, we have

$$y(x) \sim y(1) \exp\left[-\frac{\mu a}{2\epsilon}\left(1 - \sqrt{1 + \frac{4\epsilon b}{\mu^2 a^2}}\right)(x - 1)\right]$$
$$+ y(0) \exp\left[-\frac{\mu a}{2\epsilon}\left(1 + \sqrt{1 + \frac{4\epsilon b}{\mu^2 a^2}}\right)x\right],$$

which converges to $u(x) \equiv 0$ as $\mu \to 0$ on any interval $[\delta, 1 - \delta]$, $\delta > 0$, but converges nonuniformly near $x = 0$ and $x = 1$ unless $y(0) = 0$ and $y(1) = 0$. Further, note that the boundary layer behavior at $x = 0$ is determined by that root

$$-\frac{\mu a}{2\epsilon}\left(1 + \sqrt{1 + \frac{4\epsilon b}{\mu^2 a^2}}\right) = -\frac{\mu}{\epsilon}\left(a + \frac{\epsilon}{\mu^2}\frac{b}{a} - \cdots\right)$$

of the auxiliary equation which approaches $-\infty$ as $\mu \to 0$ while the boundary layer at $x = 1$ is associated with the root

$$-\frac{\mu a}{2\epsilon}\left(1 - \sqrt{1 + \frac{4\epsilon b}{\mu^2 a^2}}\right) = \frac{1}{\mu}\left(\frac{b}{a} - \frac{\epsilon}{\mu^2}\frac{b^2}{a^3} - \cdots\right)$$

which approaches $+\infty$ as $\mu \to 0$. Likewise, for Case 2, where $\mu^2/\epsilon \to 0$ as $\epsilon \to 0$,

$$y(x) \sim y(1) \exp\left[-\sqrt{\frac{b}{\epsilon}}\left(\sqrt{\frac{\mu^2 a^2}{4\epsilon b}} - \sqrt{1 + \frac{\mu^2 a^2}{4\epsilon b}}\right)(x - 1)\right]$$
$$+ y(0) \exp\left[-\sqrt{\frac{b}{\epsilon}}\left(\sqrt{\frac{\mu^2 a^2}{4\epsilon b}} + \sqrt{1 + \frac{\mu^2 a^2}{4\epsilon b}}\right)x\right]$$

and features boundary layer behavior near $x = 0$ and $x = 1$. In particular, if $\mu = o(\epsilon)$,

$$y(x) \sim y(1) \exp\left[\sqrt{\frac{b}{\epsilon}}\,(x-1)\right] + y(0) \exp\left[-\sqrt{\frac{b}{\epsilon}}\,x\right]$$

—the solution of the "semi-reduced" boundary value problem

$$\epsilon z'' - bz = 0,$$

$$z(0) = y(0), \qquad z(1) = y(1).$$

Examine, now, the variable coefficient problem

$$\begin{cases} \epsilon y'' + \mu a(x)\, y' - b(x)\, y = 0, \\ y(0), \quad y(1) \qquad \text{prescribed}, \end{cases} \tag{4.4}$$

for $x \in [0, 1]$ and $a(x)$ and $b(x)$ positive, with auxiliary equation

$$\epsilon D^2 + \mu a(x)\, D - b(x) = 0. \tag{4.5}$$

For Case 1, $\epsilon/\mu^2 \to 0$ as $\mu \to 0$, introduce

$$\frac{d_1\left(x, \frac{\epsilon}{\mu^2}\right)}{\mu} = -\frac{\mu a(x)}{2\epsilon}\left(1 - \sqrt{1 + \frac{4\epsilon b(x)}{\mu^2 a^2(x)}}\right)$$

and

$$-\frac{\mu}{\epsilon}\, d_2\left(x, \frac{\epsilon}{\mu^2}\right) = -\frac{\mu a(x)}{2\epsilon}\left(1 + \sqrt{1 + \frac{4\epsilon b(x)}{\mu^2 a^2(x)}}\right)$$

as the roots of (4.5), with d_1 and d_2 having positive limiting values. Proceeding as usual, we can prove the following.

Theorem 4.3. *Consider the boundary value problem*

$$\epsilon y'' + \mu a(x)\, y' - b(x)\, y = 0,$$

$$y(0), \quad y(1) \qquad \textit{prescribed}$$

for $x \in [0, 1]$ where $a(x)$ and $b(x)$ are positive and for $\epsilon/\mu^2 \to 0$ as $\mu \to 0$. The boundary value problem has a unique solution $y(x)$. Further, *coefficients $a_{rs}(x)$ and $b_{rs}(x)$ may be successively obtained formally such that, for each $N \geqslant 1$,*

$$y(x) = \left(\sum_{\substack{r,s\geqslant 0\\ r+s\leqslant N}} a_{rs}(x)\left(\frac{\epsilon}{\mu}\right)^r\left(\frac{\epsilon}{\mu^2}\right)^s\right)\exp\left[\frac{1}{\mu}\int_1^x d_1\left(s, \frac{\epsilon}{\mu^2}\right) ds\right]$$

$$+ \left(\sum_{\substack{r,s\geqslant 0\\ r+s\leqslant N}} b_{rs}(x)\left(\frac{\epsilon}{\mu}\right)^r\left(\frac{\epsilon}{\mu^2}\right)^s\right)\exp\left[-\frac{\mu}{\epsilon}\int_0^x d_2\left(s, \frac{\epsilon}{\mu^2}\right) ds\right]$$

$$+ O\left(\mu\left(\frac{\epsilon}{\mu^2}\right)^{N+1}\right) \qquad \textit{uniformly for} \quad x \in [0, 1].$$

Likewise, for Case 2, we have the following.

Theorem 4.4. *Consider the boundary-value problem*

$$\epsilon y'' + \mu a(x)\, y' - b(x)\, y = 0,$$

$$y(0), \quad y(1) \qquad \textit{prescribed},$$

for $x \in [0, 1]$ where $b(x)$ is positive and for $\mu^2/\epsilon \to 0$ as $\epsilon \to 0$.
The boundary value problem has a unique solution. Further, let

$$-\frac{d_1(x, \mu/\tau)}{\tau} \qquad \textit{and} \qquad \frac{d_2(x, \mu/\tau)}{\tau}$$

be the roots of the auxiliary equation

$$\epsilon D^2 + \mu a(x)\, D - b(x) = 0$$

where $\tau = \sqrt{\epsilon}$ and d_1 and d_2 have positive limiting values. Coefficients $a_{rs}(x)$ and $b_{rs}(x)$ may be successively defined, in the usual manner, such that for each $N \geqslant 1$,

$$y(x) = \left(\sum_{\substack{r,s\geqslant 0\\ r+s\leqslant N}} a_{rs}(x)\,\tau^r\left(\frac{\mu}{\tau}\right)^s\right)\exp\left[-\frac{1}{\tau}\int_0^x d_1\left(s, \left(\frac{\mu}{\tau}\right)\right) ds\right]$$

$$+ \left[\sum_{\substack{r,s\geqslant 0\\ r+s\leqslant N}} b_{rs}(x)\,\tau^r\left(\frac{\mu}{\tau}\right)^s\right]\exp\left[\frac{1}{\tau}\int_1^x d_2\left(s, \left(\frac{\mu}{\tau}\right)\right) ds\right]$$

$$+ O(\tau\lambda^{N+1}) \qquad \textit{uniformly for} \quad x \in [0, 1] \quad \textit{where} \quad \lambda = \max(\tau, \mu/\tau).$$

Note.

(1) In proving the previous two theorems, the following maximum-minimum principle is useful:

Lemma. *Suppose z satisfies*

$$\epsilon z'' + \mu a(x)\, z' - b(x)\, z = f(x)$$

for $\epsilon \geqslant 0$ on the x-interval $[0, 1]$ *where $b(x)$ is positive.*
Then

$$|z(x)| \leqslant \operatorname{Max} \left\{ |z(0)|, |z(1)|, \operatorname*{Max}_{x \in [0,1]} \left| \frac{f(x)}{b(x)} \right| \right\}.$$

(2) For Case 2, the coefficient $a(x)$ need not be positive.

(3) The boundary value problem

$$\epsilon y'' - \mu a(x)\, y' - b(x)\, y = 0,$$

$$y(0), \quad y(1) \qquad \text{prescribed},$$

for $x \in [0, 1]$ and $a(x)$ and $b(x)$ positive can likewise be solved in the same manner in the two cases: $\epsilon/\mu^2 \to 0$ as $\mu \to 0$ and $\mu^2/\epsilon \to 0$ as $\epsilon \to 0$.

C. *Other Boundary Value Problems*

The general theory of linear ordinary differential equations implies that for every $\epsilon > 0$, the second-order homogeneous equations considered have two linearly independent solutions and that every solution is a linear combination of these two solutions as well as of every pair of linearly independent solutions, and further, that the solution of the nonhomogeneous equations are obtainable by the method of variation of parameters. This does not imply, however, that any boundary value problem has a unique solution, nor does it imply that the solution of any such boundary value problem has a well-defined limiting behavior as μ and ϵ simultaneously approach zero.

Consider, again, the homogeneous equation

$$\epsilon y'' + \mu a(x)\, y' - b(x)\, y = 0 \tag{4.6}$$

for $x \in [0, 1]$ where $a(x)$ and $b(x)$ are positive and $\epsilon/\mu^2 \to 0$ as $\mu \to 0$. Further, let $y_1(x)$ and $y_2(x)$ be linearly independent solutions of this equation with boundary conditions $y_1(0) = 0, y_1(1) = 1, y_2(0) = 1$, and $y_2(1) = 0$, and note that the asymptotic expansions of y_1 and y_2 follow immediately from Theorem 4.3. Restricting attention to the specific boundary value problem

$$\epsilon y'' + \mu a(x)\, y' - b(x)\, y = 0,$$
$$y'(0), \quad y'(1) \qquad \text{prescribed},$$

we completely determine its asymptotic solution by setting

$$y(x) = C_1(\mu)\, y_1(x) + C_2(\mu)\, y_2(x) \tag{4.7}$$

and obtaining asymptotic expansions for $C_1(\mu)$ and $C_2(\mu)$ by using the known expansions for y_1 and y_2.

Alternately, by *a priori* considerations, one could predict—without calculating—that $C_1(\mu) = O(\mu)$ and is expandable as a double power series in μ and ϵ/μ^2, while $C_2(\mu) = O(\epsilon/\mu)$ and is asymptotically a double series in ϵ/μ and ϵ/μ^2. Thus, knowing that $y(x)$ is given by (4.7) and knowing the form of the expansions for $y_1(x)$ and $y_2(x)$, we may set

$$y(x) \sim \mu A\left(x, \mu, \frac{\epsilon}{\mu^2}\right) \exp\left[\frac{1}{\mu} \int_1^x d_1\left(s, \frac{\epsilon}{\mu^2}\right) ds\right]$$
$$+ \frac{\epsilon}{\mu} B\left(x, \frac{\epsilon}{\mu}, \frac{\epsilon}{\mu^2}\right) \exp\left[-\frac{\mu}{\epsilon} \int_0^x d_2\left(s, \frac{\epsilon}{\mu^2}\right) ds\right].$$

We then proceed to determine A and B in the usual formal manner as double series in the indicated variables for d_1/μ and $-\mu/\epsilon\, d_2$ roots of the auxiliary polynomial (4.5). Clearly, this approach is more efficient than first obtaining the expansions for $y_1(x)$, $y_2(x)$, $C_1(\mu)$, and $C_2(\mu)$ and multiplying the appropriate expansions.

Considering, instead, the initial value problem for Eq. (4.6) where $y(0)$ and $y'(0)$ are prescribed, we find it impossible to construct a solution $y(x)$ which is bounded throughout $[0, 1]$ as $\mu \to 0$. Here, however, the root d_1/μ of the auxiliary polynomial (4.5) approaches $+\infty$ as $\mu \to 0$, so we intuitively expect it to be associated with a boundary layer at $x = 1$, not at $x = 0$.

It is apparent that we need some theorems stating *à priori* which boundary value problems have converging solutions as ϵ and μ simultaneously approach zero and, further, when the limiting behavior is convergent, some *à priori* estimates on the orders of the coefficients $C_i(\mu)$. Restricting attention to the boundary value problems for

$$\epsilon y'' + \mu a(x) y' + b(x) y = 0,$$

on $0 \leqslant x \leqslant 1$ where $a(x) \neq 0 \neq b(x)$ and for $\epsilon/\mu^2 \to 0$ as $\mu \to 0$, we find that the solutions of

(1) the initial-value problem when $a(x)$ and $b(x)$ are positive,
(2) the terminal-value problem when $a(x)$ and $b(x)$ are negative, and
(3) any problem with separated (or unmixed) boundary conditions when $a(x)$ and $b(x)$ have opposite signs

will all converge nonuniformly on $[0, 1]$ (as $\epsilon, \mu \to 0$) to zero (the solution of the reduced equation). Further, complete asymptotic solutions are obtainable. For other boundary conditions, however, the solution of the boundary value problem will, in general, fail to converge (as $\epsilon, \mu \to 0$) as simple, constant coefficient examples easily demonstrate.

D. *Nonhomogeneous Boundary Value Problem*

Consider, now, the nonhomogeneous problem

$$\begin{cases} \epsilon y'' + \mu a(x) y' - b(x) y = R(x), \\ y(0), \quad y(1) \quad \text{prescribed}, \end{cases} \tag{4.8}$$

for $x \in [0, 1]$ where $a(x)$ and $b(x)$ are positive and $\epsilon/\mu^2 \to 0$ as $\mu \to 0$.

Let $y_1(x)$ and $y_2(x)$ be solutions of the homogeneous equation satisfying the boundary conditions $y_1(0) = 0$, $y_1(1) = 1$, $y_2(0) = 1$, and $y_2(1) = 0$ and let

$$J[y_2, y_1] = \epsilon \exp\left[\frac{\mu}{\epsilon}\int_0^x a(s)\, ds\right] [y_1'(x)\, y_2(x) - y_2'(x)\, y_1(x)].$$

This "conjoint" of y_1 and y_2 is a constant and we use it to represent the solution $y(x)$ of (4.8) by the Green's function formula

$$\begin{aligned} y(x) = &- \frac{y_2(x)}{J[y_2, y_1]} \int_0^x y_1(t) \exp\left[\frac{\mu}{\epsilon}\int_0^t a(s)\, ds\right] R(t)\, dt \\ &- \frac{y_1(x)}{J[y_2, y_1]} \int_x^1 y_2(t) \exp\left[\frac{\mu}{\epsilon}\int_0^t a(s)\, ds\right] R(t)\, dt \\ &+ y(1)\, y_1(x) + y(0)\, y_2(x) \end{aligned}$$

(see, e.g., Friedman [32], Chapter 3).

Formally substituting the expansions for $y_1(x)$ and $y_2(x)$ into this expression (see O'Malley [82]), we obtain

$$y(x) \sim z\left(x, \frac{\epsilon}{\mu}, \frac{\epsilon}{\mu^2}\right) + \tilde{A}\left(x, \frac{\epsilon}{\mu}, \frac{\epsilon}{\mu^2}\right) \exp\left[\frac{1}{\mu}\int_1^x d_1\left(s, \frac{\epsilon}{\mu^2}\right) ds\right]$$

$$+ \tilde{B}\left(x, \frac{\epsilon}{\mu}, \frac{\epsilon}{\mu^2}\right) \exp\left[-\frac{\mu}{\epsilon}\int_0^x d_2\left(s, \frac{\epsilon}{\mu^2}\right) ds\right], \tag{4.10}$$

where $z(x, \epsilon/\mu, \epsilon/\mu^2)$ is an asymptotic solution of the nonhomogeneous equation and a constant multiple of

$$B\left(x, \frac{\epsilon}{\mu}, \frac{\epsilon}{\mu^2}\right) \int_0^x A\left(t, \frac{\epsilon}{\mu}, \frac{\epsilon}{\mu^2}\right) \exp\left[-\frac{\mu}{\epsilon}\int_t^x d_2\left(s, \frac{\epsilon}{\mu^2}\right) ds\right] R(t)\, dt$$

$$+ A\left(x, \frac{\epsilon}{\mu}, \frac{\epsilon}{\mu^2}\right) \int_1^x B\left(t, \frac{\epsilon}{\mu}, \frac{\epsilon}{\mu^2}\right) \exp\left[-\frac{1}{\mu}\int_x^t d_1\left(s, \frac{\epsilon}{\mu^2}\right) ds\right] R(t)\, dt$$

where $\tilde{A}(1, \epsilon/\mu, \epsilon/\mu^2)$ and $\tilde{B}(0, \epsilon/\mu, \epsilon/\mu^2)$ are determined so that the sum asymptotically solves the boundary value problem (4.8). The nonhomogeneous problem has thus been asymptotically solved by the Green's function approach. It should be noted, however, that an asymptotic solution of the nonhomogeneous equation could also be obtained as a regular perturbation (in powers of ϵ and μ) of $-R(x)/b(x)$, the solution of the reduced equation, and that the nonhomogeneous problem could then also be solved by adding an appropriate asymptotic solution of the homogeneous equation.

Higher Order Problems

Now consider boundary value problems for the equation

$$\epsilon My + \mu Ny + Ly = 0 \tag{4.11}$$

on the x-interval $[0, 1]$ where M, N, and L are linear ordinary differential operators having orders m, n, and l, respectively, where $m > n > l \geqslant 0$, and ϵ and μ simultaneously approach zero in an interrelated way. Specifically, let

$$My = y^{(m)} + \alpha_1(x)\, y^{(m-1)} + \cdots + \alpha_k(x)\, y^{(m-k)},$$

$$Ny = \beta(x)[y^{(n)} + \beta_1(x)\, y^{(n-1)} + \cdots + \beta_j(x)\, y^{(n-j)}],$$

and

$$Ly = \gamma(x)[y^{(l)} + \gamma_1(x)\, y^{(l-1)} + \cdots + \gamma_{l-1}(x)\, y' + \gamma_l(x)\, y],$$

where $\beta(x) \neq 0 \neq \gamma(x)$. Further, let us consider (4.11) subject to the boundary conditions

$$\begin{cases} y^{(\lambda_i)}(0) = l_i\,, & i = 1, 2,..., r, \\ y^{(\tau_i)}(1) = l_{r+i}\,, & i = 1, 2,..., m - r, \end{cases} \tag{4.12}$$

where

$$m > \lambda_1 > \lambda_2 > \cdots > \lambda_r \geqslant 0,$$

and

$$m > \tau_1 > \tau_2 > \cdots > \tau_{m-r} \geqslant 0.$$

As usual, we associate with the differential equation (4.11) its auxiliary equation

$$\epsilon\left[D^m + \sum_{i=1}^{k} \alpha_i(x)\, D^{m-i}\right] + \mu\beta(x)\left[D^n + \sum_{i=1}^{j} \beta_i(x)\, D^{n-i}\right]$$
$$+ \gamma(x)\left[D^l + \sum_{i=1}^{l} \gamma_i(x)\, D^{l-i}\right] = 0 \tag{4.13}$$

and use its singular (as $\epsilon, \mu \to 0$) roots to obtain $m - l$ asymptotic solutions of (4.11). Further, we call equation (4.11) exceptional if this set of $(m - l)$ singular roots of (4.13) contains members with purely imaginary singular parts. Otherwise, (4.11) is called nonexceptional and $m - l = p + q$, where p is the number of singular roots whose real parts have negative limiting values and q is the number of singular roots whose real parts have positive limiting values.

The limiting behavior of the roots of (4.13) as well as the limiting behavior of the solution of the boundary value problem (4.11), (4.12) are completely different in the two cases

$$(1)\quad \frac{\epsilon}{\mu^{((m-l)/(n-l))}} \to 0 \quad \text{as} \quad \mu \to 0, \qquad \text{and} \qquad (2)\quad \frac{\mu^{((m-l)/(n-l))}}{\epsilon} \to 0$$

as $\epsilon \to 0$. Thus, they are considered separately below. A more complete analysis, including coverage of the problem when (4.11) is exceptional and when $\epsilon = \mu^{(m-l)/(n-l)}$, is given in O'Malley [83].

Case 1: $\epsilon/\mu^{((m-l)/(n-l))} \to 0$ *as* $\mu \to 0$.

Introduce two new parameters $\sigma = \mu^{1/(n-l)}$ and

$$\nu = \left(\frac{\epsilon}{\mu}\right)^{1/(m-n)}, \quad \text{so that} \quad \frac{\nu}{\sigma} = \left(\frac{\epsilon}{\mu^{(m-l)/(n-l)}}\right)^{1/(m-n)} \to 0 \quad \text{as} \quad \mu \to 0.$$

For Case 1, then, $(n - l)$ singular solutions of (4.13) are of the form

$$D\left(x, \sigma, \left(\frac{\nu}{\sigma}\right)^{m-n}\right) = \frac{d\left(x, \sigma, \left(\frac{\nu}{\sigma}\right)^{m-n}\right)}{\sigma}$$

$$= \frac{1}{\sigma}\left[\omega_1{}^r g(x) + \sigma\left(\frac{\gamma_1(x) - \beta_1(x)}{n - l}\right) - \left(\frac{\nu}{\sigma}\right)^{m-n}\left(\frac{(g(x)\,\omega_1{}^r)^{m-n+1}}{(n - l)\,\beta(x)}\right) + \cdots\right], \quad (4.14)$$

where $g(x)$ is positive, r is an integer, and ω_1 is a complex number of modulus one such that

$$(\omega_1{}^r g(x))^{n-l} = -\gamma(x)/\beta(x)$$

and each of the $(n - l)$ determinations $d(x, \sigma, (\nu/\sigma)^{m-n})$ can be expanded as a double power series in σ and $(\nu/\sigma)^{m-n}$ with variable coefficients which may be successively determined by substitution into (4.13). (For details, see O'Malley [81], Appendix A.) Let p_1 be the number of these roots whose real parts approach $-\infty$ as $\mu \to 0$ and let q_1 be the number of these roots whose real parts approach $+\infty$ as $\mu \to 0$. Further, $m - n$ singular roots of (4.13) are of the form

$$D\left(x, \nu, \left(\frac{\nu}{\sigma}\right)^{n-l}\right) = \frac{d\left(x, \nu, \left(\frac{\nu}{\sigma}\right)^{n-l}\right)}{\nu}$$

$$= \frac{1}{\nu}\left[\omega_2{}^s c(x) + \nu\left(\frac{\beta_1(x) - \alpha_1(x)}{m - n}\right) + \left(\frac{\nu}{\sigma}\right)^{n-l}\left(\frac{\gamma(x)(\omega_2{}^s c(x))^{1-(n-l)}}{(m - n)\,\beta(x)}\right) + \cdots\right], \quad (4.15)$$

where $c(x)$ is positive, s is an integer, ω_2 is a complex number of modulus one such that

$$(\omega_2{}^s c(x))^{m-n} = -\beta(x)$$

and each $d(x, \nu, (\nu/\sigma)^{n-l})$ has a double series expansion in σ and $(\nu/\sigma)^{m-n}$ whose terms can be obtained successively. Define p_2 to be the number of these roots whose real parts approach $-\infty$ as $\mu \to 0$ and q_2 to be the number whose real parts approach $+\infty$.

Corresponding to each of these $m - l$ solutions of (4.13), it is frequently convenient to associate its singular part. That is, suppose there exists (least) positive integers K_1 and K_2 such that

$$\frac{1}{\sigma}\left(\frac{\nu}{\sigma}\right)^{K_1(m-n)} = o(1) = \frac{1}{\nu}\left(\frac{\nu}{\sigma}\right)^{K_2(n-l)} \tag{4.16}$$

as $\mu \to 0$. (Hereafter, we refer to these conditions as hypothesis (4.16).) Then, to $d(x, \sigma, (\nu/\sigma)^{m-n})$ and $d(x, \nu, (\nu/\sigma)^{n-l})$, we associate the corresponding finite expansions $\tilde{d}(x, \sigma, (\nu/\sigma)^{m-n})$ and $\tilde{d}(x, \nu, (\nu/\sigma)^{n-l})$ consisting of terms which make singular contributions in the expressions (4.14) and (4.15).

To each of the $n - l$ complex roots (4.14), we let correspond a formal complex solution of (4.11), namely,

$$y = \frac{A\left(x, \sigma, \left(\frac{\nu}{\sigma}\right)^{m-n}\right)}{(g(x))^{(n+l-1)/2}} \exp\left[\frac{1}{\sigma}\int_t^x d\left(s, \sigma, \left(\frac{\nu}{\sigma}\right)^{m-n}\right) ds\right], \tag{4.17}$$

and to the $m - n$ roots (4.15), the formal solutions

$$y = \frac{B\left(x, \nu, \left(\frac{\nu}{\sigma}\right)^{n-l}\right)}{(c(x))^{(m+n-1)/2}} \exp\left[\frac{1}{\nu}\int_t^x d\left(s, \nu, \left(\frac{\nu}{\sigma}\right)^{n-l}\right) ds\right]. \tag{4.18}$$

Moreover, when hypothesis (4.16) holds, we replace

$$d\left(s, \sigma, \left(\frac{\nu}{\sigma}\right)^{m-n}\right) \quad \text{and} \quad d\left(s, \nu, \left(\frac{\nu}{\sigma}\right)^{n-l}\right)$$

by the related finite sums

$$\tilde{d}\left(s, \sigma, \left(\frac{\nu}{\sigma}\right)^{m-n}\right) \quad \text{and} \quad \tilde{d}\left(s, \nu, \left(\frac{\nu}{\sigma}\right)^{n-l}\right).$$

In both cases, t is either 0 or 1, and A and B possess asymptotic double power series in σ and $(\nu/\sigma)^{m-n}$, and ν and $(\nu/\sigma)^{n-l}$, respectively, both with constant leading term and such that the variable higher order terms may be successively determined by Eq. (4.11) to within additive constants (see O'Malley [81], Appendix B).

Becoming more explicit, we define q_1 linearly independent solutions $y_1, y_2, \ldots, y_{q_1}$ of the equation (4.11) of the form (4.17) with $t = 1$

corresponding to the q_1 roots (4.14) whose real part approaches $+\infty$ as $\mu \to 0$, and to the p_1 roots whose real part approaches $-\infty$, we define linearly independent solutions $y_{q_1+1}, y_{q_1+2}, \ldots, y_{q_1+p_1}$ of the form (4.17), all with $t = 0$. Likewise, to the q_2 roots (4.15) whose real part approaches $+\infty$ as $\mu \to 0$, we define linearly independent solutions $y_{n-l+1}, y_{n-l+2}, \ldots, y_{n-l+q_2}$ of the form (4.18), all with $t = 1$, and to the p_2 roots whose real part approaches $-\infty$, we define linearly independent solutions $y_{n-l+q_2+1}, \ldots, y_{n-l+q_2+p_2}$, all with $t = 0$. Clearly, $y_1, y_2, \ldots, y_{q_1}$ and $y_{n-l+1}, y_{n-l+2}, \ldots, y_{n-l+q_2}$ and their derivatives are all exponentially small away from $x = 1$, while $y_{q_1+1}, \ldots, y_{q_1+p_1}$ and $y_{n-l+q_2+1}, \ldots, y_{n-l+q_2+p_2}$ exhibit this boundary layer behavior near $x = 0$.

In order to simplify the exposition, we shall restrict attention here to the nonexceptional case where $p_1 + q_1 = n - l$ and $p_2 + q_2 = m - n$.

A remaining set of l linearly independent asymptotic solutions of (4.11) can be obtained as regular perturbations of any set $z_1, z_2, \ldots, z_l$ of linearly independent solutions of the reduced equation $Lz = 0$, i.e., we set

$$y_{m-l+i} \sim \sum_{r,s=0}^{\infty} y_{irs}(x)\, \mu^r \epsilon^s, \qquad i = 1, 2, \ldots, l,$$

where $y_{i00} = z_i$, and further terms are determined successsively from the differential equation

$$Ly_{irs} = -My_{i,r,s-1} - Ny_{i,r-1,s}$$

obtained by substituting the expansion for y_{m-l+i} into Eq. (4.11) and formally equating coefficients. (This nonhomogeneous equation can be solved by variation of parameters since the right-hand side is known at each step.) The results of Turrittin [102] show that $y_1, \ldots, y_m$ form a fundamental system for (4.11).

Altogether then, we have constructed m linearly independent asymptotic solutions of Eq. (4.11) and therefore expect to write the limiting solution of the boundary value problem, if it exists, as a linear combination of this asymptotic fundamental system. Further, in event of convergence, we expect the limiting solution to converge uniformly in any closed interval outside the boundary layer to the solution of a reduced boundary value problem, consisting of the reduced equation $Lz = 0$ plus l boundary conditions. To determine which boundary conditions are canceled from the set (4.12), we must obtain a cancellation law.

Cancellation Law (Nonexceptional Case). *Cancel $p = p_1 + p_2$ boundary conditions at $x = 0$ and $q = q_1 + q_2$ boundary conditions at $x = 1$, starting from those involving the highest-order derivatives.*

Continuing in the manner of Wasow [109] as detailed in O'Malley [83], we obtain the following,

Theorem 4.5. *Consider the differential equation*

$$\epsilon My + \mu Ny + Ly = 0 \tag{4.11}$$

on the x-interval $[0, 1]$ *subject to the boundary conditions*

$$y^{(\lambda_i)}(0) = l_i, \qquad i = 1, 2, \ldots, r,$$
$$y^{(\tau_i)}(1) = l_{r+i}, \qquad i = 1, 2, \ldots, m - r,$$

where

$$\beta(x) \neq 0 \neq \gamma(x), \qquad m > n > l \geqslant 0,$$

and

$$m > \lambda_1 > \lambda_2 > \cdots > \lambda_r \geqslant 0$$

and

$$m > \tau_1 > \tau_2 > \cdots > \tau_{m-r} \geqslant 0.$$

Further, let

$$\frac{\epsilon}{\mu^{(m-l)/(n-l)}} \to 0 \quad as \quad \mu \to 0,$$

let the differential equation be nonexceptional on the x-interval $[0, 1]$, *and let $z(x)$ satisfy the reduced boundary value problem*

$$Lz = 0,$$
$$z^{(\lambda_i)}(0) = l_i, \qquad i = p + 1, p + 2, \ldots, r,$$
$$z^{(\tau_i)}(1) = l_{r+i}, \qquad i = q + 1, q + 2, \ldots, m - r,$$

where $p \leqslant r$ and $q \leqslant m - r$.

Suppose

1. *the reduced problem has a unique solution, and*
2. (a) $\lambda_1, \lambda_2, \ldots, \lambda_{p_2}$ *are distinct modulo $m - n$,*
   (b) $\lambda_{p_2+1}, \lambda_{p_2+2}, \ldots, \lambda_p$ *are distinct modulo $n - l$,*
   (c) $\tau_1, \tau_2, \ldots, \tau_{q_2}$ *are distinct modulo $m - n$, and*
   (d) $\tau_{q_2+1}, \tau_{q_2+2}, \ldots, \tau_q$ *are distinct modulo $n - l$.*

Then $y(x)$ *has a well-defined limiting behavior on the interval* $0 \leqslant x \leqslant 1$ *as* $\mu \to 0$ *given by*

$$y(x) = \frac{1}{(g(x))^{(n+l-1)/2}} \left[\sigma^{\tau_q} \sum_{k=1}^{q_1} u_k \left(x, \sigma, \frac{\nu}{\sigma}\right) + \sigma^{\lambda_p} \sum_{k=1}^{p_1} u_{k+q_1} \left(x, \sigma, \frac{\nu}{\sigma}\right) \right]$$

$$+ \frac{1}{(c(x))^{(m+n-1)/2}} \left[\sigma^{\tau_q} \left(\frac{\nu}{\sigma}\right)^{\tau_{q_2}} \sum_{k=1}^{q_2} u_{n-l+k} \left(x, \sigma, \frac{\nu}{\sigma}\right) \right.$$

$$\left. + \sigma^{\lambda_p} \left(\frac{\nu}{\sigma}\right)^{\lambda_{p_2}} \sum_{k=1}^{p_2} u_{n-l+q_2+k} \left(x, \sigma, \frac{\nu}{\sigma}\right) \right] + u\left(x, \sigma, \frac{\nu}{\sigma}\right), \tag{4.19}$$

where $u(x, \sigma, \nu/\sigma)$ *is a regular perturbation of* $z(x)$ *in powers of* σ *and* ν/σ, *and*

$$u_k\left(x, \sigma, \frac{\nu}{\sigma}\right) = \hat{A}_k\left(x, \sigma, \frac{\nu}{\sigma}\right) \exp\left[\frac{1}{\sigma} \int_t^x d\left(s, \sigma, \left(\frac{\nu}{\sigma}\right)^{m-n}\right) ds\right],$$

$$k = 1, 2, \ldots, n - l$$

and

$$u_k\left(x, \sigma, \frac{\nu}{\sigma}\right) = \hat{B}_k\left(x, \sigma, \frac{\nu}{\sigma}\right) \exp\left[\frac{1}{\nu} \int_t^x d\left(s, \nu, \left(\frac{\nu}{\sigma}\right)^{n-l}\right) ds\right]$$

$$k = n - l + 1, \ldots, m - l,$$

with the determinations of t *and* d *(or* $\tilde{d}$*) the same as that used in the expression for the corresponding* y_k, *and where* $\hat{A}_k$ *and* $\hat{B}_k$ *have asymptotic double power series in* σ *and* ν/σ *with constant leading term and higher-order terms determined uniquely from the differential equation and the boundary conditions. In particular, then,*

$$y(x) \sim z(x) \quad on \quad 0 < x < 1.$$

Note. (1) That the hypotheses 1 and 2, and the conditions $p \leqslant r$, $q \leqslant m - r$ are essentially necessary is seen by considering a series of constant coefficient boundary value problems whose solutions, in general, diverge as $\epsilon, \mu \to 0$. See O'Malley [81] for a list of examples.

(2) If in the expansion for some $u_k(x, \sigma, \nu/\sigma)$,

$$d\left(s, \sigma, \left(\frac{\nu}{\sigma}\right)^{m-n}\right) \quad \text{or} \quad d\left(s, \nu, \left(\frac{\nu}{\sigma}\right)^{n-l}\right)$$

is represented by any finite partial sum $\hat{d}$ of its double series expansion, the resulting error in the function u_k has the same order as the difference $d - \hat{d}$.

Further results are reported in O'Malley [81]. There, e.g., the more complicated problem when equation (4.1) is exceptional is considered and a detailed determination of an asymptotic solution for the illustrative problem

$$\mu^2 y^{(7)} + \mu\beta(x)\, y^{(5)} + \gamma(x)\, y^{(2)} = 0$$

with

$$\beta(x) < 0 < \gamma(x) \qquad \text{on} \qquad [0, 1],$$

and

$$y^{(4)}(0), y^{(3)}(0), y^{(2)}(0), y'(0), y(0), y^{(2)}(1),$$

and

$$y'(1) \qquad \text{prescribed}$$

is given.

Case 2: $\mu^{((m-l)/(n-l))}/\epsilon \to 0$ *as* $\epsilon \to 0$.

In this case, we introduce $\kappa = \epsilon^{1/(m-l)} > 0$ [so that μ/κ^{n-l} is also a small parameter] and find $m - l$ singular roots of the auxiliary equation (4.13) of the form

$$\begin{aligned} D\left(x, \kappa, \frac{\mu}{\kappa^{n-l}}\right) &= \frac{d\left(x, \kappa, \frac{\mu}{\kappa^{n-l}}\right)}{\kappa} \\ &= \frac{1}{\kappa}\left[\omega^i h(x) + \kappa\left(\frac{\gamma_1(x) - \alpha_1(x)}{m - l}\right)\right. \\ &\qquad \left. - \frac{\mu}{\kappa^{n-l}}\left(\frac{\beta(x)(g(x))^{1-(m-n)}}{m - l}\right) + \cdots\right], \end{aligned} \tag{4.20}$$

where $h(x)$ is positive, i is an integer, ω is a complex number of modulus one such that

$$(\omega^i h(x))^{m-l} = -\gamma(x),$$

and $d(x, \kappa, \mu/\kappa^{n-l})$ can be expanded as a double power series in κ and μ/κ^{n-l} such that each coefficient may be uniquely determined successively from Eq. (4.13). Further, to each of these $m - l$ singular roots, we associate formal solutions of (4.11), namely

$$y = \frac{E(x, \kappa, \mu/\kappa^{n-l})}{(h(x))^{(m+l-1)/2}} \exp\left[\frac{1}{\kappa}\int_t^x d(s, \kappa, \mu/\kappa^{n-l})\, ds\right],$$

where E has a double power series expansion in κ and μ/κ^{n-l} with constant leading term and higher-order terms obtainable successively from the differential equation (4.11) up to additive constants.

Again, we restrict attention to the nonexceptional case where $m - l = p + q$ for p the number of roots (4.20) whose real part approaches $-\infty$ as $\mu \to 0$ and q, the number whose real part approaches $+\infty$. We obtain the following results:

Cancellation Law (Nonexceptional Case). *Cancel p boundary conditions at $x = 0$ and q boundary conditions at $x = 1$, starting from those containing the highest derivatives.*

Theorem 4.6. *Consider the differential equation*

$$\epsilon My + \mu Ny + Ly = 0$$

on the x-interval $[0, 1]$ *subject to the boundary conditions*

$$y^{(\lambda_i)}(0) = l_i, \qquad i = 1, 2,..., r,$$

$$y^{(\tau_i)}(1) = l_{r+i}, \qquad i = 1, 2,..., m - r,$$

where

$$\gamma(x) \neq 0, \qquad m > n > l \geqslant 0,$$

and

$$m > \lambda_1 > \lambda_2 > \cdots > \lambda_r \geqslant 0,$$

and

$$m > \tau_1 > \tau_2 > \cdots > \tau_{m-r} \geqslant 0.$$

Further, let

$$\frac{\mu^{(m-l)/(n-l)}}{\epsilon} \to 0 \quad as \quad \epsilon \to 0,$$

let the differential equation be nonexceptional, and let $z(x)$ solve the reduced boundary value problem

$$Lz = 0,$$

$$z^{(\lambda_i)}(0) = l_i, \qquad i = p + 1,..., r,$$

$$z^{(\tau_i)}(1) = l_{r+i}, \qquad i = q + 1,..., m - r$$

where $p \leqslant r$ and $q \leqslant m - r$.

Suppose

1. *the reduced problem has a unique solution, and*
2. (a) $\lambda_1, \lambda_2, \ldots, \lambda_p$ *are distinct modulo* $m - l$,
   (b) $\tau_1, \tau_2, \ldots, \tau_q$ *are distinct modulo* $m - l$.

Then $y(x)$ *has a well-defined limiting behavior as* $\epsilon \to 0$ *given by*

$$y(x) = \frac{1}{(h(x))^{(m+l-1)/2}} \left[\kappa^{\tau_q} \sum_{k=1}^{q} u_k \left(x, \kappa, \frac{\mu}{\kappa^{n-l}} \right) \right.$$

$$\left. + \kappa^{\lambda_p} \sum_{k=1}^{p} u_{q+k} \left(x, \kappa, \frac{\mu}{\kappa^{n-l}} \right) \right] + u(x, \kappa, \mu/\kappa^{n-l}), \tag{4.21}$$

where $u(x, \kappa, \mu/\kappa^{n-l})$ *is a regular perturbation of* $z(x)$, *and the linearly independent functions*

$$u_k \left(x, \kappa, \frac{\mu}{\kappa^{n-l}} \right) = A_k \left(x, \kappa, \frac{\mu}{\kappa^{n-l}} \right) \exp \left[\frac{1}{\kappa} \int_t^x d_k \left(s, \kappa, \frac{\mu}{\kappa^{n-l}} \right) ds \right],$$

$$k = 1, 2, \ldots, m - l$$

are determined formally, in the usual manner, where

$t = 1$ *and* Re d_k *has positive limiting values for* $k = 1, 2, \ldots, q$

while

$t = 0$ *and* Re d_k *has negative limiting values for* $k = q + 1, \ldots, m - l$.

Again,

$$y(x) \sim z(x) \qquad on \qquad 0 < x < 1.$$

Note. (1) In the special (one-parameter) case where $\mu \equiv 0$, this theorem coincides with the conclusion in Wasow [109].

(2) For Case 2, the condition $\beta(x) \neq 0$ on $[0, 1]$ is not required in the nonexceptional case.

Comment. It seems feasible to also consider singular perturbation problems for systems of first-order equations involving many parameters. Thus, this author has analyzed systems of the form

$$\Omega(\epsilon)(dy/dx) = A(x, \epsilon)\, y \tag{4.19}$$

where $\Omega(\epsilon) = \operatorname{diag}(\epsilon_1 I_{m_1} : \epsilon_2 I_{m_2} : \cdots : \epsilon_s I_{m_s})$ (for I_{m_j}, the identity matrix of order m_j) and $A(x, \epsilon)$ is holomorphic in the small parameters ϵ_1/ϵ_2, $\epsilon_2/\epsilon_3, \ldots, \epsilon_{s-1}/\epsilon_s$, and ϵ_s. The system can (under appropriate conditions) be diagonalized in the manner of Wasow [115] and Harris [47] and the component problems can be analyzed by using the techniques of Turrittin [102] or Sibuya [96] (in the case of turning point problems).

Consider, e.g., boundary value problems for the equation

$$\epsilon z'' + \mu a(x) z' + b(x) z = 0$$

on the interval [0, 1] where $a(x) \neq 0$ when $\epsilon/\mu^2 \to 0$ as $\mu \to 0$. Introducing $y_1 = z'$ and $y_2 = z/\mu$, a system of the form (4.19) results with

$$\Omega(\epsilon) = \begin{pmatrix} \epsilon/\mu & 0 \\ 0 & \mu \end{pmatrix} \quad \text{and} \quad A(x, \epsilon) = \begin{pmatrix} -a(x) & -b(x) \\ 1 & 0 \end{pmatrix}.$$

Moreover, the results obtained above can be recovered using the plan outlined.

5. Relaxation Oscillations

The phenomenon known as relaxation oscillations (see Wasow [112]) and Friedman [33] arises in considering differential equations of the form

$$\epsilon \ddot{x} - F(\dot{x}) + x = 0 \tag{5.1}$$

or, equivalently, nonlinear autonomous systems

$$\begin{aligned} \dot{x} &= y, \\ \epsilon \dot{y} &= F(y) - x \end{aligned} \tag{5.2}$$

in the x–y phase plane for ϵ a small positive parameter. Differentiating (5.1) with respect to the independent variable t yields

$$\epsilon \ddot{y} - F'(y) \dot{y} + y = 0 \tag{5.3}$$

for $y = \dot{x}$. Note that (5.3) has a turning point (see, e.g., Wasow [116]) where $F'(y) = 0$. Moreover, in the special case where $F(y) = y - \frac{1}{3} y^3$, (5.3) is equivalent to the van der Pol equation

$$\frac{d^2 y}{d\tau^2} - \lambda(1 - y^2) \frac{dy}{d\tau} + y = 0, \tag{5.4}$$

where $\lambda \equiv \epsilon^{-1/2}$ is a large positive parameter and $\tau \equiv \lambda t$,

For $F(y) = y - \frac{1}{3}y^3$, define the "fundamental curve" $\Gamma : x = F(y)$ in the phase plane, noting that the reduced system ((5.2) with $\epsilon = 0$) is

$$\begin{cases} \dot{x} = y \\ 0 = F(y) - x. \end{cases}$$

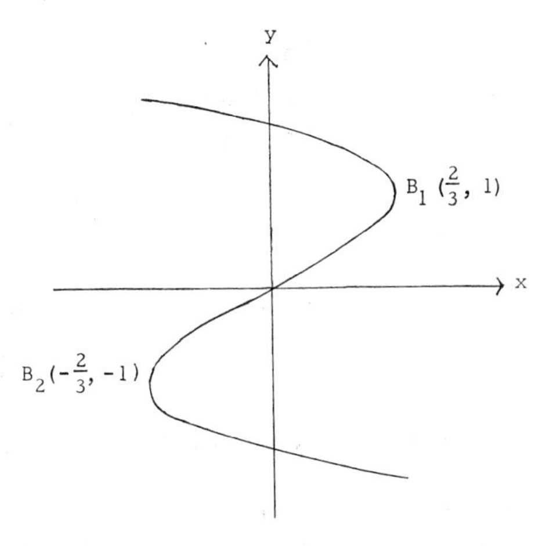

FIG. 1. The fundamental curve $\Gamma : x = F(y) = y - \frac{1}{3}y^3$.

The differential equation of the trajectories of (5.2) is

$$\frac{dy}{dx} = \frac{F(y) - x}{\epsilon y}. \tag{5.4}$$

Thus since this selection of $F(y)$ is an odd function, (5.4) implies that the vector field dy/dx will be symmetric with respect to the origin. Moreover, analyzing the field dy/dx determines the paths of (5.2). Specifically, let a trajectory start at a point P above the upper arc of Γ and to the left of $B_1 = (\frac{2}{3}, 1)$. As t increases, the trajectory must drop very rapidly, almost vertically, toward the fundamental curve Γ, then it will move to the right remaining close to, but above, Γ until it passes the point B_1, when it will drop rapidly, almost vertically, toward the lower arc of Γ. Crossing Γ, the trajectory will move to the left, remaining close to, but below, Γ until it passes $B_2 \equiv (-\frac{2}{3}, -1)$, when it will rise rapidly, almost vertically, to the upper arc of Γ. Crossing Γ, the trajectory will continue indefinitely in nearly the same orbit.

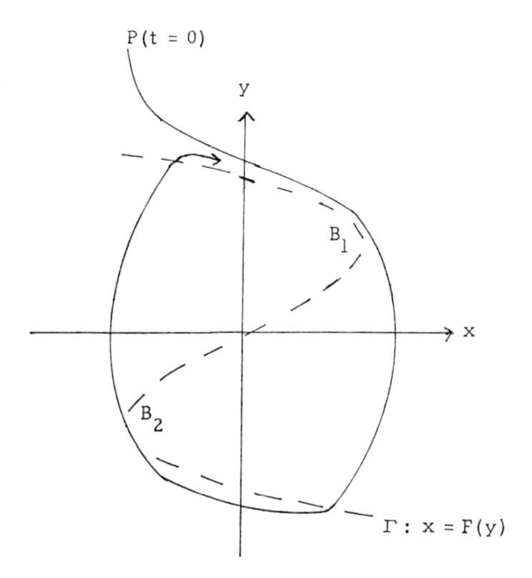

FIG. 2. Trajectory for (5.2) with $F(y) = y - \frac{1}{3}y^3$.

Liénard [69] showed that for each $\epsilon > 0$, (5.2) with $F(y) = y - \frac{1}{3}y^3$ has a unique (up to parameterization) limit cycle $C(\epsilon)$ (i.e., a periodic trajectory which other trajectories approach in a spiral-like manner as $t \to \infty$). That these limit cycles $C(\epsilon)$ actually have a limit as $\epsilon \to 0$ was proved in Flanders and Stoker [30]. This limit is the closed curve D formed by the outer arcs of the characteristic curve Γ and its two vertical tangents.

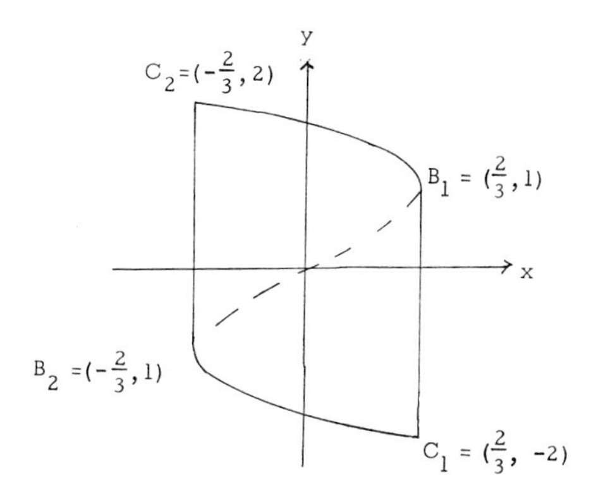

FIG. 3. The closed curve D.

Note that although D is a continuous curve in the phase plane, its tangent has discontinuities at $C_1 : (\frac{2}{3}, -2)$ and $C_2 : (-\frac{2}{3}, 2)$. Note, too, that (5.3), the scalar differential equation for y, has turning points at $y = \pm 1$, i.e., at B_1 and B_2. Further, only part of the limit curve D coincides with a path of a solution of the reduced system (5.2), and although D is a closed curve, this reduced system has no nontrivial periodic solutions and, hence, no closed trajectories.

Oscillations corresponding to the periodic trajectories of (5.3) are called relaxation oscillations and are characterized by jerky, almost instantaneous (in time, t), periodic jumps in $y = \dot{x}$. Expressing this physically, Stoker [98] states that these oscillations "exhibit two distinct and characteristic phases: one during which energy is stored up slowly (in a spring or condenser) and another in which the energy is discharged nearly instantaneously when a certain threshold potential is attained." This nonuniform convergence (in time) of the relaxation oscillations $C(\epsilon) : (x(t, \epsilon), y(t, \epsilon))$ of (5.2) as $\epsilon \to 0$ is, clearly, related to the boundary layer phenomenon discussed previously.

One expects that the limiting value of the period of the relaxation oscillations $C(\epsilon)$ as $\epsilon \to 0$ is given by

$$\lim_{\epsilon \to 0} \int_{C(\epsilon)} \frac{dx}{y} = 2 \int_2^1 \frac{dF(y)}{y} = 3 - 2 \log 2,$$

since $dt = dx/y$, $D = \lim_{\epsilon \to 0} C(\epsilon)$, $dx = 0$ on the vertical arcs of D, and $dx = dF(y)$ on Γ. Indeed, Dorodnicyn's analysis (see Dorodnicyn [23], Urabe [103], Urabe [104], and Ponzo and Wax [90]) gives the period $T(\epsilon)$ as

$$T(\epsilon) = (3 - 2 \log 2) + 3\alpha\epsilon^{2/3} + \tfrac{1}{3}\epsilon \log \epsilon + c\epsilon + O(\epsilon^{7/6})$$

where $-\alpha \doteq -2.338$ is the algebraically largest zero of the Airy function $Ai(z)$ and c is a given transcendental constant. Likewise, Dorodnicyn calculated the amplitude a of the y-oscillation, namely,

$$a = 2 + \frac{\alpha}{3} \epsilon^{2/3} + \frac{8}{27} \epsilon \log \epsilon + d\epsilon + O(\epsilon^{4/3}),$$

where d is another given transcendental constant. Roughly speaking, we could say that the "solution" D of the reduced system has y-amplitude 2 and "period" $3 - 2 \log 2$. Cartwright [13] and Haag [38], [39] both obtain similar results by different methods.

Stoker ([98], pp. 141–147) considers the illuminating, yet tractable problem of the relaxation oscillations for

$$\epsilon\ddot{x} - F(\dot{x}) + x = 0 \tag{5.1}$$

where F is the piecewise linear function

$$F(y) = \begin{cases} -2 - y, & y < -1, \\ \quad y \quad, & -1 \leqslant y \leqslant 1, \\ 2 - y \ , & y > 1. \end{cases}$$

In this case, one must simply solve two linear, constant coefficient, second-order differential equations subject to matching and periodicity requirements. Proceeding in the obvious manner, the asymptotic solution for the relaxation oscillations is completely obtainable. In particular, the period $T(\epsilon)$ is given by

$$T(\epsilon) = 2[\log 3 - \tfrac{4}{3}\epsilon \log \epsilon + \epsilon(\tfrac{4}{3} - \log 3) + \tfrac{1}{9}\epsilon^2 \log^2 \epsilon] + O(\epsilon^2 \log \epsilon).$$

Comparing this expansion with the determination when $F(y) = y - \frac{1}{3}y^3$, is is clear that the form of the characteristic F markedly affects the nature of the asymptotic solution. In particular, we note that this polygonal characteristic F has discontinuous tangents at $y = \pm 1$.

These results were extended to more general fundamental curves by Mishchenko [75], and Mishchenko and Pontryagin [76]. Mishchenko considers the first-order system

$$\begin{cases} \dot{x} = f(x, y), \\ \epsilon\dot{y} = g(x, y), \end{cases}$$

where f and g are sufficiently smooth. Then if the reduced system has a "stable, discontinuous, periodic solution" $\mathfrak{F}_0$ with "period" T_0 (we shall define these terms precisely, below), the original (or full) system has a unique limit cycle $\mathfrak{F}_\epsilon$ near $\mathfrak{F}_0$, tending to $\mathfrak{F}_0$ as $\epsilon \to 0$, with period $T(\epsilon)$ given by the formula

$$T(\epsilon) = T_0 + \epsilon^{2/3}Q_1 + \epsilon \log \epsilon Q_2 + \epsilon Q_3 + O(\epsilon^{7/6})$$

for Q_1, Q_2, and Q_3 numerical constants depending only on the values of f, g, and their derivatives. [$\mathfrak{F}_0$ and T_0 are defined to include the van der Pol case.]

For a discussion of relaxation oscillation problems with time-dependent forcing term, see Cartwright [12], Wendell [117], and Brock [9].

THE TWO-TERM EXPANSION FOR THE AMPLITUDE AND PERIOD OF THE RELAXATION OSCILLATIONS FOR VAN DER POL'S EQUATION

Returning to van der Pol's equation (5.4), or equivalently the system (5.2) with $F(y) = y - \frac{1}{3}y^3$, we study the relaxation oscillation with periodic trajectory $C(\epsilon)$ passing through the points $A_i = (x_i, y_i)$, $i = 1, 2, \ldots, 7$.

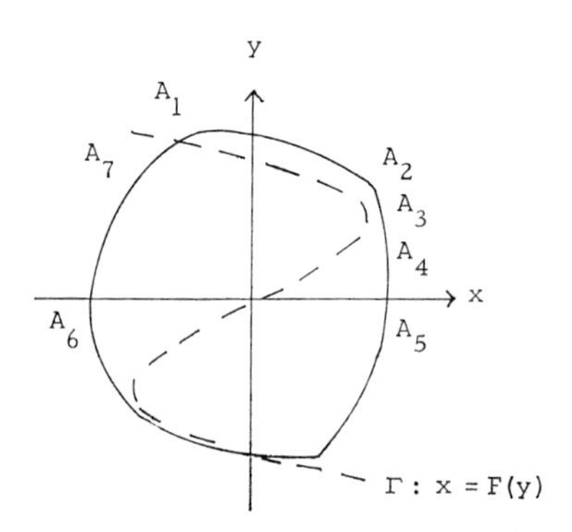

FIG. 4. Trajectory $C(\epsilon)$.

Let the trajectory $C(\epsilon)$ begin at some point A_1 on the fundamental curve $\Gamma : x = F(y)$, let $y_2 = 1 + \epsilon^r$ for $0 < r < \frac{1}{3}$, $y_3 = 1$, $y_4 = 1 - \epsilon^r$, $y_5 = y_6 = 0$, and $y_7 = y_1 - \epsilon$, and let $C(\epsilon)$ be symmetric with respect to the origin. Introduce

$$w = x - F(y) = x - y + \tfrac{1}{3}y^3,$$

the horizontal deviation of $C(\epsilon)$ from Γ, so that

$$\frac{dw}{dy} = -\frac{\epsilon y}{w} - F'(y) = \frac{y^2 - 1}{w}\left(w - \frac{\epsilon y}{y^2 - 1}\right).$$

Thus

$$0 < w(y) < \frac{\epsilon}{y - 1} \tag{5.5}$$

on $(1, y_1)$ (between A_1 and A_3) since either

$$w(y) \leqslant w(y^*) = \frac{\epsilon y^*}{(y^*)^2 - 1} < \frac{\epsilon}{y^* - 1}, \qquad y_1 > y^* > 1,$$

with y decreasing, or w is nonincreasing and

$$w(y) \leqslant \frac{\epsilon y}{y^2 - 1} < \frac{\epsilon}{y - 1}.$$

Near A_3, let

$$x = \tfrac{2}{3} + \epsilon^{2/3}u, \qquad y = 1 + \epsilon^{1/3}v$$

so that

$$0 \leqslant w(y) = \epsilon^{2/3}\left(u + v^2 + \epsilon^{1/3}\frac{v^3}{3}\right) < \frac{\epsilon^{2/3}}{v} \tag{5.6}$$

on $(1, y_1)$. Thus

$$v_2 = \epsilon^{r-1/3} \to \infty, \quad w(y_2) \to 0, \qquad \text{and} \qquad u_2 \to -\infty$$

as $\epsilon \to 0$. Further,

$$(u + v^2)\frac{du}{dv} + 1 = -\epsilon^{1/3}\left(\frac{v^3}{3}\frac{du}{dv} + v\right) \tag{5.7}$$

with the corresponding reduced equation

$$d\tilde{v}/du = -(\tilde{v}^2 + u) \tag{5.8}$$

having the general solution

$$\tilde{v}(u) = -\left(\frac{a_1 Ai'(-u) + a_2 Bi'(-u)}{a_1 Ai(-u) + a_2 Bi(-u)}\right)$$

where $Ai(z)$ and $Bi(z)$ are Airy functions and a_1 and a_2 are arbitrary constants (see Miller [73] and Copson [19] for the definitions and the relevant expansions of these functions). Assuming that the required solution v of (5.7) is a regular perturbation of some solution $\tilde{v}$ of (5.8), we seek a solution $\tilde{v}$ which approaches $+\infty$ as u approaches $-\infty$. Expanding $Ai(z)$ and $Bi(z)$ for large positive argument leads us to select $a_2 = 0$, so that $\tilde{v}(u) = -Ai'(-u)/Ai(-u)$. That $v = \tilde{v} + o(1)$ for $|v| \leqslant \epsilon^{r-1/3}$, r not too small, for this boundary condition, is shown in the appendix to Wasow [112], following the earlier argument by Haag. Further, $\tilde{v}'(u)$ is negative and uniformly bounded on $(-\infty, u_3)$ so that $u + \tilde{v}^2$ is bounded and

$$\tilde{v}(u) \leqslant \tilde{v}(u_2) \equiv v(u_2) = \epsilon^{r-1/3}$$

there. Thus, since $v = \tilde{v} + o(1)$, (5.6) implies that

$$w = O(\epsilon^{2/3})$$

on A_2A_3 for $\frac{2}{9} < r < \frac{1}{3}$.

Letting $y_4 = 1 - \epsilon^r$, $v_4 = -\epsilon^{r-1/3}$, so that $v_4 \to -\infty$ and u_4 must approach α as $\epsilon \to 0$ (using the expression for $\tilde{v}(u)$) where $-\alpha$ is the first zero of $Ai(z)$, i.e.,

$$x_4 = \tfrac{2}{3} + \epsilon^{2/3}\alpha + o(\epsilon^{2/3}).$$

Expanding about $(\frac{2}{3}, 1)$,

$$w = (y-1)^2 + \tfrac{1}{3}(y-1)^3 + (x - \tfrac{2}{3}) > (y-1)^2[1 - \tfrac{1}{3}(1-y)] > \tfrac{2}{3}(y-1)^2$$

on A_4A_5, so that

$$x_5 - x_4 = \epsilon \int_0^{y_4} \frac{y\,dy}{w} < \frac{3\epsilon}{2} \int_0^{1-\epsilon^r} (y-1)^{-2}\,dy$$
$$= O(\epsilon^{1-r}) = o(\epsilon^{2/3}).$$

Likewise,

$$\begin{aligned} -w &= -(x - x_1) - \tfrac{1}{3}(y - y_1)^3 - y_1(y - y_1)^2 + (1 - y_1^2)(y - y_1) \\ &> -(y - y_1)[\tfrac{1}{3}(y - y_1)^2 + y_1(y - y_1) - (1 - y_1^2)] \\ &> -(y - y_1)\left[\frac{y_1^3}{3} - 1\right] > -\delta(y - y_1) \end{aligned}$$

on A_6A_7 provided $y_1^2 > 3(1 + \delta)$, $\delta > 0$. Thus,

$$x_7 - x_6 = \epsilon \int_0^{y_7} \frac{y\,dy}{-w} \leqslant -\frac{\epsilon y_1}{\delta} \int_0^{y_1 - \epsilon} \frac{dy}{y - y_1} = O(\epsilon \log \epsilon).$$

Further, since

$$-w(dw/dy) = -w(y^2 - 1) + \epsilon y \geqslant \epsilon y$$

on A_7A_1,

$$w^2 \geqslant \epsilon(y_1^2 - y^2) > \epsilon y_7(y_1 - y),$$

and

$$x_1 - x_7 \leqslant \epsilon y_1 \int_{y_1-\epsilon}^{y_1} \frac{dy}{-w} \leqslant \frac{\epsilon^{1/2} y_1}{\sqrt{y_7}} \int_{y_1-\epsilon}^{y_1} \frac{dy}{\sqrt{y_1 - y}} = O(\epsilon).$$

Putting together these estimates,

$$x_1 = x_1 - x_7 + x_7 - x_6 - (x_5 - x_4) - x_4$$

since $x_6 = -x_5$, and therefore

$$x_1 = -\tfrac{2}{3} - \epsilon^{2/3}\alpha + o(\epsilon^{2/3}).$$

Further, since A_1 lies on Γ,

$$y_1 - 2 = \frac{-3(x_1 + \frac{2}{3})}{(y_1 + 1)^2} = \frac{\epsilon^{2/3}\alpha}{3} + o(\epsilon^{2/3}),$$

and the amplitude a of the relaxation oscillations for van der Pol's equation (5.4) is therefore

$$a = 2 + \tfrac{1}{3}(\epsilon^{2/3}\alpha) + o(\epsilon^{2/3})$$

as $\epsilon \to 0$ for $\alpha \doteq 2.338$ and $\epsilon = \lambda^{-2}$ since y_1 is the maximum value of y on the periodic trajectory $C(\epsilon)$.

Proceeding analogously, and using the preceding estimates, we can estimate the time T_{ij} required for traversing the arc A_iA_j. Specifically, we obtain the period T.

$$\begin{aligned} T &= 2(T_{12} + T_{23} + T_{34} + T_{45} + T_{67} + T_{71}) \\ &= 2\Big[\Big(\int_{y_1}^{y_3} \frac{dF(y)}{y} + o(\epsilon^{2/3})\Big) + \epsilon^{2/3}(u_3 + o(1)) \\ &\quad + \epsilon^{2/3}(\alpha - u_3 + o(1)) + o(\epsilon^{2/3}) + O(\epsilon \log \epsilon) + O(\epsilon)\Big] \\ &= 3 - 2 \log 2 + 3\epsilon^{2/3}\alpha + o(\epsilon^{2/3}). \end{aligned}$$

6. Initial-Value Problems for Nonlinear Systems of Equations

Perturbations of Discontinuous Solutions

Levinson [65], [67] considers the nonlinear system

$$\begin{cases} \dot{x} = f\dot{y} + r, \\ \epsilon\ddot{y} + g\dot{y} + h = 0 \end{cases} \tag{6.1}$$

for x an n-vector, y a scalar, and ϵ a small, positive parameter where the vectors f and r and the scalars g and h are functions of x, y, and the independent variable t.

Note that this general system includes systems of the form

$$\frac{dx}{dt} = f(x, y, t),$$

$$\epsilon \frac{dy}{dt} = g(x, y, t)$$

[by differentiating the last equation, or by introducing $w = \int^t y(s)\, ds$], and relaxation oscillation problems given by scalar equations of the form

$$\epsilon \ddot{x} - F(\dot{x}) + x = 0$$

(by differentiating and introducing $y = \dot{x}$).

Corresponding to (6.1), we define the reduced (or degenerate) system

$$\begin{cases} \dot{u} = f\dot{v} + r, \\ g\dot{v} + h = 0. \end{cases} \tag{6.2}$$

Note that dv/dt may become singular when g vanishes.

Recall the special case of relaxation oscillations where the limit of the phase plane limit cycles consists of "regular arcs" on which the reduced system is satisfied and of "jump arcs" "traversed in zero time" (i.e., nearby trajectories which are traversed nearly instantaneously). For the more complicated examples encompassed by (6.1), however, the path of the degenerate trajectory in E_{n+1} (i.e., the path of the limiting solution, if its exists) is not *à priori* clear. Levinson defines the solution of the reduced system (6.2) as a curve C_0 in E_{n+2} with coordinates (u, v, t) in such a manner that these questions are clarified and most potent results are obtained.

The vector $(u(t), v(t), t) \in E_{n+2}$, $0 \leqslant t \leqslant 1$, is defined to be a solution of the reduced system (6.2) if

(1) $(u(t), v(t))$ is continuous for $0 \leqslant t \leqslant 1$ except for a finite number of values t_j, $0 < t_1 < t_2 < \cdots < t_m < 1$. Moreover, the limits $(u(t_j \pm 0), v(t_j \pm 0))$ exist for each j, $j = 1, 2, \ldots, m$.

(2) $(u(t), v(t))$ satisfies (6.2) except for $t = t_j$, $j = 1, 2, \ldots, m$.

(3) $g(u(t), v(t), t)$ is positive for all $t \in [0, 1]$ except possibly for the values $t = t_j$. Moreover, for each j, $g(u(t), v(t), t)$ tends to zero as $t \to t_j^-$ while g remains positive as $t \to t_j^+$.

(4) At each point,

$$(u_-, v_-, t_j) \equiv (u(t_j - 0), v(t_j - 0), t_j)),$$

we require that

$$I \equiv \sum_{i=1}^{n} \frac{\partial g}{\partial u_i} f_i + \frac{\partial g}{\partial v} \neq 0$$

and further that

$$h_- \equiv h(u_-, v_-, t_j) \neq 0.$$

(5) From each coordinate u_-, u satisfies the initial-value problem

$$\frac{du}{dv} = f(u, v, t_j)$$

$$u(v_-) = u_-,$$

where v increases if h_- is negative and decreases otherwise, until the first value of $v \neq v_-$ is reached for which

$$\int_{v_-}^{v} g(u(v), v, t_j)\, dv = 0.$$

This value of v is denoted by $v(t_j + 0)$ and the associated value of u by $u(t_j + 0)$.

Thus C_0, as constructed, is continuous as a curve in E_{n+1} (but not as a function of t), but it has a discontinuous tangent, in general, where $t = t_j$, $j = 1, 2, \ldots, m$.

Note. (1) The construction of the solution C_0 of (6.2) places specific requirements on the functions $f, r, g,$ and h. These conditions are sufficient for the theorems which follow. They are not necessary, e.g., Levinson states that if I changes sign where g becomes zero, no jumps will occur.

(2) In relaxation oscillation problems, the solution (u, v) of the reduced system (6.2) features jumps in the scalar v only, since the function f is then identically zero.

With this definition of C_0, Levinson proves the following.

Theorem 6.1. *Consider the nonlinear system*

$$\dot{x} = f\dot{y} + r,$$

$$\epsilon\ddot{y} + g\dot{y} + h = 0,$$

where x is an n-vector and y is a scalar with $x(0)$ and $y(0)$ prescribed. Further, let $(u(t), v(t))$ solve the corresponding reduced system

$$\dot{u} = f\dot{v} + r,$$

$$g\dot{v} + h = 0,$$

for $t \in [0, 1]$ with initial conditions $(u(0), v(0))$ such that

$$\| x(0) - u(0)\| + | y(0) - v(0)| \leqslant \delta_1$$

and

$$\left| \frac{dy}{dt}(0) - \frac{dv}{dt}(0) \right| \leqslant \frac{\delta_2}{\epsilon}.$$

Then the full system has a solution $(x(t), y(t))$ for $t \in [0, 1]$ provided ϵ, δ_1, and δ_2 are all sufficiently small. Moreover, as ϵ, δ_1, and δ_2 tend to zero, the solution curve $(x(t), y(t), t)$ in E_{n+2} tends to $C_0 : (u(t), v(t), t)$. In particular, for any fixed $\delta > 0$,

$$\| x(t) - u(t)\|, \qquad | y(t) - v(t)|,$$

and

$$\left| \frac{dy}{dt} - \frac{dv}{dt} \right|$$

all tend uniformly to zero over the intervals

$$0 < \delta \leqslant t \leqslant t_1 - \delta, \qquad t_1 + \delta \leqslant t \leqslant t_2 - \delta, \ldots,$$

$$t_{m-1} + \delta \leqslant t \leqslant t_m - \delta, \qquad t_m + \delta \leqslant t \leqslant 1 - \delta$$

as ϵ, δ_1, and δ_2 tend to zero.

The existence of a solution C_0, even though discontinuous, of the reduced system (6.2), then, implies the existence of nearby solutions of the full system (6.1). Additional theorems discuss the analogous convergence of the derivatives of $x(t)$ and $y(t)$ with respect to initial conditions. In particular, when f, r, g, and h of (6.1) do not depend directly on t or in case they are periodic in t of period $\tilde{T}$, if the reduced system (6.2) has a solution of period $\tilde{T}$ and if the Jacobian associated with the determination of this solution by varying initial conditions is different from zero, then the full system (6.1) will also have a nearby periodic solution, say of period $T \sim \tilde{T}$ as $\epsilon, \delta_1, \delta_2 \to 0$. For then, the equations

$$\varphi_1(T) \equiv x(T) - x(0) = 0,$$
$$\varphi_2(T) \equiv y(T) - y(0) = 0,$$
$$\varphi_3(T) = \frac{\partial y}{\partial t}(T) - \frac{\partial y}{\partial t}(0) = 0$$

must hold for small ϵ, δ_1, and δ_2 which requires that the Jacobian

$$\frac{\partial(\varphi_1(T), \varphi_2(T), \varphi_3(T))}{\partial\left(x(0), y(0), \frac{\partial y}{\partial t}(0)\right)}$$

be continuous and nonvanishing as ϵ, δ_1, and $\delta_2 \to 0$. The convergence results, however, show that this will be satisfied provided only that the Jacobian

$$\frac{\partial(\psi_1(\tilde{T}), \psi_2(\tilde{T}))}{\partial(u(0), v(0))}$$

is nonvanishing for $\psi_1(\tilde{T}) = u(\tilde{T}) - u(0)$ and $\psi_2(\tilde{T}) = v(\tilde{T}) - v(0)$.

Asymptotic Solutions for Initial-Value Problems

Vasil'eva (see, especially the survey article, Vasil'eva [106], where the list of her previous papers can be found) considers the system of nonlinear equations

$$\begin{cases} dx/dt = f(x, y, t), \\ \epsilon(dy/dt) = g(x, y, t), \end{cases} \tag{6.3}$$

for ϵ a small positive parameter, x and y vectors, and f and g sufficiently smooth functions. Introducing

$$z(t) = \begin{pmatrix} x(t) \\ y(t) \end{pmatrix},$$

let

$$z(0) = \begin{pmatrix} x(0) \\ y(0) \end{pmatrix}$$

be prescribed.

Since the reduced system

$$\begin{cases} dx/dt = f(x, y, t), \\ \quad 0 = g(x, y, t), \end{cases} \tag{6.4}$$

corresponding to (6.3) is of lower order than the full system (6.3), it cannot, in general, satisfy all the initial conditions prescribed for $z(t)$. Obvious questions therefore arise. Does the solution of the initial value problem (6.3) have a well-defined limiting solution as $\epsilon \to 0$? If so, does this limiting solution satisfy the reduced system? In that case, does the limiting solution assume the initial condition $x(0)$, and further, which solution of the system $g(x, y, t) = 0$ determines the limiting behavior of y? Before analyzing these questions, it is necessary to introduce further terminology.

Let $y = \varphi(x, t)$ be any solution of the system of equations $g(x, y, t) = 0$ defined on a closed bounded set D. We shall denote the solution of the reduced system (6.4) corresponding to the root $y = \varphi(x, t)$ with $x(0)$ prescribed by

$$\bar{z}(t) = \begin{pmatrix} \bar{x}(t) \\ \varphi(\bar{x}(t), t) \end{pmatrix}.$$

Further, we shall call the root $y = \varphi(x, t)$ isolated on D if there exists some positive number γ such that $g(x, y, t) = 0$ has no root other than $y = \varphi(x, t)$ for $| y - \varphi(x, t)| < \gamma$.

For the full system (6.3), y is rapidly-varying compared to x and t when ϵ is small and g is positive. This observation leads us to consider the boundary layer (or adjoined) system

$$dy/d\tau = g(x^*, y, t^*), \tag{6.5}$$

where $\tau = t/\epsilon$ and x^* and t^* are fixed parameters.

An isolated root $y = \varphi(x, t)$ will be called stable in D if, for all points $(x^*, y^*) \in D$, the points $y^* = \varphi(x^*, t^*)$ are asymptotically stable critical points of the boundary layer equation (6.5) as $\tau \to \infty$, uniformly for $(x^*, t^*) \in D$; i.e.:

1. $g(x^*, y^*, t^*) = 0$;

2. for each $\mu > 0$, there exists $\delta(\mu) > 0$ independent of x^* and t^* such that

$$\| y(0) - y^* \| < \delta(\mu)$$

implies that a solution $y(\tau)$ of (6.5) exists and satisfies the inequality

$$\| y(\tau) - y^* \| < \mu$$

for all $\tau > 0$ as well as the asymptotic requirement

$$\lim_{\tau \to \infty} y(\tau) = y^* = \varphi(x^*, t^*).$$

(For an elementary introduction to stability theory, see LaSalle and Lefschetz [62].)

Lastly, a point (x^*, y^*, t^*) is said to be in the domain of influence of the stable root $y = \varphi(x, t)$ if the solution of the boundary layer system (6.5) satisfying the initial condition $y \mid_{\tau=0} = y^*$ tends to the value $\varphi(x^*, t^*)$ as $\tau \to \infty$. [Here, $y^* \neq \varphi(x^*, t^*)$, in general.] Using this terminology, we are able to state the basic conclusion of Tikhonov [99]:

Theorem 6.2. *Suppose*

(1) *the root $y = \varphi(x, t)$ of the system of equations $g(x, y, t) = 0$ is continuous, isolated, and stable in some closed, bounded domain D,*

(2) *the point $(x(0), y(0), 0)$ belongs to the domain of influence of this root,*

and

(3) *the solution $\bar{x}(t)$ of the reduced initial-value problem*

$$dx/dt = f(x, y, t),$$
$$y = \varphi(x, t),$$
$$x(0) \qquad \textit{prescribed}$$

is unique and lies in D for $0 \leqslant t \leqslant T$.
Then the solution of the initial-value problem for the full system

$$dx/dt = f(x, y, t)$$
$$\epsilon(dy/dt) = g(x, y, t)$$
$$x(0), \quad y(0) \qquad \textit{prescribed}$$

converges to the solution $(\bar{x}(t), \bar{y}(t))$ of the reduced system as $\epsilon \to 0$ in such a manner that

$$\lim_{\epsilon \to 0} x(t) = \bar{x}(t), \qquad 0 \leqslant t \leqslant T_0 < T,$$
$$\lim_{\epsilon \to 0} y(t) = \bar{y}(t) = \varphi(\bar{x}(t), t), \qquad 0 < t \leqslant T_0 < T$$

where T_0 is arbitrarily close to, but less than, T.

Note. (1) As Wasow [116] points out, this formulation of the Tikhonov theorem implicitly assumes the existence of a unique solution of the full initial-value problem on $0 \leqslant t \leqslant T$. Note, however, that since we have asked that f and g be sufficiently smooth, the existence is assured.

(2) The limiting solution of the full initial-value problem is discontinuous at $t = 0$ unless, by chance, $y(0) = \varphi(x(0), 0)$. In a small neighborhood of $t = 0$, then, a boundary layer phenomenon occurs as the solution passes rapidly from the initial point of the full system $(x(0), y(0), 0)$ into the neighborhood of the initial point $(x(0), \varphi(x, 0), 0)$ of the reduced system.

(3) In Tikhonov's definition of a stable root $y = \varphi(x, t)$, the asymptotic stability is not required to be uniform for $(x^*, y^*) \in D$. That this is incorrect follows by a counterexample given in Hoppensteadt [49].

(4) The analogous theorem for t lying in the semi-infinite interval $[0, \infty)$ is proved in Hoppensteadt [48]. In particular, this paper contains a new proof of the Tikhonov theorem. These results may also be extended to initial-value problems involving several parameters.

(5) Murphy [78] reports an interesting method for solving initial-value problems of the form (6.3) numerically.

An Illustrative Example

Consider the nonlinear autonomous system

$$\frac{dx}{dt} = e^{x^3} \sin y + e^{xy^2} \sin x - y^2 - xy - 1,$$

$$\epsilon(dy/dt) = y^2 - x^2,$$

with the initial conditions $x(0) = \alpha$ and $y(0) = \beta$ prescribed. Tikhonov's theorem implies that the solution of this problem converges (as $\epsilon \to 0$) to the solution of the reduced initial-value problem

$$dx/dt = -1,$$

$$y = -x, \qquad x(0) = \alpha,$$

provided $\alpha > 0$ for all values of t on any closed subinterval of $(0, \infty)$.

For $\alpha < 0$, however, the solution converges to the solution of the nonlinear reduced initial-value problem

$$\frac{dx}{dt} = 2(e^{x^3}\sin x - x^2) - 1,$$

$$y = x, \qquad x(0) = \alpha,$$

for t on any closed (and bounded) subinterval of $(0, \infty)$.

For this example, $g(x, y, t) \equiv y^2 - x^2$ so that the roots of the equation $g(x, y, t) = 0$ are $y = \pm x$. Each of these roots is isolated in any closed, bounded subset D of the $x - y$ plane which does not intersect the line $x = 0$.

The boundary layer equation

$$dy/d\tau = y^2 - x^{*2}$$

where $\tau = t/\epsilon$ and x^* is a fixed parameter has the general solution

$$y(\tau) = -x^* \left(\frac{a_1 e^{x^*\tau} + a_2 e^{-x^*\tau}}{a_1 e^{x^*\tau} - a_2 e^{-x^*\tau}} \right)$$

for a_1 and a_2 arbitrary constants, while the initial-value problem is solved by

$$y(\tau) = -x^* \left[\frac{(y(0) - x^*)\, e^{x^*\tau} + (y(0) + x^*)\, e^{-x^*\tau}}{(y(0) - x^*)\, e^{x^*\tau} - (y(0) + x^*)\, e^{-x^*\tau}} \right].$$

For x^* lying in any closed, bounded subset D_1 of the right half-plane (we take the half-planes to be open sets),

$$| y(0) - (-x^*)| < \tfrac{1}{2}\mu$$

implies that $y(\tau)$ exists for all $\tau > 0$ and that

$$| y(\tau) - (-x^*)| = \left| \frac{2x^*(y(0) + x^*)\, e^{-2x^*\tau}}{(y(0) - x^*) - (y(0) + x^*)\, e^{-2x^*\tau}} \right| < \mu$$

for μ sufficiently small. In D_1, then, $y(\tau) \to -x^*$, so that the root $y = -x$ is stable and the root $y = x$ is not stable. Analogously, for any closed, bounded subset D_2 of the left half-plane, the root $y = x$ is stable while the root $y = -x$ is not stable.

Moreover, any point in the right half-plane lies in the domain of influence of the root $y = -x$, and any point in the left half-plane is in the domain of influence of the root $y = x$.

If $x(0) = \alpha > 0$, the solution of the initial-value problem

$$dx/dt = -1, \qquad x(0) = \alpha$$

will remain positive for $t \in [0, \alpha)$. Likewise, the solution of the initial-value problem

$$\frac{dx}{dt} = 2(e^{x^3} \sin x - x^2) - 1, \qquad x(0) = \alpha$$

for $\alpha < 0$ will remain negative as t increases since $2(e^{x^3} \sin x - x^2) - 1 < 0$ for $x < 0$. Applying Theorem 6.2, then, the stated results follow immediately.

The Construction Procedure

To construct the asymptotic solutions desired, Vasil'eva asks somewhat more than the requirement that the root $y = \varphi(x, t)$ of the system $g(x, y, t) = 0$ be uniformly asymptotically stable. Instead, she asks that the real parts of the characteristic roots of the matrix

$$(\partial g/\partial y)(x, \varphi(x, t), t) \tag{6.6}$$

be negative in D. Hereafter, we refer to this condition as hypothesis (6.6).

Following Vasil'eva, we first construct an "inner" solution of the initial-value problem valid in the boundary layer near $t = 0$. The initial-value problem for (6.3) is equivalent to the initial-value problem for the system

$$\begin{cases} dx/d\tau = \epsilon f(x, y, \epsilon\tau), \\ dy/d\tau = g(x, y, \epsilon\tau) \end{cases} \tag{6.7}$$

for $\tau = t/\epsilon$. Proceeding, we let $z(\tau) = \binom{x(\tau)}{y(\tau)}$ and set

$$z(\tau) = \sum_{k=0}^{\infty} z_k(\tau)\, \epsilon^k$$

(with $z(0)$ prescribed),

$$f(x(\tau), y(\tau), \epsilon\tau) = \sum_{k=0}^{\infty} \epsilon^k f_k(\tau),$$

and

$$g(x(\tau), y(\tau), \epsilon\tau) = \sum_{k=0}^{\infty} \epsilon^k g_k(\tau).$$

Equating coefficients of each power of ϵ in (6.7) to zero yields

$$\begin{cases} dx_k/d\tau = f_{k-1}(\tau), \\ dy_k/d\tau = g_k(\tau), \end{cases} \tag{6.8}$$

for each $k \geqslant 0$ where $f_{-1}(\tau) \equiv 0$, $z_0(0) = z(0)$, and $z_k(0) = 0$ for $k > 0$. Integrating then, for $k = 0$,

$$x_0(\tau) = x(0)$$

$$\frac{dy_0}{d\tau} = g(x_0(\tau), y_0(\tau), 0), \qquad y_0(0) = y(0). \tag{6.9a}$$

By the hypotheses of Theorem 6.2, however, we are guaranteed that (6.9a) has a solution which converges to $\varphi(x(0), 0)$ as $\tau \to \infty$ (since (6.9a) coincides with the boundary layer equation). Likewise,

$$dx_1/d\tau = f(x_0(\tau), y_0(\tau), 0), \qquad x_1(0) = 0$$

implies that

$$x_1(\tau) = \int_0^\tau f(x_0(s), y_0(s), 0)\, ds$$

and

$$\begin{aligned} \frac{dy_1}{d\tau} &= g_x(x_0(\tau), y_0(\tau), 0)\, x_1(\tau) \\ &\quad + g_y(x_0(\tau), y_0(\tau), 0)\, y_1(\tau) \\ &\quad + g_t(x_0(\tau), y_0(\tau), 0)\, \tau \\ &\equiv g_y(x_0(\tau), y_0(\tau), 0)\, y_1(\tau) + \tilde{q}_1(\tau), \end{aligned} \tag{6.8b}$$

with $y_1(0) = 0$ can also be integrated. In general, for $k > 1$,

$$\begin{aligned} f_k(\tau) &= f_x(x_0(\tau), y_0(\tau), 0)\, x_k(\tau) \\ &\quad + f_y(x_0(\tau), y_0(\tau), 0)\, y_k(\tau) + p_{k-1}(\tau) \end{aligned}$$

and

$$\begin{aligned} g_k(x) &= g_x(x_0(\tau), y_0(\tau), 0)\, x_k(x) \\ &\quad + g_y(x_0(\tau), y_0(\tau), 0)\, y_k(x) + q_{k-1}(\tau), \end{aligned}$$

where the $p_{k-1}(\tau)$ and $q_{k-1}(\tau)$ depend on τ, and the $x_j(\tau)$ and $y_j(\tau)$ for $j \leqslant k - 1$. Continuing, then, the $x_k(\tau)$ and $y_k(\tau)$ may be uniquely obtained successively, in order, from the linear equations (6.8) and the condition that $z_k(0) = 0$ for $k > 0$.

By Tikhonov's theorem, however, the solution of the initial-value problem for (6.3) is asymptotically equal to a solution of the reduced problem for (6.4) with $x(0)$ prescribed outside the boundary layer. Thus, following Vasil'eva, we develop $z(t)$ in an "outer solution" about $\epsilon = 0$, i.e., we let

$$\bar{z}(t) = \sum_{k=0}^{\infty} \bar{z}_k(t)\, \epsilon^k,$$

$$f(\bar{x}(t), \bar{y}(t), t) = \sum_{k=0}^{\infty} \bar{f}_k(t)\, \epsilon^k,$$

and

$$g(\bar{x}(t), \bar{y}(t), t) = \sum_{k=0}^{\infty} \bar{g}_k(t)\, \epsilon^k.$$

(*Note.* $\bar{x}(t)$ and $\bar{y}(t)$ are not the same here as the functions so denoted in the statement of Theorem 6.1.) Then, formally equating powers of ϵ in (6.3), we obtain

$$\begin{cases} d\bar{x}_k/dt = \bar{f}_k(t), \\ \bar{g}_k(t) = d\bar{y}_{k-1}/dt \end{cases} \tag{6.10}$$

for $k \geqslant 0$ where $\bar{y}_{-1} \equiv 0$. In particular, when $k = 0$,

$$\bar{g}_0(t) \equiv g(\bar{x}_0(t), \bar{y}_0(t), t) = 0$$

and we select $\bar{y}_0(t) = \varphi(\bar{x}_0(t), t)$ which, in turn, requires that $\bar{x}_0$ must satisfy the reduced system corresponding to the root $\varphi(x, t)$, i.e.,

$$dx/dt = f(x, y, t),$$

$$y = \varphi(x, t),$$

with $x(0)$ prescribed. The hypotheses of Theorem 6.2, however, assume that this nonlinear problem is solvable for $0 \leqslant t \leqslant T$. For $k = 1$,

$$d\bar{x}_1/dt = f_x(\bar{x}_0(t), \bar{y}_0(t), t)\, \bar{x}_1(t) + f_y(\bar{x}_0(t), \bar{y}_0(t), t)\, \bar{y}_1(t)$$

and

$$\bar{g}_1(t) = g_x(\bar{x}_0(t), \bar{y}_0(t), t)\, \bar{x}_1(t) + g_y(\bar{x}_0(t), \bar{y}_0(t), t)\, \bar{y}_1(t) = d\bar{y}_0/dt.$$

Since $g_y(\bar{x}_0(t), \bar{y}_0(t), t)$ is invertible, however, $\bar{y}_1(t)$ is a well-defined function of $\bar{x}_1(t)$, and $\bar{x}_1(t)$ satisfies a simple, linear, nonhomogeneous system of equations. Thus $\bar{x}_1(t)$ and $\bar{y}_1(t)$ are completely determined on $0 \leqslant t \leqslant T$ up to the selection of the initial condition $\bar{x}_1(0)$. (Note that Tikhonov's theorem does not yield higher-order approximate solutions, and therefore gives no information regarding $\bar{x}_k(0)$ for $k > 0$.) Likewise, for $k > 1$,

$$\bar{f}_k(t) = f_x(\bar{x}_0(t), \bar{y}_0(t), t)\,\bar{x}_k + f_y(\bar{x}_0(t), \bar{y}_0(t), t)\,\bar{y}_k + \bar{p}_{k-1}(t)$$

and

$$\bar{g}_k(t) = g_x(\bar{x}_0(t), \bar{y}_0(t), t)\,\bar{x}_k + g_y(\bar{x}_0(t), \bar{y}_0(t), t)\,\bar{y}_k + \bar{q}_{k-1}(t),$$

where $\bar{p}_{k-1}(t)$ and $\bar{q}_{k-1}(t)$ depend on t and $\bar{x}_j(t)$ and $\bar{y}_j(t)$ for $j < k$. Solving for $\bar{y}_k(t)$ in terms of $\bar{x}_k(t)$ and integrating, the $\bar{x}_k(t)$ and $\bar{y}_k(t)$ can be successively determined on the interval $[0, T]$ up to the selection of $\bar{x}_k(0)$.

Intuitively, we expect to obtain these unknown constants by, somehow, matching the inner expansion $z(\tau)$ (which can be completely determined to any number of terms) with the outer expansion $\bar{z}(t)$ (which can be determined up to these constants through any number of terms) for large values of τ and small values of t. Thus, we formally expand the outer solution $\bar{z}(t)$ about $t = 0$, i.e., we set

$$\tilde{z} = \sum_{m=0}^{\infty} \sum_{j=0}^{m} z_{j,m-j} t^j \epsilon^{m-j}$$

where the z_{jl}'s are defined by the expansions

$$\bar{z}_l(t) = \sum_{j=0}^{\infty} z_{jl} t^j.$$

Further, we let $(\tilde{z})_N$ represent the partial sum of $\tilde{z}$ through all Nth-order terms, in the two variables t and ϵ. Expressing $(\tilde{z})_N$ in terms of $\tau = t/\epsilon$, we have

$$(\tilde{z})_N = \sum_{m=0}^{N} \tilde{z}_m(\tau)\,\epsilon^m$$

for

$$\tilde{z}_m(\tau) = \sum_{j=0}^{m} z_{j,m-j}\tau^j.$$

Letting $(z)_N$ be the N-term inner expansion, the difference $(z)_N - (\tilde{z})_N$ satisfies

$$(z)_N - (\tilde{z})_N = \sum_{m=0}^{N} \prod_m (z, \tau)\, \epsilon^m$$

for the "boundary layer functions"

$$\prod_m (z, \tau) = z_m(\tau) - \tilde{z}_m(\tau).$$

Note that

$$\prod_m (z, 0) = \begin{cases} 0\;\;\;\;\;, & m = 0, \\ -\tilde{z}_m(0), & m > 0. \end{cases}$$

Since

$$(d/d\tau) \prod_m (x, \tau) = f_{m-1}(\tau) - \tilde{f}_{m-1}(\tau)$$

where

$$\tilde{f}_{m-1}(\tau) = \sum_{i=0}^{n-1} f_{i,m-1-i}\tau^i$$

and

$$f_k(\tau) = \sum_{l=0}^{\infty} f_{lk}t^l,$$

$$\prod_m (x, \tau) = \left[\prod_m (x, 0) + \int_0^\infty [f_{m-1}(\tau) - \tilde{f}_{m-1}(\tau)]\, d\tau\right]$$

$$- \int_\tau^\infty [f_{m-1}(s) - \tilde{f}_{m-1}(s)]\, ds.$$

Thus $\prod_m(x, \tau) \to 0$ as $\tau \to \infty$ (i.e., at the matching point) provided

$$\tilde{x}_m(0) = \int_0^\infty [f_{m-1}(\tau) - \tilde{f}_{m-1}(\tau)]\, d\tau \tag{6.11}$$

for each $m > 0$.

In general, we expect that $(z)_N - (\tilde{z})_N$ is negligible outside the boundary layer, while $(\bar{z})_N - (\tilde{z})_N$ for $(\bar{z})_N = \sum_{k=0}^N \bar{z}_k(t)\epsilon^k$ is negligible within the boundary layer. Further, we expect that the solution z of the initial-value problem will have the asymptotic solution $z(\tau)$ within the boundary layer and $\bar{z}(t)$ outside the boundary layer. Expressing this formally, Vasil'eva proves

Theorem 6.3. *Let $z(t, \epsilon)$ solve the nonlinear initial-value problem for the system of equations* (6.3). *Further, let the hypotheses of Theorem* 6.2 *hold as well as hypothesis* (6.6), *and let the expansions $(\bar{z})_N$, $(z)_N$, and $(\tilde{z})_N$ be those formally constructed above. Then for each integer $N \geqslant 0$,*

$$z(t, \epsilon) = Z_N + O(\epsilon^{N+1})$$

uniformly on $0 \leqslant t \leqslant T_0$ for all sufficiently small ϵ where

$$Z_N = (z)_N + (\bar{z})_N - (\tilde{z})_N .$$

Note. The coefficients in the expansions $(\tilde{z})_N$ and z_N are functions of the order N of the approximation. This was not the case for the expansions obtained in Sections 2–4.

A Second Example

To illustrate the method of Vasil'eva, consider the nonhomogeneous, linear, constant coefficient system

$$dx/dt = a_1 x + a_2 y,$$

$$\epsilon(dy/dt) = b_1 x + b_2 y.$$

where b_2 is negative.

The corresponding reduced system

$$dx/dt = a_1 x + a_2 y,$$

$$0 = b_1 x + b_2 y$$

has $y = -(b_1/b_2)x$ as the only root of the equation $g(x\ y, t) \equiv b_1 x + b_2 y = 0$, so this root is *à fortiori* isolated. For the initial-value problem where $x(0)$ and $y(0)$ are prescribed for the full system and t is increasing, the boundary layer equation is

$$dy/d\tau = b_1 x^* + b_2 y,$$

where x^* has any arbitrary (fixed) value and $\tau = t/\epsilon$. Since $\partial g/\partial y \equiv b_2$ is negative, the root $y = -(b_1/b_2)x$ is stable. Furthermore, any initial point $(x(0), y(0), 0)$ will then lie in the domain of influence of this root.

Tikhonov's theorem, then, implies that the limiting solution, except for the y-limit at $t = 0$, will coincide with the solution of the reduced initial value problem, i.e.,

$$\left(x(0) \exp \left[\left(a_1 - \frac{b_1}{b_2} a_2\right) t\right], -\frac{b_1}{b_2} x(0) \exp \left[\left(a_1 - \frac{b_1 a_2}{b_2}\right) t\right]\right)$$

on any finite t interval. Indeed, if $y(0) \neq -(b_1/b_2)x(0)$, a boundary layer phenomenon (nonuniform convergence) must occur at $t = 0$.

The inner solution $z(\tau) = \sum_{k=0}^{\infty} z_k(\tau)\epsilon^k$ is constructed to satisfy

$$dx/d\tau = \epsilon(a_1 x + a_2 y),$$
$$dy/d\tau = b_1 x + b_2 y,$$

with $z(0)$ prescribed. Equating coefficients, we ask that

$$dx_k/d\tau = a_1 x_{k-1} + a_2 y_{k-1} ,$$
$$dy_k/d\tau = b_1 x_k + b_2 y_k ,$$

for each $k \geqslant 0$ where $x_{-1} = y_{-1} \equiv 0$, $z_0(0) = z(0)$ and $z_k(0) = 0$ for $k > 0$. Integrating, then,

$$x_0(\tau) = x(0),$$

$$y_0(\tau) = \left(y(0) + \frac{b_1}{b_2} x(0)\right) e^{b_2 \tau} - \frac{b_1}{b_2} x(0),$$

$$x_1(\tau) = \left(a_1 - a_2 \frac{b_1}{b_2}\right) x(0)\, \tau + \frac{a_2}{b_2} \left(y(0) + \frac{b_1}{b_2} x(0)\right) (e^{b_2 \tau} - 1),$$

and

$$y_1(\tau) = -\frac{b_1}{b_2^2} \left(a_1 - \frac{a_2 b_1}{b_2}\right) x(0)(1 + b_2 \tau - e^{b_2 \tau})$$
$$+ \frac{b_1 a_2}{b_2^2} \left(y(0) + \frac{b_1}{b_2} x(0)\right) (1 + (b_2 \tau - 1)\, e^{b_2 \tau}).$$

Likewise, higher-order terms are uniquely obtainable by integrating successively, in order, for $x_k(\tau)$ and $y_k(\tau)$.

The outer solution

$$\bar{z}(t) = \sum_{k=0}^{\infty} \bar{z}_k(t)\, \epsilon^k$$

is constructed to formally satisfy the equations

$$d\bar{x}_k/dt = a_1\bar{x}_k + a_2\bar{y}_k\,,$$
$$b_1\bar{x}_k + b_2\bar{y}_k = d\bar{y}_{k-1}/dt$$

for each integer $k \geqslant 0$ where $\bar{y}_{-1} \equiv 0$ and $\bar{x}_0(0) = x(0)$. Thus,

$$\bar{x}_0(t) = x(0)\exp\left[\left(a_1 - \frac{a_2b_1}{b_2}\right)t\right],$$

$$\bar{y}_0(t) = -\frac{b_1x(0)}{b_2}\exp\left[\left(a_1 - \frac{a_2b_1}{b_2}\right)t\right],$$

$$\bar{x}_1(t) = \left[\bar{x}_1(0) - \frac{a_2b_1}{b_2{}^2}\left(a_1 - \frac{a_2b_1}{b_2}\right)x(0)\,t\right]\exp\left[\left(a_1 - \frac{a_2b_1}{b_2}\right)t\right],$$

and

$$\bar{y}_1(t) = \left[-\frac{b_1}{b_2}\,\bar{x}_1(0) + \frac{a_2b_1{}^2}{b_2{}^3}\left(a_1 - \frac{a_2b_1}{b_2}\right)x(0)\,t\right.$$
$$\left. - \frac{b_1x(0)}{b_2{}^2}\left(a_1 - \frac{a_2b_1}{b_2}\right)\right]\exp\left[\left(a_1 - \frac{a_2b_1}{b_2}\right)t\right].$$

Higher-order terms may be analogously determined up to the selection of the initial values $\bar{x}_k(0)$.

Expanding the coefficients of the outer solution $\bar{z}(t)$ for small values of t, i.e.,

$$\bar{z}_k(t) = \sum_{l=0}^{\infty} z_{lk}t^l,$$

and inserting these expansions into $\bar{z}(t)$, we define $(\tilde{z})_N$ as the finite sum obtained by collecting all terms of order $\leqslant N$ in the two variables ϵ and t. Thus, for example,

$$\bar{x}_1(t) = \bar{x}_1(0) + \left[\bar{x}_1(0)\left(a_1 - \frac{a_2b_1}{b_2}\right) - \frac{a_2b_1}{b_2{}^2}\left(a_1 - \frac{a_2b_1}{b_2}\right)x(0)\right]t + \cdots$$

and

$$(\tilde{x})_0 = x(0),$$
$$(\tilde{y})_0 = -(b_1/b_2)\,x(0),$$
$$(\tilde{x})_1 = x(0) + tx(0)\left(a_1 - \frac{a_2b_1}{b_2}\right) + \epsilon\bar{x}_1(0),$$

and

$$(\tilde{y})_1 = -\frac{b_1 x(0)}{b_2} - t\frac{b_1}{b_2}x(0)\left(a_1 - \frac{a_2 b_1}{b_2}\right) - \frac{\epsilon b_1}{b_2}\bar{x}_1(0)$$

$$-\frac{\epsilon b_1 x(0)}{b_2{}^2}\left(a_1 - \frac{a_2 b_1}{b_2}\right).$$

Since b_2 is negative, $(\tilde{z})_0$ and $z_0(\tau)$ will have the same limit as $\tau \to \infty$. Further, $(\tilde{z})_1$ and $(z)_1 = z_0(\tau) + \epsilon z_1(\tau)$ will have the same limit as $\tau \to \infty$ (since $t = \tau\epsilon$) provided

$$\bar{x}_1(0) = -\frac{a_2}{b_2}\left(y(0) + \frac{b_1}{b_2}x(0)\right).$$

Note, however, that this is the value determined by formula (6.11) since, here,

$$f_0(\tau) = \left(a_1 - \frac{b_1 a_2}{b_2}\right)x(0) + a_2\left(y(0) + \frac{b_1}{b_2}x(0)\right)e^{b_2\tau}$$

and

$$\tilde{f}_0(\tau) = \left(a_1 - \frac{b_1 a_2}{b_2}\right)x(0).$$

Higher-order initial values $\bar{x}_k(0)$ are successively obtained by matching $(\tilde{z})_k$ and $(z)_k$ as $\tau \to \infty$.

Introducing $(\bar{z})_N = \sum_{k=1}^{N} \bar{z}_k(t)\epsilon^k$, we obtain a uniformly valid solution to the initial-value problem on any finite interval $[0, T]$ by constructing

$$Z_N = (z)_N + (\bar{z})_N - (\tilde{z})_N\,,$$

i.e., for each $N \geqslant 0$, the solution z satisfies

$$z = Z_N + O(\epsilon^{N+1}) \qquad \text{uniformly as} \quad \epsilon \to 0.$$

Hence

$$X_0 = x(0)\exp\left[\left(a_1 - \frac{a_2 b_1}{b_2}\right)t\right],$$

$$Y_0 = -\frac{b_1}{b_2}x(0)\exp\left[\left(a_1 - \frac{a_2 b_1}{b_2}\right)t\right] + \left(y(0) + \frac{b_1}{b_2}x(0)\right)e^{b_2 t/\epsilon},$$

$$X_1 = \left[x(0) - \epsilon\left(\frac{a_2}{b_2}\left(y(0) + \frac{b_1}{b_2}x(0)\right)\right.\right.$$

$$\left.\left.+\frac{a_2 b_1}{b_2{}^2}\left(a_1 - \frac{a_2 b_1}{b_2}\right)x(0)\,t\right)\right]\exp\left[\left(a_1 - \frac{a_2 b_1}{b_2}\right)t\right]$$

$$+\,\epsilon\frac{a_2}{b_2}\left(y(0) + \frac{b_1}{b_2}x(0)\right)e^{b_2 t/\epsilon},$$

and

$$
\begin{aligned}
Y_1 = \Big[&-\frac{b_1}{b_2}\,x(0) + \epsilon\Big(\frac{b_1 a_2}{b_2{}^2}\Big(y(0) + \frac{b_1}{b_2}\,x(0)\Big) \\
&+ \frac{b_1 x(0)}{b_2{}^2}\Big(a_1 - \frac{a_2 b_1}{b_2}\Big)\Big(a_2\,\frac{b_1}{b_2}\,t - 1\Big)\Big)\Big] \exp\Big[\Big(a_1 - \frac{a_2 b_1}{b_2}\Big)\,t\Big] \\
&+ \Big[\Big(y(0) + \frac{b_1}{b_2}\,x(0)\Big)\Big(1 + \frac{a_2 b_1 t}{b_2}\Big) \\
&+ \epsilon\Big(\frac{b_1}{b_2{}^2}\Big(a_1 - \frac{a_2 b_1}{b_2}\Big)\,x(0) - \frac{b_1 a_2}{b_2{}^2}\Big(y(0) + \frac{b_1}{b_2}\,x(0)\Big)\Big)\Big]\, e^{b_2 t/\epsilon}.
\end{aligned}
$$

Moreover these expansions may be verified, albeit with much labor, since such constant coefficient systems may be straightforwardly integrated.

7. Singular Perturbation Problems for Partial Differential Equations

In this section, we restrict attention to several representative problems. The literature abounds with many additional problems. In addition to the references cited below, the reader is urged to note, e.g., Birkhoff [3], Levin [64], Bobisud [4], Kisynski [59], and Smoller [97].

Examples where the Reduced Equation is of Lower Order

Levinson [66] considers the first boundary value problem for the linear second-order elliptic partial differential equation

$$
\epsilon\Delta u + A(x, y)\, u_x + B(x, y)\, u_y + C(x, y)\, u = D(x, y) \tag{7.1}
$$

where ϵ is a small positive parameter and Δ is the Laplacian operator $\partial^2/\partial x^2 + \partial^2/\partial y^2$. Here, we will not present Levinson's results in their complete generality, but instead assume that the coefficients A, B, C, and D are sufficiently smooth in some open region R_0 of the x–y plane which contains a region R whose boundary S consists of a finite number of simple, closed curves. Further, let x and y be smooth functions of arc length s along each closed curve of S and let smooth boundary values u independent of ϵ be assigned along each such curve. Further, we ask that $A^2(x, y) + B^2(x, y) > 0$ in R_0 and that either R_0 be simply connected or that $C(x, y) < 0$ in R_0. Either hypothesis suffices to establish a maximum principle in $R \cup S$, i.e., a bound for $|u|$ on S determines a bound for $|u|$ in R. Under these hypotheses, the first boundary value problem for (7.1) will have a unique solution within $R \cup S$ for each $\epsilon > 0$.

As $\epsilon \to 0$, under quite general circumstances, we would expect the solutions of boundary value problems for (7.1) to converge to solutions of the reduced equation

$$A(x, y)\, u_x + B(x, y)\, u_y + C(x, y)\, u = D(x, y) \tag{7.2}$$

where the characteristic curves (or more simply, the characteristics) of this first-order equation satisfy

$$\frac{dx}{A(x, y)} = \frac{dy}{B(x, y)}.$$

The hypotheses listed above imply that this differential equation for the characteristics has no singular points in R_0.

Let us define the positive direction of the boundary curve S to be that direction on a tangent line to S from which a counterclockwise 90° rotation yields the interior normal, and let the arc length s of S be increasing in the positive direction. Let S_1 be a segment of one of the curves of S such that none of its tangents is a characteristic of the reduced equation (7.2), and let S_2 be that segment of a curve of S intersected by the characteristics of (7.2) passing through S_1, and conversely. Requiring that S_2 be nowhere characteristic,

$$B\, dx - A\, dy \neq 0$$

on both S_1 and S_2, so without loss of generality, let

$$B\frac{dx}{ds} - A\frac{dy}{ds} < 0 \quad \text{on} \quad S_1 \qquad \text{and} \qquad B\frac{dx}{ds} - A\frac{dy}{ds} > 0 \quad \text{on} \quad S_2 .$$

A closed simply-connected region in $R \cup S$ is formed by S_1 and S_2 and the two-characteristics of (7.2) joining the endpoints of S_1 and S_2. Hereafter, we refer to such a closed region as a "regular quadrilateral" Q.

Formal solutions of Eq. (7.1) can easily be constructed. For example, let $U(x, y, \epsilon)$ be an expansion of the form

$$U(x, y, \epsilon) = \sum_{k=0}^{\infty} U_k(x, y)\, \epsilon^k,$$

where U assumes the boundary values prescribed for u on S_1. Substituting into (7.1) and proceeding formally, we ask that the U_k satisfy

$$A(x, y)\, U_{kx}(x, y) + B(x, y)\, U_{ky} + C(x, y)\, U_k(x, y)$$

$$= \begin{cases} D(x, y) & \text{when} \quad k = 0, \\ -\Delta U_{k-1}(x, y) & \text{for} \quad k \geqslant 1, \end{cases}$$

with $U_0 = u$ and $U_k = 0$ for $k \geqslant 1$ on S_1. These linear equations may be uniquely solved successively since S_1 is a noncharacteristic curve for (7.2) and therefore for each of these equations.

In the event that the solution u of the boundary value problem for (7.1) converges as $\epsilon \to 0$ to the solution U_0 of the reduced boundary value problem for (7.2), a boundary layer phenomena can be expected along S_2 since the boundary values for U_0 on S_2 will not, in general, coincide with the boundary values prescribed there for u. Thus, we construct a formal boundary-layer-type solution of the homogeneous equation corresponding to (7.1) valid near S_2. Specifically, we set

$$z = \exp\left(-\frac{g(x, y)}{\epsilon}\right) h(x, y, \epsilon)$$

where $g = 0$ and $h = u - U$ on S_2 and g is positive away from S_2. Substituting z into the differential equation, we obtain

$$e^{-g/\epsilon}\Big[\frac{h}{\epsilon}(g_x{}^2 + g_y{}^2 - Ag_x - Bg_y)$$
$$+ ((A - 2g_x)\, h_x + (B - 2g_y)\, h_y + (C - \Delta g)\, h) + \epsilon\Delta h\Big] = 0.$$

Thus, we ask that g solve the nonlinear equation

$$g_x{}^2 + g_y{}^2 - Ag_x - Bg_y = 0.$$

Introducing the parameter t and setting $p = g_x$ and $q = g_y$, this equation is equivalent to the first-order system:

$$\frac{dx}{dt} = 2p - A,$$

$$\frac{dy}{dt} = 2q - B,$$

$$\frac{dg}{dt} = p(2p - A) + q(2q - B) = p^2 + q^2,$$

$$\frac{dp}{dt} = A_x p + B_x q,$$

and

$$\frac{dq}{dt} = A_y p + B_y q.$$

(The reader who is unfamiliar with the details of this procedure should see, e.g., Courant and Hilbert [21], pp. 75–84.) Letting $t = 0$ along S_2, the initial data is expressible as

$$x(s, 0) = x(s), \qquad y(s, 0) = y(s), \qquad g(s, 0) = 0$$

and since

$$\frac{dg}{ds} = p\frac{dx}{ds} + q\frac{dy}{ds} = 0 \qquad \text{and} \qquad dx^2 + dy^2 = ds^2$$

along S_2, the nontrivial determinations of p and q there are

$$p = -\left(B\frac{dx}{ds} - A\frac{dy}{ds}\right)\frac{dy}{ds}$$

and

$$q = \left(B\frac{dx}{ds} - A\frac{dy}{ds}\right)\frac{dx}{ds}.$$

Thus, $p(s, 0)$ and $q(s, 0)$ are completely specified. Moreover, since

$$\frac{dx}{ds}\frac{dy}{dt} - \frac{dy}{ds}\frac{dx}{dt} = B\frac{dx}{ds} - A\frac{dy}{ds} \neq 0$$

on S_2, g exists and is uniquely determined for this selection of p and q in a neighborhood of the noncharacteristic initial data curve S_2.

Knowing g, at least locally, we ask that $h(x, y, \epsilon)$ satisfy

$$(A - 2g_x)\, h_x + (B - 2g_y)\, h_y + (C - \Delta g)\, h = -\epsilon \Delta h.$$

Expanding, we let

$$h(x, y, \epsilon) = \sum_{k=0}^{\infty} h_k(x, y)\, \epsilon^k,$$

where each h_k satisfies the linear first-order equation

$$(A - 2g_x)\, h_{kx} + (B - 2g_y)\, h_{ky} + (C - \Delta g)\, h_k = -\Delta h_{k-1}$$

with

$$h_0 = u - U_0 \quad \text{on} \quad S_2,$$

while $h_k = -U_k$ for $k \geqslant 1$. For these equations, too, S_2 is noncharacteristic, so that the formal expansion for h can be uniquely obtained through any number of terms in the neighborhood of S_2 where g exists.

Using these formal results, Levinson proves the following.

Theorem 7.1. *Let $u(x, y)$ solve the first boundary value problem for*

$$\epsilon \Delta u + A(x, y)\, u_x + B(x, y)\, u_y + C(x, y)\, u = D(x, y)$$

in $R \cup S$ where a maximum principle holds. Then for each integer $N \geqslant 0$,

$$u(x, y) = \sum_{k=0}^{N} U_k(x, y)\, \epsilon^k + z_N(x, y, \epsilon) + O(\epsilon^{N+1/2})$$

uniformly in any regular quadrilateral Q where

$$z_N(x, y, \epsilon) = \begin{cases} e^{-g(x,y)/\epsilon} \sum_{k=0}^{N} h_k(x, y)\, \epsilon^k, & \textit{near } S_2 \\ e^{-\delta/\epsilon} \quad \textit{for some} \quad \delta > 0, & \textit{elsewhere in } Q \end{cases}$$

Note. (1) Eckhaus and de Jager [24] considerably extend Levinson's results and those of Visik and Lyusternik [107]. In particular, they obtain limited results at corner points and when the boundary S of the region R contains curves that are characteristics of the reduced equation.

(2) Oleinik [80] has obtained corresponding results for the second boundary value problem. Bobisud [6], [7] considers initial-boundary value problems for linear second-order parabolic equations with a small parameter multiplying the highest-order derivatives.

(4) In a series of papers, D. Huet (see, e.g., Huet [50]–[53], has established convergence theorems for singular perturbation problems for certain higher order elliptic partial differential equations in various Hilbert spaces. Weaker results for certain fourth-order equations have also been obtained by Davis [22].

Examples Illustrating Levinson's Theorem

Specializing, we consider the first boundary value problem for

$$\epsilon \Delta u + u_x + u_y = D(x, y) \tag{7.3}$$

in the unit square T with smooth boundary values

$$\begin{cases} u(0, y) = a(y), \\ u(1, y) = b(y), \\ u(x, 0) = c(x), \\ u(x, 1) = d(x), \end{cases} \tag{7.4}$$

prescribed on the boundary S of T. The characteristics of the reduced equation

$$u_x + u_y = D(x, y)$$

will, then, coincide with no tangent to S. Let S_1 and S_2 represent those edges of S along the lines $y = 1$ and $x = 0$, respectively. Then, we formally construct power series $U(x, y, \epsilon)$ and $H(x, y, \epsilon)$ such that an asymptotic solution $u(x, y)$ of the boundary value problem considered is given by

$$u(x, y) = U(x, y, \epsilon) + \exp\left(-\frac{g(x, y)}{\epsilon}\right) H(x, y, \epsilon)$$

where $y \geqslant x$, $U(x, 1, \epsilon) = d(x)$, $g(x, 1) = 0$, $g > 0$ elsewhere in T, and $H(0, y, \epsilon) = a(y) - U(0, y, \epsilon)$. Thus, we ask that U, g, and H satisfy the equations

$$U_x + U_y = D(x, y) - \epsilon \Delta U,$$

$$g_x^2 + g_y^2 - g_x - g_y = 0,$$

and

$$-2(H_x g_x + H_y g_y) + H_x + H_y - H\Delta g = -\epsilon \Delta H.$$

First, letting $U(x, y, \epsilon) = \sum_{k=0}^{\infty} U_k(x, y)\epsilon^k$, and proceeding formally, we obtain

$$U_0(x, y) = d(1 + x - y) + \int_1^y D(r - y + x, r)\, dr$$

and, successively, for $k \geqslant 1$,

$$U_k(x, y) = -\int_1^y \Delta U_{k-1}(r - y + x, r)\, dr,$$

where $\Delta U_j(r - y + x, r)$ represents $\Delta U_j(x, y)$ evaluated at $x = r - y + x$ and $y = r$.

Since $x = 0$ is not a characteristic of the nonlinear equation for g, its unique solution is $g = x$. Letting $H(x, y, \epsilon) = \sum_{k=0}^{\infty} h_k(x, y)\epsilon^k$, then, we ask that

$$-H_x + H_y = -\epsilon \Delta H$$

and obtain

$$h_0(x, y) = a(x + y) - U_0(0, x + y)$$

and, successively, for $k \geqslant 1$,

$$h_k(x, y) = -U_k(0, x + y) + \int_0^x \Delta h_{k-1}(t, x + y - t)\, dt.$$

Levinson's theorem, then, implies

Corollary 1. *Let $u(x, y)$ solve the first boundary value problem for*

$$\epsilon \Delta u + u_x + u_y = D(x, y)$$

in the unit square T. For each $N \geqslant 1$,

$$u(x, y) = \sum_{k=0}^{N} U_k(x, y)\, \epsilon^k + e^{-x/\epsilon} \sum_{k=0}^{N} h_k(x, y)\, \epsilon^k + O(\epsilon^{N+1/2})$$

uniformly in each regular quadrilateral

$$Q = \{(x, y) : (x, y) \in R \cup S, \quad 0 < \delta_1 \leqslant y - x \leqslant 1 - \delta_2 < 1\}$$

where the coefficients $U_k(x, y)$ and $h_k(x, y)$ are determined above.

Note. The analogous procedure can be used to construct the asymptotic solution in any regular quadrilateral

$$\tilde{Q} = \{(x, y) : (x, y) \in R \cup S, \quad 0 < \delta_1 \leqslant x - y \leqslant 1 - \delta_2 < 1\}$$

below the line $x = y$. Behavior along this line, except at $x = 0$ and $x = 1$, can also be obtained (see Eckhaus and de Jager [24]).

A second example is the equation

$$\epsilon \Delta u + u_x = D(x, y) \tag{7.5}$$

in the unit square T with smooth boundary values (7.4) prescribed. This problem was first considered, using different methods, in Wasow [110]. Here, the reduced equation,

$$u_x = D(x, y)$$

has the set of horizontal lines as characteristic curves. In particular, the boundary curves along $y = 0$ and $y = 1$ are characteristics.

Proceeding as usual, we obtain

Corollary 2. *Let $u(x, y)$ solve the equation*

$$\epsilon \Delta u + u_x = D(x, y)$$

in the unit square T with smooth boundary values

$$u(0, y) = a(y),$$
$$u(1, y) = b(y),$$
$$u(x, 0) = c(x), \qquad \text{and}$$
$$u(x, 1) = d(x).$$

For each $N \geqslant 0$, then,

$$u(x, y) = \sum_{k=0}^{N} U_k(x, y)\, \epsilon^k + e^{-x/\epsilon} \sum_{k=0}^{N} h_k(x, y)\, \epsilon^k + O(\epsilon^{\ +1/2})$$

uniformly in any regular quadrilateral

$$Q = \{(x, y):\ 0 \leqslant x \leqslant 1,\ \ 0 < \delta_1 \leqslant y \leqslant 1 - \delta_2 < 1\}$$

where

$$U_0(x, y) = b(y) + \int_1^x D(s, y)\, ds,$$
$$h_0(x, y) = a(y) - U_0(0, y),$$

and, successively, for $k \geqslant 1$

$$U_k(x, y) = -\int_1^x \Delta U_{k-1}(s, y)\, ds$$

and

$$h_k(x, y) = -U_k(0, y) + \int_0^x \Delta h_{k-1}(s, y)\, ds.$$

Thus the asymptotic solution $u(x, y)$ is completely determined throughout the interior of T and along those portions of the boundary S which are not characteristics of the reduced equation. The complete asymptotic behavior along the lines $y = 0$ and $y = 1$ remains unknown,

AN EXTENSION OF LEVINSON'S RESULTS TO TWO-PARAMETER PROBLEMS

In attempting to extend the methods outlined in Levinson [66], we obtain, for example,

Theorem 7.2. *Let $u(x, y)$ solve the first boundary value problem for*

$$\epsilon \Delta u - \mu A(x, y)\, u_x - C(x, y)\, u = 0 \tag{7.6}$$

in the unit square T where A and C are positive and ϵ and μ are small, positive, interrelated parameters. Further, let A, C, and the boundary values prescribed be sufficiently smooth and let $\epsilon/\mu^2 \to 0$ as $\mu \to 0$.

Then

$$u(x, y) \sim z_1\left(x, y, \frac{\epsilon}{\mu}, \frac{\epsilon}{\mu^2}\right) + z_2\left(x, y, \frac{\epsilon}{\mu}, \frac{\epsilon}{\mu^2}\right)$$

uniformly in any horizontal strip

$$Q : \{(x, y)\colon\ 0 \leqslant x \leqslant 1,\ \ 0 < \delta_1 \leqslant y \leqslant 1 - \delta_2 < 1\},$$

where

$$z_1\left(x, y, \frac{\epsilon}{\mu}, \frac{\epsilon}{\mu^2}\right) = h_1\left(x, y, \frac{\epsilon}{\mu}, \frac{\epsilon}{\mu^2}\right) \exp\left[-\frac{1}{\mu} g_1\left(x, y, \frac{\epsilon}{\mu^2}\right)\right]$$

and

$$z_2\left(x, y, \frac{\epsilon}{\mu}, \frac{\epsilon}{\mu^2}\right) = \begin{cases} h_2\left(x, y, \frac{\epsilon}{\mu}, \frac{\epsilon}{\mu^2}\right) \exp\left[-\frac{\mu}{\epsilon} g_2\left(x, y, \frac{\epsilon}{\mu^2}\right)\right] \\ \qquad\qquad \text{in a neighborhood of } x = 1 \\ O(e^{-\mu\delta/\epsilon}) \quad \text{for some } \delta > 0 \text{ elsewhere in } Q. \end{cases}$$

The functions g_1 and g_2 are represented by power series in ϵ/μ^2, while h_1 and h_2 are represented by double power series in ϵ/μ and ϵ/μ^2. Their variable coefficients can be calculated successively by formal substitution into Eq. (7.6) with the additional conditions

$$h_1\left(0, y, \frac{\epsilon}{\mu}, \frac{\epsilon}{\mu^2}\right) = u(0, y), \quad g_1\left(0, y, \frac{\epsilon}{\mu^2}\right) = 0, \quad g_1 > 0$$

elsewhere in T,

$$h_2\left(1, y, \frac{\epsilon}{\mu}, \frac{\epsilon}{\mu^2}\right) = u(1, y), \quad g_2\left(1, y, \frac{\epsilon}{\mu^2}\right) = 0, \quad \text{and} \quad g_2 > 0$$

elsewhere.

Specifically, if we define z_{1N} and z_{2N} by terminating the expansions for h_1 and h_2, respectively, after all terms of order $\leqslant N$,

$$u(x, y) = z_{1N}\left(x, y, \frac{\epsilon}{\mu}, \frac{\epsilon}{\mu^2}\right) + z_{2N}\left(x, y, \frac{\epsilon}{\mu}, \frac{\epsilon}{\mu^2}\right) + O\left(\mu\left(\frac{\epsilon}{\mu^2}\right)^{N+1}\right)$$

uniformly in Q.

Note. (1) We do not attempt a boundary layer analysis in the neighborhoods of the lines $y = 0$ and $y = 1$ since they are characteristics of the "semi-reduced" or "intermediate" equation

$$\mu A(x, y)\, u_x - C(x, y)\, u = 0.$$

(2) General results for equations of the form

$$\epsilon(\Delta u + a(x, y)\, u_x + b(x, y)\, u_y + c(x, y)\, u) + \mu(e(x, y)\, u_x + f(x, y)\, u_y + g(x, y)\, u) + h(x, y)\, u = r(x, y)$$

for h negative and ϵ and μ small, positive, interrelated parameters simultaneously approaching zero are reported in O'Malley [84] for the Dirichlet problem and O'Malley [87] for the Robin problem.

An Example where the Reduced Equation is of a Different Type

Latta [63] considers the first boundary value problem for the equation

$$\epsilon u_{yy} + u_{xx} - u_y = 0 \tag{7.7}$$

in fairly general domains of the x–y plane. Note that this differential equation is elliptic for $\epsilon > 0$ and its first boundary value problem then admits a unique solution, but the reduced equation

$$u_{xx} - u_y = 0$$

is parabolic. Restricting attention to the unit square T, let smooth boundary values be prescribed. In the event that the solution of the boundary value problem converges to a solution of the reduced equation, we can anticipate boundary layer behavior in the vicinity of the upper edge of T since the solution of the reduced equation is uniquely determined throughout T by the boundary values assumed on the other three edges.

Thus, we attempt to determine an asymptotic solution $u(x, y, \epsilon)$ of the form

$$u(x, y, \epsilon) = A(x, y, \epsilon) + B(x, y, \epsilon) \exp\left(-\frac{g(x, y)}{\epsilon}\right)$$

where A and B represent power series in ϵ with variable coefficients and $g(x, y)$ has the value zero along the line $y = 1$, but is positive elsewhere in T. Formally substituting this expression into (7.7), we have

$$\begin{aligned}\epsilon A_{yy} &+ A_{xx} - A_y \\ &+ \frac{\exp(-g(x, y)/\epsilon)}{\epsilon^2}[Bg_x^{\,2} + \epsilon(-2B_x g_x - Bg_{xx} + Bg_y^{\,2} + Bg_y) \\ &+ \epsilon^2(B_{xx} - 2B_y g_y - Bg_{yy} - B_y) + \epsilon^3 B_{yy}] = 0.\end{aligned}$$

Setting $A(x, y, \epsilon) = \sum_{k=0}^{\infty} a_k(x, y)\epsilon^k$, then, we ask that the coefficients a_k satisfy

$$a_{kxx} - a_{ky} = -a_{k-1,yy}$$

where $a_{-1} \equiv 0$ and the boundary values $a_0(x, y) = u(x, y)$ and $a_k(x, y) = 0$ for $k \geqslant 1$ are prescribed on the lower base and the vertical sides of T. (This assumes that bounded derivatives $a_{k-1,yy}$ exist throughout T.) Thus, the a_k's can be uniquely determined successively, being solutions of well-posed boundary value problems for nonhomogeneous parabolic equations.

Likewise, asking that

$$Bg_x^{\,2} = 0$$

for nonzero B, we set $g_x = 0$. Then setting

$$-2B_x g_x - Bg_{xx} + Bg_y^{\,2} + Bg_y = Bg_y(g_y + 1) = 0,$$

with $g(x, 1) = 0$, we obtain

$$g(x, y) = 1 - y.$$

Proceeding, we set

$$B_{xx} - 2B_y g_y - Bg_{yy} - B_y + \epsilon B_{yy} = B_{xx} + \epsilon B_{yy} + B_y = 0$$

where

$$B(x, y, \epsilon) = \sum_{k=0}^{\infty} b_k(x, y)\,\epsilon^k.$$

The coefficients b_k can be uniquely obtained by solving the following well-posed boundary value problems successively:

$$b_{0xx} + b_{0y} = 0$$

with

$$b_0(x, 1) = u(x, 1) - a_0(x, 1)$$

and

$$b_0(x, y) = 0 \quad \text{on the vertical sides of } T$$

and for $k \geqslant 1$, assuming that $b_{k-1,yy}$ exists and is bounded throughout T,

$$b_{kxx} + b_{ky} = -b_{k-1,yy}$$

with

$$b_k(x, 1) = -a_k(x, 1)$$

and

$$b_k(x, y) = 0 \quad \text{on the vertical sides of } T.$$

Applying an appropriate maximum principle argument, Latta is able to prove:

Theorem 7.3. *Let $u(x, y)$ solve the first boundary value problem for the equation*

$$\epsilon u_{yy} + u_{xx} - u_y = 0$$

in the unit square T. The, for each $N \geqslant 1$,

$$u(x, y) = \sum_{k=0}^{N} a_k(x, y)\, \epsilon^k + \left(\sum_{k=0}^{N} b_k(x, y)\, \epsilon^k\right) \exp\left(-\frac{(1-y)}{\epsilon}\right) + O(\epsilon^{N+1})$$

as $\epsilon \to 0$ uniformly in T where $a_k(x, y)$ and $b_k(x, y)$ are the coefficients determined above.

Note. (1) In Latta [63], the error term for the general problems considered is proved to be $O(\epsilon^{N+1-\delta})$ for any $\delta > 0$. Later, in unpublished work, Latta has shown that the error is, in fact, $O(\epsilon^{N+1})$ for the problem considered here.

(2) Zlamal [119] considers more general parabolic equations as the limiting case of certain elliptic equations. He carefully analyzes difficulties arising because the existence of bounded derivatives of the formally constructed asymptotic solution is not assured.

(3) Long ago, Hadamard [40] showed how to obtain the fundamental solution of the heat equation as the limit of the Riemann function for a hyperbolic equation involving a small parameter in such a manner that the heat equation is its limiting form. Likewise, Zlamal [120] (and certain previous papers) treats more general parabolic equations as the limit of certain hyperbolic equations, using the Fourier transform method to show asymptotic convergence. Bobisud [5] considers analogous problems for systems of partial differential equations of the first order in the "time" variable t. Recently, several authors (see, e.g., Lions [70], [71], and Kohn and Nirenberg [60]) have utilized results for singularly perturbed elliptic equations as a smoothing technique. This method is known as elliptic regularization.

A Problem for Oseen Flow

Latta [63] also considers Oseen flow for a viscous incompressible fluid past a semi-infinite flat plate at zero angle of attack. Letting the plate coincide with the positive x-axis, $0 \leqslant x < \infty$, in the x–y plane, the equations of motion are

$$\begin{cases} u_x + p_x = \nu \Delta u, \\ v_x + p_y = \nu \Delta v, \\ u_x + v_y = 0, \end{cases} \tag{7.8}$$

where ν is the coefficient of viscosity, assumed to be small and positive, and u and v represent the perturbation components of the flow velocity. (In these equations, the constant free-stream velocity, or, equivalently, the flow velocity at upstream infinity, and the density are both normalized to have the value one.) The boundary conditions are that

$$u = u_0 \quad \text{and} \quad v = 0 \quad \text{along the plate,}$$

while

$$u = v = p = 0 \quad \text{at upstream infinity.}$$

Thus, u_0 represents the relative velocity of the plate with respect to the constant free stream velocity. For a fixed plate, then, $u_0 = -1$.

Making the change to parabolic coordinates (the "optimal coordinates," as later defined in Kaplan [54]), we set $z = x + iy = w^2 = (\xi + i\eta)^2$, so that the z-plane corresponds to $\eta \geqslant 0$ and the plate lies along the

ξ-axis, $\eta = 0$. In (ξ, η) coordinates, then, the equations of motion are rewritten:

$$\nu \Delta u - 2\xi u_\xi + 2\eta u_\eta - 2\xi p_\xi + 2\eta p_\eta = 0,$$
$$\nu \Delta v - 2\xi v_\xi + 2\eta v_\eta - 2\xi p_\eta - 2\eta p_\xi = 0,$$
$$\xi u_\xi - \eta u_\eta + \xi v_\eta + \eta v_\xi = 0.$$

The classical boundary layer analysis (see, e.g., Goldstein [37]) predicts on physical grounds the boundary layer thickness to be of order $\sqrt{\nu}$. Introducing ξ and $\sigma = \eta/\sqrt{\nu}$, then, as the independent variables in the boundary layer, the resulting boundary layer equations have the asymptotic solutions

$$u \sim u_0 \operatorname{erfc} \sigma,$$
$$v \sim 0,$$
$$p \sim 0,$$

as $\nu \to 0$ where

$$\operatorname{erfc} \sigma = \frac{2}{\sqrt{\pi}} \int_\sigma^\infty e^{-t^2}\, dt.$$

Thus, in an attempt to obtain a uniformly valid asymptotic solution of the equations of motion for all ξ and η, we introduce the boundary layer stretching coordinate $g(\xi, \eta)/\nu$ where g is zero along the plate, i.e., when $\eta = 0$, but g is positive elsewhere. Then, we formally set

$$u = u_1(\xi, \eta) \operatorname{erfc} \sqrt{\frac{g(\xi, \eta)}{\nu}},$$
$$+ \sqrt{\nu}\left(u_2(\xi, \eta, \sqrt{\nu}) + u_3(\xi, \eta, \sqrt{\nu}) \exp\left(-\frac{g(\xi, \eta)}{\nu}\right)\right),$$
$$v = \sqrt{\nu}\left(v_2(\xi, \eta, \sqrt{\nu}) + v_3(\xi, \eta, \sqrt{\nu}) \exp\left(-\frac{g(\xi, \eta)}{\nu}\right)\right),$$

and

$$p = \sqrt{\nu}\left(p_2(\xi, \eta, \sqrt{\nu}) + p_3(\xi, \eta, \sqrt{\nu}) \exp\left(-\frac{g(\xi, \eta)}{\nu}\right)\right),$$

where, as usual, u_2, u_3, v_2, v_3, p_2 and p_3 represent power series in $\sqrt{\nu}$ with variable coefficients. Note that $\operatorname{erfc} \sqrt{g(\xi, \eta)/\nu}$ and $\exp(-g(\xi, \eta)/\nu)$ are both exponentially small as $\nu \to 0$ away from the boundary $\eta = 0$.

Proceeding formally and applying the boundary conditions, we obtain

$$g(\xi, \eta) = \eta^2$$

and the solutions

$$u = u_0 \operatorname{erfc} \frac{\eta}{\sqrt{\nu}} + \sqrt{\nu} \left(\frac{u_0 \eta}{\sqrt{\pi} (\xi^2 + \eta^2)} \right) (1 - e^{-\eta^2/\nu}),$$

$$v = - \sqrt{\nu} \left(\frac{u_0 \xi}{\sqrt{\pi} (\xi^2 + \eta^2)} \right) (1 - e^{-\eta^2/\nu}),$$

$$p = - \sqrt{\nu} \frac{u_0 \eta}{\sqrt{\pi} (\xi^2 + \eta^2)} .$$

These formally determined solutions are exact for all values of ν, so, *à fortiori*, are also the asymptotic solutions desired as $\nu \to 0$.

Note. (1) This solution was also obtained in Lewis and Carrier [68] using Wiener–Hopf techniques.

(2) The form of solution suggests that one could have proceeded, without the benefit of physical reasoning, by formally expanding u, v, and p in the form

$$\sum_{k=1}^{\infty} f(\xi, \eta, \zeta) \, \nu^{k/2}$$

where $\zeta = g(\xi, \eta)/\nu$ is treated as a new independent variable, in the manner of Cochran [14] and O'Malley [85].

References

1. R. Bellman, "Perturbation Techniques in Mathematics, Physics, and Engineering." Holt, Rinehart and Winston, New York, 1964.
2. G. D. Birkhoff, On the asymptotic character of the solutions of certain linear differential equations containing a parameter. *Trans. Am. Math. Soc.* **9** (1908), 219–231.
3. G. D. Birkhoff, Quantum mechanics and asymptotic series. *Bull. Am. Math. Soc.* **39** (1933), 681–700.
4. L. Bobisud, On the single first-order partial differential equation with a small parameter. *SIAM Rev.* **8** (1966), 479–493.
5. L. Bobisud, Degeneration of the solutions of certain well-posed systems of partial differential equations depending on a small parameter. *J. Math. Anal. Appl.* **16** (1966), 419–454.
6. L. Bobisud, Second-order linear parabolic equations with a small parameter. *Arch. Rat. Mech. Anal.* **27** (1968), 385–397.

7. L. Bobisud, The second initial-boundary value problem for a linear parabolic equation with a small parameter (to be published).
8. W. E. Boyce and G. H. Handelman, Vibrations of rotating beams with tip mass. *Z. Angew. Math. Phys.* **12** (1961), 369–392.
9. P. Brock, The nature of solutions of a Rayleigh type forced vibration equation with a large coefficient of damping. *J. Appl. Phys.* **24** (1953), 1004–1007.
10. G. F. Carrier, Boundary Layer Problems in Applied Mechanics, *in* "Advances in Applied Mechanics III," pp. 1–19. Academic Press, New York, 1953.
11. G. F. Carrier, Boundary layer problems in applied mathematics. *Commun. Pure Appl. Math.* **7** (1954), 11–17.
12. M. L. Cartwright, Forced oscillations in nonlinear systems. *Contrib. Theory Nonlinear Oscillations* **1** (1950), 149–241.
13. M. L. Cartwright, Van der Pol's equation for relaxation oscillations. *Contrib. Theory Nonlinear Oscillations* **2** (1952), 3–8.
14. J. A. Cochran, Problems in Singular Perturbation Theory, unpublished Doctoral dissertation, Stanford University, 1962.
15. J. A. Cochran, On the uniqueness of solutions of linear differential equations *J. Math. Anal. Appl.* **22** (1968), 418–426.
16. E. A. Coddington and N. Levinson, A boundary value problem for a nonlinear differential equation with a small parameter. *Proc. Amer. Math. Soc.* **3** (1952), 73–81.
17. E. A. Coddington and N. Levinson, "Theory of Ordinary Differential Equations." McGraw-Hill, New York, 1955.
18. J. D. Cole and J. Kevorkian, Uniformly Valid Asymptotic Expansions for Certain Nonlinear Differential Equations, *in* "Nonlinear Differential Equations and Nonlinear Mechanics," pp. 113–120. Academic Press, New York, 1963.
19. E. T. Copson, "Asymptotic Expansions." Cambridge University Press, London, 1965.
20. J. G. van der Corput, Asymptotic developments I. Fundamental theorems of asymptotics. *J. Anal. Math.* **4** (1956), 341–418.
21. R. Courant and D. Hilbert, "Methods of Mathematical Physics II, Partial Differential Equations." Interscience, New York, 1962.
22. R. B. Davis, Asymptotic solutions of the first boundary value problem for a fourth-order elliptic partial differential equation. *J. Ratl. Mech. Anal.* **5** (1956), 605–620.
23. A. A. Dorodnicyn, Asymptotic solution of van der Pol's equation. *Prikl. Mat. Meh.* **11** (1947), 313–328. [*Am. Math. Soc. Transl.*, No. 88, 1953.]
24. W. Eckhaus and E. M. de Jager, Asymptotic solutions of singular perturbation problems for linear differential equations of elliptic type. *Arch. Rational Mech. Anal.* **23** (1966), 26–86.
25. A. Erdélyi, An expansion procedure for singular perturbations. *Atti Acad. Sci. Torino, Cl. Sci. Fis. Mat. Natur.* **95** (1961), 651–672.
26. A. Erdélyi, On a nonlinear boundary value problem involving a small parameter. *J. Australian Math. Soc.* **2** (1962), 425–439.
27. A. Erdélyi, Singular perturbations of boundary value problems involving ordinary differential equations. *J. Soc. Indust. Appl. Math.* **11** (1963), 105–116.
28. A. Erdélyi, The Integral Equations of Asymptotic Theory, *in* "Asymptotic Solutions of Differential Equations and Their Applications," pp. 211–229. Wiley, New York, 1964.

29. A. Erdélyi, Two-variable expansions for singular perturbations, *J. Inst. Math. Appl.* **4** (1968), 113–119.
30. D. A. Flanders and J. J. Stoker, The Limit Case of Relaxation Oscillations, *in* "Studies in Nonlinear Vibration Theory," pp. 50–64. New York University, New York, 1946.
31. L. E. Fraenkel, On the method of matched asymptotic expansions, Parts I–III (to be published).
32. B. Friedman, "Principles and Techniques of Applied Mathematics." Wiley, New York, 1956.
33. B. Friedman, Singular Perturbations of Ordinary Differential Equations, Mathematical Note No. 245, Mathematics Research Laboratory, Boeing Scientific Research Laboratories, Seattle, 1961.
34. B. Friedman, Singular Perturbations of Differential Equations II, Mathematical Note No. 275, Mathematics Research Laboratory, Boeing Scientific Research Laboratories, Seattle, 1962.
35. K. O. Friedrichs and W. Wasow, Singular perturbations of nonlinear oscillations. *Duke Math. J.* **13** (1946), 367–381.
36. K. O. Friedrichs, Asymptotic phenomena in mathematical physics. *Bull. Am. Math. Soc.* **61** (1955), 485–504.
37. S. Goldstein (Ed.), "Modern Developments in Fluid Dynamics," Vol. 1. Clarendon Press, Oxford, England, 1938.
38. J. Haag, Étude asymptotique des oscillations de relaxation. *Ann. Sci. École Norm. Sup.* **60** (1943), 35–64; 65–111.
39. J. Haag, Exemples concrets d'étude asymptotique d'oscillations de relaxation. *Ann. Sci. École Norm. Sup.* **61** (1944), 73–117.
40. J. Hadamard, "Lectures on Cauchy's Problem in Linear Partial Differential Equations." Reprinted: Dover, New York, 1952 (original ed. 1923).
41. G. H. Handelman and J. B. Keller, Small Vibrations of a Slightly Stiff Pendulum, *in* "Proceedings of the 4th U.S. National Congress on Applied Mechanics", Vol. 1, pp. 195–202. American Society Mechanical Engineers, New York, 1962.
42. G. H. Handelman, J. B. Keller, and R. E. O'Malley, Jr., Loss of boundary conditions in the asymptotic solution of linear ordinary differential equations I. Eigenvalue problems. *Commun. Pure Appl. Math.* **21** (1968), 243–261.
43. W. A. Harris, Jr., Singular perturbations of two-point boundary problems for systems of ordinary differential equations. *Arch. Ratl. Mech. Anal.* **5** (1960), 212–225.
44. W. A. Harris, Jr., Singular perturbations of eigenvalue problems. *Arch. Ratl. Mech. Anal.* **7** (1961), 224–241.
45. W. A. Harris, Jr., Singular perturbations of a boundary value problem for a system of differential equations. *Duke Math. J.* **29** (1962), 429–445.
46. W. A. Harris, Jr., Singular perturbations of two-point boundary problems. *J. Math. Mech.* **11** (1962), 371–382.
47. W. A. Harris, Jr., Equivalent classes of singular perturbation problems. *Rendi. Circ. Mat. Palermo* **14** (1965), 61–75.
48. F. C. Hoppensteadt, Singular perturbations on the infinite interval. *Trans. Am. Math. Soc.* **123** (1966), 521–535.
49. F. Hoppensteadt, "Stability in Systems with Parameter." *J. Math. Anal. Appl.* **18** (1967), 129–134.

50. D. HUET, Phénomènes de perturbation singulière dans les problèmes aux limites. *Ann. Inst. Fourier, Grenoble* **10** (1960), 1–96.
51. D. HUET, Perturbations singulières relatives au problème de Dirichlet dans un demi-espace. *Ann. Scuola Norm. Sup. Pisa* **18** (1964), 427–448.
52. D. HUET, Sur quelques problèmes de perturbation singulière dans les espaces L_p. *Rev. Fac. Ci. Lisboa* **11** (1965), 137–164.
53. D. HUET, Remarque sur un théorème d'Agmon et applications à quelques problèmes de perturbation singulière. *Boll. Un. Mat. Ital.* **21** (1966), 219–227.
54. S. KAPLUN, The role of coordinate systems in boundary layer theory. *Z. Angew. Math. Phys.* **5** (1954), 111–135.
55. S. KAPLUN, Low Reynolds number flow past a circular cylinder. *J. Math. Mech.* **6** (1957), 595–603.
56. S. KAPLUN, "Fluid Mechanics and Singular Perturbations" (P. A. Lagerstrom, L. N. Howard, and C. S. Liu, Eds.). Academic Press, New York, 1967.
57. S. KAPLUN AND P. A. LAGERSTROM, Asymptotic expansions of Navier–Stokes solutions for small Reynolds numbers. *J. Math. Mech.* **6** (1957), 585–593.
58. J. KEVORKIAN, The Two-Variable Expansion Procedure for the Approximate Solution of Certain Nonlinear Differential Equations, Douglas Report SM-42620. Missile and Space Systems Division, Douglas Aircraft Company, Santa Monica, California, 1962.
59. J. KISYŃSKI, Sur les équations hyperboliques avec petit paramètre. *Colloq. Math.* **10** (1963), 331–343.
60. J. J. KOHN AND L. NIRENBERG, Noncoercive boundary value problems. *Commun. Pure Appl. Math.* **18** (1965), 443–492.
61. P. A. LAGERSTROM, Note on the preceding two papers. *J. Math. Mech.* **6** (1957), 605–606.
62. J. P. LASALLE AND S. LEFSCHETZ, "Stability by Liapunov's Direct Method, with Applications." Academic Press, New York, 1961.
63. G. E. LATTA, Singular Perturbation Problems, unpublished Doctoral dissertation, California Institute of Technology, Pasadena, 1951.
64. J. J. LEVIN, First order partial differential equations containing a small parameter. *J. Rational Mech. Anal.* **4** (1955), 481–501.
65. N. LEVINSON, Perturbations of discontinuous solutions of nonlinear systems of differential equations. *Proc. Natl. Acad. Sci. U.S.* **33** (1947), 214–218.
66. N. LEVINSON, The first boundary value problem for $\epsilon\Delta u + A(x, y)u_x + B(x, y)u_y + C(x, y)u = D(x, y)$ for small ϵ. *Ann. Math.* **51** (1950), 428–445.
67. N. LEVINSON, "Perturbations of Discontinuous Solutions of Nonlinear Systems of Differential Equations." *Acta Math.* **82** (1951), 71–106.
68. J. A. LEWIS AND G. F. CARRIER, Some remarks on the flat plate boundary layer. *Quart. Appl. Math.* **7** (1949), 228–234.
69. A. LIÉNARD, Étude des oscillations entretenues. *Rev. Gen. Élec.* **23** (1928), 901–912; 946–954.
70. J. L. LIONS, Singular Perturbations and some Nonlinear Boundary Value Problems. Tech. Summary Rept. No. 421, Mathematics Research Center, U.S. Army, University of Wisconsin, October 1963.
71. J. L. LIONS, Sur certaines équations paraboliques non linéaires. *Bull. Soc. Math. France* **93** (1965), 155–175.
72. J. W. MACKI, Singular perturbations of a boundary value problem for a system of

nonlinear ordinary differential equations. *Arch. Ratl. Mech. Anal.* **24** (1967), 219–232.

73. J. C. P. Miller, "The Airy Integral" [British Association for Advancement of Science, Mathematical Tables, Part-Volume B]. Cambridge University Press, Cambridge, England, 1946.
74. W. L. Miranker, Singular perturbation eigenvalues by a method of undetermined coefficients. *J. Math. Phys.* **42** (1963), 47–58.
75. E. F. Mishchenko, Asymptotic calculation of periodic solutions of systems of differential equations containing parameters in the derivatives. *Izv. Akad. Nauk SSSR Ser. Mat.* **21** (1957), 627–654. [*Am. Math. Soc. Transl., Ser. 2* **18** (1961), 199–230.]
76. E. F. Mishchenko and L. S. Pontryagin, Differential equations with a small parameter attached to the highest derivatives and some problems in the theory of oscillation. *IRE Trans. Circuit Theory* **7** (1960), 527–535.
77. J. Moser, Singular perturbation of eigenvalue problems for linear differential equations of even order." *Commun. Pure Appl. Math.* **8** (1955), 251–278.
78. W. D. Murphy, Numerical Analysis of Boundary Layer Problems. Courant Institute of Mathematical Sciences, AEC Computing and Applied Mathematics Center, TID-4500, NYO-1480-63, 1966.
79. P. Noaillon, Développements asymptotiques dans les équations différentielles linéaires à paramètre variable. *Mém. Soc. Sci. Liége, Ser. 3* **9** (1912), 197.
80. O. A. Oleinik, On equations of elliptic type with a small parameter in the highest derivatives. *Mat. Sb.* **31** (1952), 104–117.
81. R. E. O'Malley, Jr., Two-Parameter Singular Perturbation Problems, unpublished Doctoral dissertation, Stanford University, Stanford, California, 1965.
82. R. E. O'Malley, Jr., Two-parameter singular perturbation problems for second order equations. *J. Math. Mech.* **16** (1967), 1143–1164.
83. R. E. O'Malley, Jr., Singular perturbations of boundary value problems for linear ordinary differential equations involving two parameters. *J. Math. Anal. Appl.* **19** (1967), 291–308.
84. R. E. O'Malley, Jr., The first boundary value problem for certain linear elliptic differential equations involving two small parameters. *Arch. Ratl. Mech. Anal.* **26** (1967), 68–82.
85. R. E. O'Malley, Jr., A boundary value problem for certain nonlinear second order differential equations with a small parameter. *Arch. Rational Mech. Anal.* **29** (1968), 66–74.
86. R. E. O'Malley, Jr., A nonlinear singular perturbation problem arising in the study of chemical flow reactors (to be published).
87. R. E. O'Malley, Jr., The third boundary value problem for certain linear elliptic differential equations involving two small parameters. *Arch. Rational Mech. Anal.* (to appear).
88. R. E. O'Malley, Jr. and J. B. Keller, Loss of boundary conditions in the asymptotic solution of linear ordinary differential equations, II, boundary value problems *Commun. Pure Appl. Math.* **21** (1968), 263–270.
89. L. S. Pontryagin, "Asymptotic Behavior of the Solutions of Differential Equations With a Small Parameter in the Higher Derivatives." *Izv. Akad. Nauk SSSR Ser. Mat.* **21** (1959), 605–626. [*Am. Math. Soc. Transl., Ser.* 2 **18** (1961), 295–319].

90. P. J. Ponzo and N. Wax, On certain relaxation oscillations: asymptotic solutions. *J. Soc. Indust. Appl. Math.* **13** (1965), 740–766.
91. L. Prandtl, Über Flüssigkeits-bewegung bei kleiner Reibung, *Verhandlungen des III. Internationalen Mathematiker-Kongresses*, 484–491. Tuebner, Leipzig, 1905.
92. I. Proudman and J. R. A. Pearson, Expansions at small reynolds numbers for the flow past a sphere and a circular cylinder. *J. Fluid Mech.* **2** (1957), 237–262.
93. Lord Rayleigh (J. W. Strutt), "The Theory of Sound," Vol. 1, (2nd ed.). Reprinted: Dover, New York, 1945 (original ed. 1894).
94. F. Rellich, Störungstheorie der Spectralzerlegung, Article in five parts appearing in *Math. Ann.* (1937–1942) as follows: Part I **113**, 600–619; Part II **113**, 677–685; Part III **116**, 555–570; Part IV **117**, 356–382; Part V **118**, 462–484.
95. L. A. Segel, The importance of asymptotic analysis in applied mathematics. *Am. Math. Monthly* **73** (1966), 7–14.
96. Y. Sibuya, Sur réduction analytique d'un système d'équations différentielles ordinaires linéaires contenant un paramètre. *J. Fac. Sci. Univ. Tokyo, Sec. I* **7** (1958), 527–540.
97. J. A. Smoller, Singular perturbations of Cauchy's problem. *Commun. Pure Appl. Math.* **18** (1965), 665–677.
98. J. J. Stoker, "Nonlinear Vibrations in Mechanical and Electrical Systems." Interscience, New York, 1950.
99. A. N. Tikhonov, Systems of differential equations containing small parameters in the derivatives. *Mat. Sb.* **31** (1952), 575–586.
100. F. G. Tricomi, "Integral Equations." Interscience, New York, 1957.
101. H. L. Turrittin, Asymptotic solutions of certain ordinary differential equations associated with multiple roots of the characteristic equation. *Am. J. Math.* **58** (1936), 364–378.
102. H. L. Turrittin, Asymptotic expansions of solutions of systems of ordinary linear differential equations containing a parameter. *Contrib. Theory Nonlinear Oscillations* **2** (1952), 81–116.
103. M. Urabe, Periodic solutions of van der Pol's equation with damping coefficient $\lambda = 0 \sim 10$. *IRE, Trans. Circuit Theory* **7** (1960), 382–386.
104. M. Urabe, Numerical Study of Periodic Solutions of the van der Pol Equation, *in* "International Symposium on Nonlinear Differential Equations and Nonlinear Mechanics," pp. 184–192. Academic Press, New York, 1963.
105. M. Van Dyke, "Perturbation Methods in Fluid Dynamics." Academic Press, New York, 1964.
106. A. B. Vasil'eva, Asymptotic behavior of solutions to certain problems involving nonlinear differential equations containing a small parameter multiplying the highest derivatives." *Uspekhi Mat. Nauk* **18** (1963), 15–86. [*Russian Math. Surveys* **18** (1963), 13–84.]
107. M. I. Visik and L. A. Lyusternik, Regular degeneration and boundary layer for linear differential equations with small parameter. *Uspekhi Mat. Nauk* **12** (1957), 3–122. [*Am. Math. Soc. Transl., Ser. 2* **20** (1961), 239–364.]
108. M. I. Visik and L. A. Lyusternik, Solution of some perturbation problems in the case of matrices and self-adjoint or non-self-adjoint differential equations I. *Uspekhi Mat. Nauk* **15** (1960), 3–80. [*Russian Math. Survey* **15** (1960), 1–73.]
109. W. Wasow, On the asymptotic solution of boundary value problems for ordinary differential equations containing a parameter. *J. Math. Phys.* **23** (1944), 173–183.

110. W. WASOW, Asymptotic solution of boundary value problems for the differential equation $\Delta U + \lambda(\partial U/\partial x) = \lambda f(x, y)$. *Duke Math. J.* **11** (1944), 405–415.
111. W. WASOW, On the construction of periodic solutions of singular perturbation problems. *Contrib. Theory Nonlinear Oscillations* **1** (1950), 313–350.
112. W. WASOW, Singular Perturbation Methods for Nonlinear Oscillations, *in* "Proceedings of the Symposium on Nonlinear Circuit Analysis," pp. 75–98. Edwards Brothers, Ann Arbor, Michigan, 1953.
113. W. WASOW, Singular perturbation of boundary value problems for nonlinear differential equations of the second order. *Commun. Pure Appl. Math.* **9** (1956), 93–113.
114. W. WASOW, Asymptotic Expansions for Ordinary Differential Equations: Trends and Problems, *in* "Asymptotic Solutions of Differential Equations and Their Applications," pp. 3–26. Wiley, New York, 1964.
115. W. WASOW, "Asymptotic Decomposition of Systems of Linear Differential Equations Having Poles with respect to a Parameter." *Rend. Circ. Mat. Palermo* **13** (1964), 329–344.
116. W. WASOW, "Asymptotic Expansions for Ordinary Differential Equations." Interscience, New York, 1965.
117. J. G. Wendell, Singular perturbations of a van der Pol equation. *Contrib. Theory Nonlinear Oscillations* **1** (1950), 243–290.
118. D. WILLETT, On a nonlinear boundary value problem with a small parameter multiplying the highest derivative. *Arch. Ratl. Mech. Anal.* **23** (1966), 276–287.
119. M. ZLÁMAL, The parabolic equation as a limiting case of a certain elliptic equation. *Ann. Mat. Pure Appl.* **4** (1962), 143–150.
120. M. ZLÁMAL, On a Singular Perturbation Problem Concerning Hyperbolic Equations, Lecture Series No. 45, Institute for Fluid Dynamics and Applied Mathematics, University of Maryland, College Park, Maryland, 1964.

ADDITIONAL BIBLIOGRAPHY ADDED IN PROOF

121. R. BELLMAN AND K. L. COOKE, On the limit of solutions of differential-difference equations as the retardation approaches zero, *Proc. Nat. Acad. Sci. U.S.* **45** (1959), 1026–1028.
122. L. BOBISUD, On the behavior of the solution of the telegraphist's equation for large velocities, *Pacific J. Math.* **22** (1967), 213–219.
123. G. F. CARRIER AND C. E. PEARSON, "Ordinary Differential Equations," Ginn-Blaisdell, Waltham, 1968.
124. K. W. CHANG, Almost periodic solutions of singularly perturbed systems of differential equations, *J. Diff. Eq.* **4** (1968), 300–307.
125. J. D. COLE, "Perturbation Methods in Applied Mathematics," Ginn-Blaisdell, Waltham, 1968.
126. C. COMSTOCK, Boundary layers for almost characteristic boundaries, *J. Math. Anal. Appl.* **22** (1968), 54–61.
127. K. L. COOKE, The condition of regular degeneration for singularly perturbed differential-difference equations, *J. Diff. Eq.* **1** (1965), 39–94.
128. K. L. COOKE AND K. R. MEYER, The condition of regular degeneration of singularly perturbed systems of linear differential-difference equations, *J. Math. Anal. Appl.* **14** (1966), 83–106.

129. W. Eckhaus, "On the Foundations of the Method of Matched Asymptotic Approximations," Mathematisch Instituut der Technische Hogeschool Delft, Nederland, July 1968.
130. A. Erdélyi, Approximate solutions of a nonlinear boundary value problem, *Arch. Rational Mech. Anal.* **29** (1968), 1–17.
131. P. C. Fife, Nonlinear deflection of thin elastic plates under tension, *Commun. Pure Appl. Math.* **14** (1961), 81–112.
132. P. C. Fife, Considerations regarding the mathematical basis for Prandtl's boundary layer theory, *Arch. Rational Mech. Anal.* **28** (1968), 184–216.
133. L. Flatto and N. Levinson, Periodic solutions of singularly perturbed systems, *J. Rational Mech. Anal.* **4** (1955), 943–950.
134. N. D. Fowkes, A singular perturbation method, *Quart. Appl. Math.* **26** (1968), 57–69 and 71–85.
135. A. Friedman, Singular perturbations for partial differential equations, *Arch. Rational Mech. Anal.* **29** (1968), 289–303.
136. W. M. Greenlee, "Rate of Convergence in Singular Perturbations," University of Kansas, Department of Mathematics, Technical Report 12, June 1967.
137. S. Haber and N. Levinson, A boundary value problem for a singularly perturbed differential equation, *Proc. Amer. Math. Soc.* **6** (1955), 866–872.
138. J. K. Hale and G. Seifert, Bounded and almost periodic solutions of singularly perturbed equations, *J. Math. Anal. Appl.* **3** (1961), 18–24.
139. F. G. Heineken, H. M. Tsuchiya, and R. Aris, On the mathematical status of the pseudo-steady state hypothesis of biochemical kinetics, *Math. Biosciences* **1** (1967), 95–113.
140. F. Hoppensteadt, On systems of ordinary differential equations with several parameters multiplying the derivatives (to be published).
141. F. Hoppensteadt, Asymptotic series solutions for nonlinear singular perturbation problems (to be published).
142. C. C. Hurd, Asymptotic theory of linear differential equations containing two parameters, *Tohoku Math. J.* **45** (1939), 58–68.
143. M. Iwano, Asymptotic solutions of Whittaker's equation as the moduli of the independent variable and two parameters tend to infinity, *Japanese J. Math.* 33 (1963), 1–92.
144. M. Iwano and Y. Sibuya, Reduction of order of a linear ordinary differential equation containing a small parameter, *Kodai Math. Sem. Rep.* **15** (1963), 1–28.
145. W. D. Kazarinoff, Asymptotic theory of second order differential equations with two simple turning points, *Arch. Rational Mech. Anal.* **2** (1958), 129–150.
146. J. B. Keller, "Perturbation Theory," lecture notes, Michigan State University, East Lansing, 1968.
147. J. K. Knowles, The Dirichlet problem for a thin rectangle, *Proc. Edinburgh Math. Soc.* **15** (1967), 315–320.
148. J. K. Knowles and R. E. Messick, On a class of singular perturbation problems, *J. Math. Anal. Appl.* **9** (1964), 42–58.
149. N. Levinson, A boundary value problem for a singularly perturbed differential equation, *Duke Math. J.* **25** (1958), 331–342.
150. C. C. Lin and A. L. Rabinstein, On the asymptotic solutions of a class of ordinary differential equations of the fourth order, *Trans. Amer. Math. Soc.* **94** (1960), 24–57.

151. D. B. MacMillan, Asymptotic methods for systems of differential equations in which some variables have very short response times, *SIAM J. Appl. Math.* **16** (1968), 704–722.
152. L. Markus and N. R. Amundson, Nonlinear boundary value problems arising in chemical reaction theory, *J. Diff. Eq.* **4** (1968), 102–113.
153. R. W. McKelvey, The solution of second order linear ordinary differential equations about a turning point of order two, *Trans. Amer. Math. Soc.* **79** (1955), 103–123.
154. R. E. Meyer, On the approximation of double limits by single limits and the Kaplun extension theorem, *J. Inst. Math. Appl.* **3** (1967), 245–249.
155. O. A. Oleinik, On the second boundary-value problem for elliptic equations with a small parameter before the highest derivative, *Dokl. Akad. Nauk SSSR* **79** (1951), 735–737.
156. R. E. O'Malley, Jr., Boundary value problems for linear systems of ordinary differential equations involving many small parameters, *J. Math. Mech.* (forthcoming).
157. R. E. O'Malley, Jr., "On a Boundary Value Problem for a Nonlinear Differential Equation with a Small Parameter," Mathematics Research Center, U. S. Army, The University of Wisconsin, TSR 831, December 1967.
158. R. E. O'Malley, Jr., "On Singular Perturbation Problems with Interior Non-uniformities," Mathematics Research Center, U. S. Army, The University of Wisconsin, TSR 864, April 1968.
159. R. E. O'Malley, Jr., "On the Asymptotic Solution of Boundary Value Problems for Nonhomogeneous Ordinary Differential Equations Containing a Parameter," Mathematics Research Center, U. S. Army, The University of Wisconsin, TSR 872, May 1968.
160. R. E. O'Malley, Jr., "On the Asymptotic Solution of Multi-Point Boundary Value Problems," Mathematics Research Center, U. S. Army, The University of Wisconsin, TSR 885, July 1968.
161. S. V. Parter, "Singular Perturbations of Second Order Differential Equations" (unpublished paper).
162. L. E. Payne and D. Sather, On singular perturbation in non-well posed problems, *Ann. Mat. Pura Appl.* (4) **75** (1967), 219–230.
163. L. M. Perko, A method of error estimation in singular perturbation problems with application to the restricted three body problem, *SIAM J. Appl. Math.* **15** (1967), 738–753.
164. P. J. Ponzo, Forced oscillations of the generalized Liénard equation, *SIAM J. Appl. Math.* **15** (1967), 75–87.
165. E. R. Rang, Periodic Solutions of Singular Perturbation Problems, in "Nonlinear Differential Equations and Nonlinear Mechanics," pp. 377–383. Academic Press, New York, (1963).
166. Y. Sibuya, Asymptotic solutions of a system of linear ordinary differential equations containing a parameter, *Funkcial. Ekvac.* **4** (1962), 83–113.
167. Y. Sibuya, "On the Convergence of Formal Solutions of Linear Ordinary Differential Equations Containing a Parameter," Mathematics Research Center, U. S. Army, The University of Wisconsin, TSR 511, September 1964.
168. C. R. Steele, On the asymptotic solution of nonhomogeneous ordinary differential equations with a large parameter, *Quart. Appl. Math.* **23** (1965), 193–201.

169. F. Stenger, Error bounds for asymptotic solutions of differential equations, *J. Res. Natl. Bur. Stds.* **70B** (1966), 167–210.
170. G. Stengle, Asymptotic solutions uniform in many asymptotic scales for Whittaker's differential equation as the independent variable and both parameters tend to infinity (to be published).
171. S. Sugiyama, Continuity properties of the retardation in the theory of difference-differential equations, *Proc. Japan Acad.* **37** (1961), 179–182.
172. K. Takahashi, Über eine Asymptotische Darstellung der Lösung eines Systems von Differentialgleichungen welche von zwei Parameter abhängen..., *Tôhoku Math. J.* **13** (1961), 1–17.
173. L. Ting and S. Chen, Perturbation solutions and asymptotic solutions in boundary layer theory, *J. Eng. Math.* **1** (1967), 327–340.
174. A. B. Vasil'eva and M. Imanaliev, The asymptotic form of the solution to the Cauchy problem for an integro-differential equation with a small parameter, *Sibirskii Mat. Z.* **7**(1966), 61–69.
175. A. B. Vasil'eva and V. M. Volosov, The work of Tikhonov and his pupils on ordinary differential equations containing a small parameter, *Uspehi Mat. Nauk* **22** (1967), No. 2, 149–168. (*Russian Math. Surveys* **22** (1967), No. 2, 124–142.)
176. A. B. Vasil'eva and V. A. Tupciev, Periodic nearly-discontinuous solutions of systems of differential equations with a small parameter in the derivatives, *Soviet Math. Dokl.* **9** (1968), No. 1, 179–183.
177. M. I. Visik and L. A. Lyusternik, On the asymptotic behavior of the solutions of boundary problems for quasi-linear differential equations, *Dokl. Akad. Nauk SSSR* **121** (1958), 778–781.
178. M. I. Visik and L. A. Lyusternik, Initial jump for nonlinear differential equations containing a small parameter, *Dokl. Akad. Nauk SSSR* **132** (1960), 1242–1245.
179. W. Wasow, "On Boundary Layer Problems in the Theory of Ordinary Differential Equations," unpublished thesis, New York University, December 1941.
180. W. Wasow, Singular perturbation problems of systems of two ordinary analytic differential equations, *Arch. Rational Mech. Anal.* **14** (1963), 61–80.
181. W. Wasow, Asymptotic simplification of self-adjoint differential equations with a parameter, *J. Diff. Eq.* **2** (1966), 378–390.
182. W. Wasow, On turning point problems for systems with almost diagonal coefficient matrix, *Funkcial. Ekvac.* **8** (1966), 143–171.
183. W. Wasow, Connection problems for asymptotic series (to be published).

Reprinted from *Advances in Mathematics* **3**, No. 4, 594–623, (1969).

Classification of Second Order Linear Differential Equations with Respect to Oscillation*

D. WILLETT

Department of Mathematics, The University of Utah, Salt Lake City, Utah

1. Introduction

Equations of the form

$$[r(t)\, y']' + q(t)\, y = 0 \tag{1.1}$$

where $r \in C^1[a, \infty)$, $r > 0$, and $q \in C[a, \infty)$, are classified by the behavior of their real solutions, as oscillatory or nonoscillatory. In the first instance, one, and thereby every, solution vanishes at an infinite number of isolated points in $[a, \infty)$; in the second instance each solution has only a finite number of zeros in $[a, \infty)$. By solution is always meant a function which is not identically zero. A special instance of nonoscillation is the disconjugate case in which every solution has at most one zero in $[a, \infty)$. Although there are many results concerning the classification of equations of the form (1.1) with respect to these properties, no completely satisfactory answer has yet been obtained. The purpose of this paper or survey is to identify the known results, to relate the new results and old

* This survey was begun while the author was at the 1967 Associated Western Universities Differential Equations Symposium at the University of Colorado. It was partially supported by NASA contract 45-003-038.

results to one another, and to unify some aspects of the known theory. For the sake of completeness, we will mention most of the results included in the excellent survey of Ráb [*73*; 1959]. There is a further justification of this duplication in that we will develop the known theory in a different manner than did Ráb.

The qualitative study of second order linear equations originated in the classic paper of Sturm [*81*; 1836]. However, the general importance and usefulness of Sturm's work was not properly recognized until the end of the 19th and the beginning of the 20th centuries. At that time the work of Bôcher [*4–7*] had a considerable influence in getting recognition of Sturm's work. For the problem of classifing the solutions of (1.1), Sturm's main result is his famous comparison theorem:

Sturm Comparison Theorem. If

$$q_1 \geqslant q_2\,,\; r_1 \leqslant r_2\,, \qquad \text{and} \qquad (r_1 y')' + q_1 y = 0$$

is nonoscillatory, then $(r_2 y')' + q_2 y = 0$ is nonoscillatory.

As an application of this theorem, let

$$l_0(t) = t, \qquad l_n(t) = \log l_{n-1}(t), \qquad n = 1, 2, \ldots, \tag{1.2}$$

and

$$L_n(t) = \prod_{k=0}^{n} l_k(t), \qquad n = 0, 1, \ldots . \tag{1.3}$$

Then $u_n = L_n^{1/2}$ is a nonoscillatory solution to the equation $u_n'' + q_n u_n = 0$, where

$$q_n(t) = \left[\sum_{k=0}^{n} L_k^{-2}(t)\right] \Big/ 4. \tag{1.4}$$

Hence, if there exists a nonnegative integer n such that

$$q(t) - q_n(t) \leqslant 0, \qquad a \leqslant t < \infty, \tag{1.5}$$

then $y'' + qy = 0$ is nonoscillatory by the Sturm Theorem. It happens that if there exist a nonnegative integer n and a number ϵ such that

$$[q(t) - q_n(t)]\, L_n^2(t) \geqslant \epsilon > 0, \qquad a \leqslant t < \infty, \tag{1.5$'$}$$

then $y'' + qy = 0$ is oscillatory. We will give a simple proof of this latter result in the next section. For $n = 0$, these propositions become $q(t) \leqslant \frac{1}{4}t^2 \rightarrow$ nonoscillation, $q(t) \geqslant (1 + \epsilon)/4t^2 \rightarrow$ oscillation, which is a result first noted by Kneser [*37*; 1893]. The general results involving (1.5, 1.5′) were essentially derived by Riemann and Weber [*75*; 1912]. They later reappeared in various forms in [*26*], [*31*], [*42*], and [*55*].

With the exception of two results in Section 5, we will not include in this survey any details of the many generalizations and ramifications of the Sturm Comparison Theorem. However, the main references in this regard are [*6*], [*7*], [*14*], [*17*], [*21*], [*24*], [*34–36*], [*38*], [*43*], [*46–49*], [*57*], [*60*], [*62*], [*64*], [*74*], [*83*], [*84*], [*94*].

Throughout this survey, $\int^t f(s)\,ds$ will denote any absolutely continuous function F with the property that $F'(t) = f(t)$. Whenever $\int^\infty$ is written, it is to be assumed that

$$\int^\infty = \lim_{t\to\infty} \int^t ,$$

and that this limit exists in the extended real numbers $[-\infty, \infty]$. Whenever equations of the form (1.1) are considered, it will be implicitly assumed that $q \in C[a, \infty)$, $r \in C^1[a, \infty)$, and $r > 0$.

2. Results originating directly from Kummer and Riccati Transformations.

Let

$$\varphi \in C^1(a, \infty), \qquad \varphi' > 0, \qquad \psi \in C^2[a, \infty), \qquad \psi(t) \neq 0, \qquad a \leqslant t < \infty. \tag{2.1}$$

The so-called Kummer transformation (cf. Kummer [*4*]; 1834])*

$$\tau = \varphi(t), \qquad y(t) = \psi(t)\, x(\tau) \tag{2.2}$$

transforms (1.1) into

$$\frac{d}{d\tau}\left[R(\tau)\,\frac{dx}{d\tau}\right] + Q(\tau)\,x = 0, \tag{2.3}$$

* Stäckel [*80*; 1893] and Lie [*51*; 1894] showed at about the same time that (2.2) is the only schlichte transformation of (t, y, y') space into (τ, x, x') space that preserves the form of (1.1). Boruvka [*11*; pp. 102–105 & 183–186] and Sansone [*78*; pp. 90–101] discuss the use and history of (2.2) in the classical theory of differential equations. Boruvka [*10*; 1962] presents a survey of results concerned with the problem of when two given equations $y'' + q(t)y = 0$ and $\ddot{x} + p(\tau)x = 0$ can be transformed into one another by a Kummer transformation.

where

$$R(\tau) = r(t)\,\varphi'(t)\,\psi^2(t),$$
$$Q(\tau) = [(r(t)\,\psi'(t))' + q(t)\,\psi(t)]\,\psi(t)[\varphi'(t)]^{-1}. \tag{2.4}$$

Equations (1.1) and (2.3) obviously have the same oscillatory behavior, because of the form of (2.2) and the assumptions on φ and ψ. Furthermore, one can always choose

$$\tau = \varphi(t) = \int^t [r(s)\,\psi^2(s)]^{-1}\,ds, \tag{2.5}$$

so that (2.3) is of the simpler form

$$\ddot{x} + p(\tau)\,x = 0, \qquad \varphi(a) \leqslant \tau < \varphi(\infty)(\cdot = d/d\tau), \tag{2.6}$$

where

$$p(\tau) = [(r(t)\,\psi'(t))' + q(t)\,\psi(t)]\,\psi^3(t)\,r(t). \tag{2.7}$$

The Kummer transformation with φ chosen as in (2.5) ($r \equiv 1$) and $\psi = u_n = L_n^{1/2}$, where L_n is defined by (1.3), produces

$$p(\tau) = [u_n''(t) + q(t)\,u_n(t)]\,u_n^{\,3}(t) = [q(t) - q_n(t)]\,L_n^{\,2}(t). \tag{2.8}$$

Hence, (1.5′) and the Sturm Comparison Theorem imply that equation (2.6) in this context, and therefore, the equation $y'' + qy = 0$, is oscillatory. This then proves the other half of the Riemann–Weber result mentioned in Section 1.

If

$$\int^\infty [r(s)\,\psi^2(s)]^{-1}\,ds = \infty, \tag{2.9}$$

then (2.5) maps the unbounded interval $[a, \infty)$ onto the unbounded interval $[\varphi(a), \infty)$. Kummer transformations with this property are often needed in oscillation theory. Condition (2.9) can be always achieved; for example:

$$\text{If} \quad \int^\infty r^{-1}(s)\,ds = \infty, \qquad \text{let} \quad \psi(t) = 1.$$
$$\text{If} \quad \int^\infty r^{-1}(s)\,ds < \infty, \qquad \text{let} \quad \psi(t) = \int_t^\infty r^{-1}(s)\,ds. \tag{2.10}$$

These choices of ψ are especially elegant, because in each case $(r\psi')' \equiv 0$. Hence, (2.7) is particularly simple.

Thus, the oscillation classification problem for equations of the form $(ry')' + qy = 0$ on an unbounded interval is equivalent to the same problem for equations of the form $y'' + py = 0$ on an unbounded interval. In this survey, we will not be consistent in whether results are stated for $(ry')' + qy = 0$ or $y'' + py = 0$.

A second useful transformation in oscillation theory is the well known Riccati transformation*:

(i) If y is a nonvanishing solution of (1.1) on an interval I, then $u = ry'y^{-1}$ is a solution of

$$u' + q + r^{-1}u^2 = 0 \tag{2.11}$$

on I.

(ii) If u is a solution of (2.11) on I, then

$$y(t) = \exp\left[\int^t u(s)\, r^{-1}(s)\, ds\right], \tag{2.12}$$

is a nonvanishing solution of (1.1) on I.

Theorem 2.1 (Bôcher [5; 1900–01]). *Equation* (1.1) *is disconjugate, if and only if, there exists* $u \in C^1(a, \infty)$ *such that*

$$u'(t) + q(t) + r^{-1}(t)\, u^2(t) \leqslant 0, \qquad a < t < \infty. \tag{2.13}$$

Proof. If (1.1) is disconjugate, then the solutions y satisfying $y(a) = 0$, $y'(a) \neq 0$ do not vanish in (a, ∞). For such a solution, $u = ry'y^{-1}$ satisfies (2.11), hence, (2.13).

Conversely, if u is a solution of (2.13), let $\psi(t)$ be defined by (2.12) for $t \geqslant b > a$ and let $\varphi(t)$ be defined by (2.5) with lower limit b. The resulting Kummer transformation (2.2) takes (1.1) into (2.6) with

$$p(\tau) = [u'(t) + r^{-1}(t)\, u^2(t) + q(t)]\, r(t)\, \psi^4(t).$$

Hence, (2.13) implies $p(\tau) \leqslant 0$, $\varphi(b) \leqslant \tau < \varphi(\infty)$. We conclude from the Sturm Comparison Theorem that $\ddot{x} + p(\tau)\, x = 0$ is disconjugate on $[\varphi(b), \varphi(\infty))$. Therefore, (1.1) is disconjugate on $[b, \infty)$. Since this is true for all $b > a$, (1.1) is disconjugate on $[a, \infty)$.

* See [85; p. 217] for a history of the Riccati transformation.

By letting $u = ry'y^{-1}$ in (2.13), we can reformulate Theorem 2.1 in the following manner: (1.1) is disconjugate, if and only if, there exist $y \in C^2[a, \infty)$, $y(t) > 0$ when $a < t < \infty$, such that

$$(ry')' + qy \leqslant 0. \tag{2.14}$$

Kondratév [*39*; 1957] has given a direct and elementary proof of this result. His proof is based upon the fact that if yw, where y satisfies (2.14), is substituted into (1.1) for y, then the resulting second order linear differential equation in w has a nonpositive coefficient of w because of (2.14). This coefficient remains nonpositive upon putting the equation in normal form (in the form (1.1)). Hence, the Sturm Comparison Theorem implies that there is a nonvanishing solution for w, and, therefore, (1.1) is disconjugate.

If we let

$$u(t) = \left[2 \int_t^\infty sp(s)\, ds + 1\right] \Big/ (2t),$$

in Theorem 2.1, then we obtain the following generalization of a result of Hille [*31*; 1948] and Hartman and Wintner [*29*; 1948]:

Corollary 2.1*. *If*

$$\left| \int_t^\infty sp(s)\, ds \right| < \infty, \qquad a \leqslant t < \infty,$$

then $y'' + py = 0$ *is nonoscillatory.*

If we apply Corollary 2.1 to (2.6) with ψ chosen as in (2.10), we obtain the following generalization of various results in [*30*; 1953], [*44*; 1949], [*58*; 1955], [*89*; 1948], and [*90*; 1949]:

Corollary 2.2. *If*

$$\int^\infty r^{-1}(s)\, ds < \infty \qquad \textit{and} \qquad \left| \int^\infty q(t) \left[\int_t^\infty r^{-1}(s)\, ds \right] dt \right| < \infty$$

or

$$\int^\infty r^{-1}(s)\, ds = \infty \qquad \textit{and} \qquad \left| \int^\infty q(t) \left[\int^t r^{-1}(s)\, ds \right] dt \right| < \infty,$$

then (1.1) *is nonoscillatory.*

* Zlámal [*95*; 1950] proved that $\int^\infty s^\sigma p(s)\, ds = \infty$ for some constant $\sigma < 1$ implies that $y'' + py = 0$ is oscillatory.

Corollary 2.2 is also an obvious consequence of the following result of Zubova [96; 1957]:

Theorem 2.2. *Equation* (1.1) *is disconjugate, if and only if, there exist positve functions* $h \in C(a, \infty)$, $f \in C^1(a, \infty)$ *such that*

$$h(t) f'(t) + \int^t f(s)\, q(s)\, ds = 0 \quad \textit{and} \quad 0 < h(t) \leqslant r(t), \quad a < t < \infty. \tag{2.15}$$

Proof. The function f, which does not vanish in (a, ∞), satisfies the equation $(hf')' + qf = 0$. Since $h \leqslant r$, the Sturm Comparison Theorem implies that no solutions of $(ry')' + qy = 0$ can vanish more than once in $[a, \infty)$. The converse is obvious.

A result similar to Theorem 2.2, but involving oscillation instead of disconjugacy, can be similarly established by using the contrapositive of the Sturm Comparison Theorem. Zubova lists such a result.

Corollary 2.3. *If*

$$(r_i y')' + q_i y = 0, \qquad i = 1,..., n,$$

are disconjugate and c_i *are nonnegative constants such that*

$$\sum_{i=1}^{n} c_i = 1,$$

then

$$\left[\left(\sum_{i=1}^{n} c_i r_i\right) y'\right]' + \left(\sum_{i=1}^{n} c_i q_i\right) y = 0 \tag{2.16}$$

is disconjugate.

Proof. Theorem 2.1 implies that there exist functions $u_i \in C^1[a, \infty)$, $i = 1,..., n$, such that in (a, ∞)

$$u_i' + q_i + r_i^{-1} u_i^2 \leqslant 0, \qquad i = 1,..., n.$$

It follows that the function

$$u = \sum_{i=1}^{n} c_i u_i\,,$$

satisfies in (a, ∞)

$$u' + \sum_{i=1}^{n} c_i q_i + \left(\sum_{i=1}^{n} c_i r_i\right)^{-1} u^2 \leqslant 0.$$

Hence, Theorem 2.1 implies that (2.16) is disconjugate.

Adamov [*1*; 1948] established Corollary 2.3 for the special case when $n = 2$, $r_1 \equiv r_2 \equiv 1$, and q_1, q_2 are periodic of the same period. Petropavlovskaya [*68*; 1955] generalized Adamov's result by removing the periodicity assumption on p_1, p_2; Markus and Moore [*54*; 1956] further removed the condition that $r_1 \equiv r_2 \equiv 1$. Finally, Kondratév [*40*; 1957] established the result for general n, but with $r_i \equiv 1$ for all values of i.

Corollary 2.4. (Hartman [*27*; 1951]). *Let $P \in C^1[a, \infty)$ be any function such that $P' = -p$. If*

$$y'' + 4P^2 y = 0 \tag{2.17}$$

is disconjugate, then $y'' + py = 0$ is disconjugate.

Proof. The disconjugacy of (2.17) implies that there exists $v \in C^1(a, \infty)$ such that $v' + v^2 + 4P^2 \leqslant 0$. But then $u = P + v/2$ satisfies $u' + u^2 + p \leqslant 0$; and so Theorem 2.1 implies that $y'' + py = 0$ is disconjugate.

Other less elegant results that can be obtained by specializing u in Theorem 2.1 have been obtained by Hartman [*26*; 1948], Kondratév [*40*, 1957], Wintner [*91*; 1951], and Zlámal [*95*; 1950].

Theorem 2.3. *The equation $y'' + py = 0$ is nonoscillatory, if and only if, there exists $\psi \in C^2[a, \infty)$, $\psi > 0$, such that*

$$\int^{\infty} \Psi(t)\,\psi(t)[\psi''(t) + p(t)\,\psi(t)]\,dt < \infty, \tag{2.18}$$

where

$$\Psi(t) = \begin{cases} \psi_1(t) & \text{when} \quad \psi_1(t) \equiv \int_t^{\infty} \psi^{-2}(s)\,ds < \infty \\ \psi_2(t) & \text{when} \quad \psi_2(t) \equiv \int^{t} \psi^{-2}(s)\,ds \to \infty \quad \text{as} \quad t \to \infty. \end{cases} \tag{2.19}$$

Proof. When (2.18) holds, Corollary 2.2, applied to equation (2.3) with $\varphi' \equiv 1$ and $r \equiv 1$, implies that $y'' + py = 0$ is nonoscillatory.

Conversely, when $y'' + py = 0$ is nonoscillatory, (2.18) is satisfied by any positive function ψ that coincides with a solution y on an interval $[b, \infty)$ in which y does not vanish. Furthermore, if y is a maximal solution, then

$$\int^{\infty} \psi^{-2}(s)\, ds < \infty, \tag{2.20}$$

and if y is a minimal solution, then

$$\int^{\infty} \psi^{-2}(s)\, ds = \infty. \tag{2.21}$$

Theorem 2.3 with ψ satisfying (2.20) in both directions was first proven by Wintner [*87*; 1948]. Levin [*50*; 1965] listed Theorem 2.3 as it is here.

3. Oscillation and the Bohl transformation.

The so-called Bohl transformation (Bohl [*8*; 1906]) can be described as follows:

(i) If $\lambda \in C^2[a, \infty)$ is a solution to the nonlinear differential equation

$$(r\lambda')' + q\lambda = (r\lambda^3)^{-1}, \tag{3.1}$$

then

$$y(t) = \lambda(t) \sin\left(\int^{t} [r(s)\, \lambda^2(s)]^{-1}\, ds\right) \tag{3.2}$$

is a solution of (1.1).

(ii) Conversely, if y_1 and y_2 are linearly independent solutions of (1.1) and have Wronskian equal to $1/r$, then

$$\lambda = (y_1{}^2 + y_2{}^2)^{1/2} \tag{3.3}$$

is a solution of (3.1).

Ráb [*73*; 1959] bases his survey of oscillation theory upon the Bohl transformation. For the sake of completeness we will give the highlights of that theory most directly related to the Bohl transformation in this section.

Theorem 3.1. (Ráb [*73*; p. 337] *Equation* (1.1) *is oscillatory, if and only if, there exists* $\lambda \in C^2[a, \infty)$, $\lambda > 0$, *such that* λ *is a solution of* (3.1) *and*

$$\int^{\infty} [r(s)\,\lambda^2(s)]^{-1}\,ds = \infty. \tag{3.4}$$

Theorem 3.2. (Ráb [*73*; p. 339]) *If for each function* $P \in C^1[a, \infty)$ *such that* $P' = -p$, *it is true that*

$$\int^{\infty} \exp\left[2 \int^{t} P(s)\,ds\right] dt = \infty, \tag{3.5}$$

then $y'' + py = 0$ *is oscillatory.*

Proof. Suppose that $y'' + py = 0$ is nonoscillatory on $[a, \infty)$. Then there exists $b \geqslant a$ and two nonvanishing linearly independent solutions y_1, y_2 of $y'' + py = 0$ on $[b, \infty)$ with the Wronskian of y_1, y_2 equal to unity. Letting y denote either y_1 or y_2, we obtain by means of a Riccati transformation that

$$y'(t)\,y^{-1}(t) \leqslant C - \int_b^t p(s)\,ds \equiv P(t), \qquad t \geqslant b,$$

where

$$C = \max_{i=1,2}\,[\,y_i'(b)\,y_i^{-1}(b)].$$

Hence,

$$y(t) \leqslant y(b) \exp\left[\int_b^t P(s)\,ds\right], \qquad t \geqslant b; \tag{3.6}$$

and (ii) implies that $\lambda = (y_1^2 + y_2^2)^{1/2}$ satisfies $\lambda'' + p\lambda = \lambda^{-3}$. Furthermore, (3.6) implies that

$$\lambda^{-2}(t) \geqslant [\,y_1^2(b) + y_2^2(b)]^{-1} \exp\left[2 \int_b^t P(s)\,ds\right], \qquad t \geqslant b.$$

Hence, (3.5) implies that

$$\int^{\infty} \lambda^{-2}(t)\,dt = \infty;$$

and so, Theorem 3.1 implies that $y'' + py = 0$ is oscillatory, which is a contradiction.

Corollary 3.1*. (Wintner [*88*; 1949[†]]) *If*

$$\int^{\infty} p(s)\, ds = \infty,$$

then $y'' + py = 0$ *is oscillatory.*

Corollary 3.2. (Moore [*58*; 1955[‡]]) *Equation* (1.1) *is oscillatory, if and only if, there exists* $\psi \in C^2[a, \infty)$, $\psi > 0$, *such that*

$$\int^{\infty} [r(s)\, \psi^2(s)]^{-1}\, ds = \infty \tag{3.7}$$

and

$$\int^{\infty} \{[r(s)\, \psi'(s)]'\, \psi(s) + q(s)\, \psi^2(s)\}\, ds = \infty. \tag{3.8}$$

Proof. Assume first that (3.7) and (3.8) hold. If (1.1) is transformed by a Kummer transformation (2.2) with φ defined by (2.5), then the resulting equation $\ddot{x} + p(\tau)\, x = 0$ has

$$\int^{\infty} p(\tau)\, d\tau = \infty,$$

because of (3.8). Hence, Corollary 3.1 implies that this equation is oscillatory, and so, (1.1) is oscillatory.

Conversely, if (1.1) is oscillatory, Theorem 3.1 implies the existence of a function λ which satisfies (3.1) and (3.4). If we let $\psi \equiv \lambda$, then (3.4) implies (3.7), and (3.1) and (3.4) imply (3.8).

If we let $\psi \equiv 1$ in Corollary 3.2, then we obtain a sufficient condition for oscillation of Leighton [*45*; 1950], namely,

$$\int^{\infty} r^{-1}(s)\, ds = \int^{\infty} q(s)\, ds = \infty. \tag{3.9}$$

Obviously one can obtain an infinite number of special sufficient conditions for oscillation by specializing ψ in Corollary 3.2. Some of the more interesting of these involve the functions l_n, L_n, and q_n, which are defined in (1.2), (1.3), and (1.4). Each of the following is a sufficient condition for (1.1) to be oscillatory:

* See the footnote attached to Corollary 2.1.

† Fite [*22*; 1918] originally proved this result with the additional assumption $p \geqslant 0$.

‡ Gagliardo [*25*; 1954] originally proved that (3.7) and (3.8) for the case $r \equiv 1$ were sufficient for oscillation.

(I) The function

$$R(t) = \int^t r^{-1}(s)\, ds \tag{3.10}$$

satisfies $R(t) \to \infty$ as $t \to \infty$, and there exist a non-negative integer n and positive number ϵ such that

$$\int^\infty q(t)\, R^2(t)[L_n(R(t))\, l_n^\epsilon(R(t))]^{-1}\, dt = \infty.$$

(II) $R(t) \to \infty$ as $t \to \infty$ and there exists a non-negative integer n such that

$$\int^\infty [q(t) - r^{-1}(t)\, q_n(R(t))]\, L_n(R(t))\, dt = \infty.$$

(III) $R(t) \to \infty$ as $t \to \infty$ and there exist a non-negative integer n and positive number ϵ such that

$$\int^\infty [q(t) - r^{-1}(t)\, q_n(R(t))]\, L_{n+1}(R(t))[l_{n+2}(R(t))]^{-(1+\epsilon)}\, dt = \infty.$$

(IV) The Function

$$\bar{R}(t) = 1\Big/\Big[\int_t^\infty r^{-1}(s)\, ds\Big], \tag{3.11}$$

is positive and there exist a non-negative integer n and positive number ϵ such that

$$\int^\infty q(t)[L_n(\bar{R}(t))\, l_n^\epsilon(\bar{R}(t))]^{-1}\, dt = \infty.$$

(V) $\bar{R} > 0$ and there exists a non-negative integer n such that

$$\int^\infty [q(t) - r^{-1}(t)\, \bar{R}^4(t)\, q_n(\bar{R}(t))]\, L_n(\bar{R}(t))\, dt = \infty.$$

The results in cases (I), (II), (IV), and (V) follow directly from Corollary 3.2 by letting ψ^2 be $R^2/L_n(R)\, l_n^\epsilon(R)$, $L_n(R)$, $1/L_n(\bar{R})\, l_n^\epsilon(\bar{R})$, and $L_n(\bar{R})$, respectively. Moore [*58*; 1955] established (I) and (IV) for the case $n = 0$. Zlámal [*95*; 1950] established (II) for the case $r \equiv 1$. Ráb [*72*; 1957], [*73*; pp. 346–351] established (III), which is somewhat different from the other four cases. Case (III) essentially follows from the following result:

Corollary 3.3. (Ráb [*73*; p. 342]) *Equation* (1.1) *is oscillatory, if and only if, there exists* $\psi \in C^2[a, \infty)$, $\psi > 0$, *such that for each function* $P \in C^1[a, \infty)$ *satisfying*

$$P' = (r\psi')'\,\psi + q\psi^2,$$

it is true that

$$\int^{\infty} [r(t)\,\psi^2(t)]^{-1} \exp\left\{2\int^{t} [r(s)\,\psi^2(s)]^{-1}\,P(s)\,ds\right\} dt = \infty. \tag{3.12}$$

Proof. If (3.12) holds for all admissible functions P, then Theorem 3.2 implies that (2.6) is oscillatory. Hence, (1.1) is oscillatory. Conversely, (3.12) is a direct consequence of Corollary 3.2.

Ráb [*73*; pp. 339–344] derives other results similar to Corollary 3.3 and supplies a good discussion of this aspect of oscillation theory. Most of the work of El'šin [*16–20*], [*32*], [*33*] is concerned with various ramifications and applications of Corollary 3.3. El'šin formulates (3.12) in terms of a function θ, where θ and ψ are related by the formula

$$\psi(t) = \exp\left[r^{1/2}(t)\int_a^t \theta(s)\,ds\right].$$

Other work related to Corollary 3.3 and reported in Ráb's survey has been done by Boruvka [*9*; 1957], Gagliardo [*25*; 1954], Kondrat'ev [*40*; 1957], Laitoch [*42*; 1955], and Zlámal [*95*; 1950].

4. Classification in terms of $\int^{\infty} p(s)\,ds$.

One of the following four cases must always occur:

(i) $\int^{\infty} p(s)\,ds = \infty$

(ii) $-\infty < \int^{\infty} p(s)\,ds < \infty$

(iii) $\int^{\infty} p(s)\,ds = -\infty$

(iv) $\limsup_{t\to\infty} \int^{t} p(s)\,ds > \liminf_{t\to\infty} \int^{t} p(s)\,ds.$

The equation $y'' + py = 0$ is always oscillatory if p satisfies (i). However, both oscillation and nonoscillation are compatible with (ii), (iii), and (iv).

The classical Euler equation

$$y'' + \alpha t^{-2} y = 0 \qquad (\alpha \text{ constant}),$$

illustrates both cases for (ii), and the classical Mathieu equation

$$y'' + (\alpha - \beta \cos t)\, y = 0 \qquad (\alpha, \beta \text{ constants}),$$

illustrates both cases for (iii). The complete classification of the Mathieu equation seems to still be an open problem. However, extensive oscillation results have been obtained by Moore [*59*; 1956] and Zubova [*97*; 1963] for the Mathieu equation and its generalization, the Hill equation. Also, Magnus and Winkler [*52*; pp. 56–78] list many results for the Hill equation, and Markus and Moore [*54*; 1956] have studied

$$y'' + [\alpha - \beta p(t)]\, y = 0 \qquad (\alpha, \beta \text{ constants}),$$

under the assumption that p is almost periodic. Yelchin [*93*; 1946] proved that $y'' + py = 0$ is oscillatory if p has a Fourier series with zero constant term. Sobol [*79*; 1951] proved that $y'' + py = 0$ is oscillatory if $\int^t p(s)\, ds$ is almost periodic and not constant. Other sufficient conditions for oscillation in the case of a periodic coefficient p have been obtained by Adamov [*1*; 1948].

The equations $y'' + y \sin t = 0$* and $y'' + yt \sin t = 0$ are examples showing that (iv) is completely compatible with oscillation. It is somewhat more difficult to show that (iv) is compatible with nonoscillation. In this regard, let $p = v' - v^2$, where $v \in C^1[a, \infty) \cap L^2[a, \infty)$. Then,

$$y = \exp\left[-\int_a^t v(s)\, ds\right],$$

is a nonvanishing solution of $y'' + py = 0$. Clearly, functions v can be found such that

$$\limsup_{t\to\infty} \int_a^t p(s)\, ds = \limsup_{t\to\infty} v(t) - \int_a^\infty v^2(s)\, ds - v(a),$$

and

$$\liminf_{t\to\infty} \int_a^t p(s)\, ds = \liminf_{t\to\infty} v(t) - \int_a^\infty v^2(s)\, ds - v(a),$$

take on any values desired in $[-\infty, \infty]$, and $[-\infty, \infty)$ respectively.

* It is interesting to note that the equation $y'' + [\sin t/(2 + \sin t)]\, y = 0$, however, is nonoscillatory.

We will next describe a method recently developed by Coles [*12*; 1968] and Willett [*86*; 1968]. One of the advantages of this method is to unify and extend the known results for cases (ii) and (iv). For these cases, a rather extensive classification of equations has been obtained.

Let

$$\mathfrak{F} = \left\{ f : f \text{ measurable on } [a, \infty), f \geqslant 0, \int^{\infty} f(s)\, ds = \infty \right\},$$

and for $f \in \mathfrak{F}$, $p \in C[a, \infty)$, let

$$A_{fp} \equiv A(s, t) = \int_t^s \left[f(\tau) \int_t^{\tau} p(\sigma)\, d\sigma \right] d\tau \Big/ \int_t^s f(\tau)\, d\tau. \tag{4.1}$$

We say that a function $p \in C[a, \infty)$ has an *averaged integral* $P \equiv P_f$ *with respect to* $\mathfrak{F}$, if there exists $f \in \mathfrak{F}$ such that, for each $t \in [a, \infty)$, $\lim A_{fp}(s, t)$, as $s \to \infty$, exists in $[-\infty, \infty]$ and

$$P(t) = \lim_{s \to \infty} A_{fp}(s, t). \tag{4.2}$$

If the limit in (4.2) exists for one value of t, then it exists for any value of t in $[a, \infty)$. In fact, an averaged integral P always satisfies the following fundamental relationship:

$$P(t) = P(b) - \int_b^t p(s)\, ds, \qquad a \leqslant b, t < \infty; \tag{4.3}$$

hence,

$$P'(t) = -p(t), \qquad a \leqslant t < \infty.$$

Furthermore, if $\int_t^{\infty} p(s)$ exists, then

$$P(t) = \int_t^{\infty} p(s)\, ds, \qquad a \leqslant t < \infty.$$

On the other hand, consider the example when $p(t) = \cos t$. Let

$$f(t) = \begin{cases} 1 & \text{when} \quad \sin t \geqslant 0 \\ 0 & \text{when} \quad \sin t < 0 \end{cases}.$$

Then, $P_f(0)$ exists and $P_f(0) = 2/\pi$. Hence,

$$P_f(t) = 2\pi^{-1} - \sin t, \qquad 0 \leqslant t < \infty.$$

For this example, $\int_t^\infty p(s)\,ds$ does not exist.

The set $\mathfrak{F}$ is too large for our purposes, and so we introduce the following two sets:

$$\mathfrak{F}_0 = \left\{ f \in \mathfrak{F} : \lim_{t\to\infty} \left[\int^t f^2(s)\,ds \right] \Big/ \left[\int^t f(s)\,ds \right]^2 = 0 \right\}.$$

$$\mathfrak{F}_1 = \left\{ f \in \mathfrak{F} : \limsup_{t\to\infty} \left(\int^t f(s)\,ds \int_t^\infty f(s) \left[\int^s f^2(\tau)\,d\tau \right]^{-1} ds \right) > 0 \right\}.$$

It is easy to show that $\mathfrak{F} \supset \mathfrak{F}_1 \supset \mathfrak{F}_0$, and if $f \in \mathfrak{F}$ and f is bounded on $[a, \infty)$, then $f \in \mathfrak{F}_0$. On the other hand, all nonnegative polynomials are in $\mathfrak{F}_0$, and so $\mathfrak{F}_0$ does contain some unbounded functions.

Theorem 4.1. *If there exists $f \in \mathfrak{F}_1$ such that the averaged integral $P_f(a) = \infty$, then $y'' + py = 0$ is oscillatory.*

For a proof of Theorem 4.1, see Willett [*86*; 1968]. Coles [*12*; 1968] proved a similar theorem using a smaller class of weight functions than $\mathfrak{F}_1$. Theorem 4.1 is no longer a true statement if $\mathfrak{F}_1$ is replaced by $\mathfrak{F}$, because if it would be, then $\limsup \int^t p(s)\,ds = \infty$ would be sufficient to imply that $y'' + py = 0$ is oscillatory. The latter is certainly not the case as some of our previous examples clearly indicated. It remains, however, an interesting question as to what is the largest class of weight functions f for which $P_f(a) = \infty$ implies oscillation.

Corollary 3.1 and the following older results are easy consequences of Theorem 4.1:

Corollary 4.1. (Olech, Opial, Wajewski [*65*; 1957]) *If*

$$\operatorname*{lim\,approx}_{t\to\infty} \int^t p(s)\,ds = \infty,$$

then $y'' + py = 0$ is oscillatory.

Corollary 4.2. (Wintner [*88*; 1949]) *If*

$$\lim_{t\to\infty} t^{-1} \int^t \left[\int^s p(\tau)\,d\tau \right] ds = \infty,$$

then $y'' + py = 0$ is oscillatory.

A general theorem similar to Theorem 4.1 but with "higher order" weighted averages has been obtained by Coles and Willett [*13*; 1968]. Rather than reproduce here the general result, which is notationally rather complicated, we will list two of the more interesting applications.

Theorem 4.2. *Let $P \in C^1[a, \infty)$ be any function such that $P' = p$. If there exists $f \in \mathfrak{F}_1$ and positive integer n such that*

$$\lim_{t_n \to \infty} \frac{\int^{t_n} f(t_{n-1}) \int^{t_{n-1}} f(t_{n-2}) \cdots \int^{t_1} f(t_0)\, P(t_0)\, dt_0 \cdots dt_{n-2}\, dt_{n-1}}{\int^{t_n} f(t_{n-1}) \int^{t_{n-1}} f(t_{n-2}) \cdots \int^{t_1} f(t_0)\, dt_0 \cdots dt_{n-2}\, dt_{n-1}} = \infty, \tag{4.4}$$

then $y'' + py = 0$ is oscillatory.

Theorem 4.3. *Let $P \in C^1[a, \infty)$ be any function such that $P' = p$. If for some positive integer n, P is not Hölder (H, n)-summable*, i.e.,

$$\lim_{t_n \to \infty} t_n^{-1} \int^{t_n} t_{n-1}^{-1} \int^{t_{n-1}} \cdots t_1^{-1} \int^{t_1} P(t_0)\, dt_0 \cdots dt_{n-2}\, dt_{n-1} = \infty,$$

then $y'' + py = 0$ is oscillatory.

The left side of (4.4) can be considered a generalized Riesz mean for P. If $f \equiv 1$, then (4.4) is the nth Cesàro sum of P.

Theorem 4.4. (Willett [*86*; 1968]) *If there exists $f \in \mathfrak{F}$, such that*

$$\liminf_{t \to \infty} A_{fp}(t, a) > -\infty, \tag{4.5}$$

then either $y'' + py = 0$ is oscillatory, or the averaged integral $P_g(t)$ exists and is finite for all $g \in \mathfrak{F}_0$ and $a \leqslant t < \infty$.

Corollary 4.3. (Hartman [*28*; 1952]) *If*

$$\limsup_{t \to \infty} t^{-1} \int^t \left[\int^s p(\tau)\, d\tau\right] ds > \liminf_{t \to \infty} t^{-1} \int^t \left[\int^s p(\tau)\, d\tau\right] ds > -\infty,$$

then $y'' + py = 0$ is oscillatory.

Theorems 4.1 and 4.4 indicate that the problem of classifying the equations $y'' + py = 0$ can be separated into two parts defined by whether an $f \in \mathfrak{F}_1$ exists such that (4.5) holds. Condition (4.5) is always satisfied if $\liminf_{t \to \infty} \int^t p(s)\, ds > -\infty$ and is never satisfied if

$\int^\infty p(s)\,ds = -\infty$. We will present results in the next section which, together with Theorems 4.1 and 4.2, will produce a reasonably complete theory for the case when (4.5) can be satisfied.

When (4.5) is not satisfied for any $f \in \mathfrak{F}_1$, few specific results seem to be known. One positive feature of this case, however, is that it can be at least theoretically eliminated by substituting $y = r^{1/2}z$, where

$$r(t) = \exp\left(-2\int^t \left[\int_a^s p(\tau)\,d\tau\right] ds\right). \tag{4.6}$$

Theorem 4.5. *The oscillatory properties of the equations $y'' + py = 0$ and $(rz')' + qz = 0$, where r is defined in (4.6) and*

$$q(t) = r(t)\left[\int_a^t p(s)\,ds\right]^2,$$

are equivalent.

Of course, the equation $(rz')' + qz = 0$ in Theorem 4.5 can be transformed to an equation of the form $w'' + \bar{p}w = 0$ by a Kummer transformation. If this is accomplished by the transformation described in (2.10), then $\bar{p} \geqslant 0$ because $q \geqslant 0$. Hence, for the function $\bar{p}$, (4.5) is satisfied by all $f \in \mathfrak{F}_1$. For some other aspects of using Kummer transformations to transform a given equation into an equation where (4.5) can be satisfied for some $f \in \mathfrak{F}_1$, see Willett [*86*; 1968].

Putnam [*71*; 1955] proved the following result, which might apply to some equations for which (4.5) cannot be satisfied by any $f \in \mathfrak{F}_1$:

Theorem 4.6. *If*

$$\limsup_{t\to\infty} t^{-1}\int^t \left[\int^s p(\tau)\,d\tau\right] ds = \infty,$$

and if there exists a constant $c > 0$ such that

$$\int^t p(s)\,ds > -e^{ct}, \qquad a \leqslant t < \infty,$$

then $y'' + py = 0$ is oscillatory.

Most of the results in this section and the next section that are stated for the equation $y'' + py = 0$ can be used to generate more general results. This can be accomplished by applying these results to equation

(2.6) to obtain a new result involving a nearly arbitrary function ψ. For example, Corollary 4.2 applied to (2.6) produces the following result of Gagliardo [*25*; 1954]:

Corollary 4.4. *If there exists* $\psi \in C^2[a, \infty)$, $\psi > 0$, *such that* (2.9) *and*

$$\lim_{t\to\infty} t^{-1} \int^t \left\{ \int^s [(r(\tau)\,\psi'(\tau))' + q(\tau)\,\psi(\tau)]\,\psi(\tau)\,d\tau \right\} ds = \infty$$

hold, then $(ry')' + qy = 0$ *is oscillatory.*

Once again we refer to Ráb [*73*; 1959] for a detailed discussion of some results of this type.

We conclude this section with the following miscellaneous results:

(i) (Potter [*69*; 1953]) If

$$p \in C^1[a, \infty), \qquad p \geqslant 0, \qquad (p')^2 p^{-3} \leqslant k < 16, \qquad \text{and} \qquad \int^\infty p^{1/2}(t)\,dt = \infty,$$

then $y'' + py = 0$ is oscillatory.

(ii) (Leighton [*46*; 1952]) If $q > 0$, $(qr)' \leqslant 0$, and $(ry')' + qy = 0$ is oscillatory, then

$$\int^\infty [q(s\ \ r^{-1}(s)]^{1/2}\,ds = \infty.$$

(iii) (Barrett [*2*; 1955]) If $q > 0$, $(qr)' \leqslant 0$, and

$$\lim_{t\to\infty} \left\{ \int^t [q(s)\,r^{-1}(s)]^{1/2}\,ds + \tfrac{1}{4}\log[q(t)\,r(t)] \right\} = \infty,$$

then $(ry')' + qy = 0$ is oscillatory.

5. Classification when p has a finite averaged integral with respect to $\mathfrak{F}_0$.

The classification of equations of the form $y'' + py = 0$ when there exists $f \in \mathfrak{F}_1$ such that

$$\liminf_{t\to\infty} A_{fp}(t, a) > -\infty,$$

has been reduced by Theorems 4.1 and 4.4 to the case when $P_g(t)$ exists and is finite for all $g \in \mathfrak{F}_0$. We now turn our attention to this case.

Theorem 5.1. (Willett [86; 1968]) *If there exist bounded functions $f, g \in \mathfrak{F}$ such that $P_f(a)$ and $P_g(a)$ exist and $P_f(a) \neq P_g(a)$, then $y'' + py = 0$ is oscillatory.*

Corollary 5.1. (Olech, Opial, Wazewski [65; 1957]) *If*

$$\lim_{t\to\infty} \text{approx sup} \int^t p(s)\, ds > \lim_{t\to\infty} \text{approx inf} \int^t p(s)\, ds,$$

then $y'' + py = 0$ is oscillatory.

Corollary 5.2. *If p is bounded on one side and*

$$\infty \geqslant \limsup_{t\to\infty} \int^t p(s)\, ds > \liminf_{t\to\infty} \int^t p(s)\, ds \geqslant -\infty, \tag{5.2}$$

then $y'' + py = 0$ is oscillatory.

Zlámal [95; 1950] proved Corollary 5.2 for the case when p is bounded and the lim sup is ∞; Petropavlovskaya [68; 1955] gave a proof for the case when p is bounded below; Moore [58; 1955] gave a proof for the case when p is bounded above and the lim sup is ∞; and Olech, Opial, Wazewski [65; 1957] proved that Corollary 5.1 implies Corollary 5.2.

Theorem 5.1 is particularly useful for the difficult problems when p is not of constant sign, or when p oscillates about some value. For example, suppose for some $\epsilon \geqslant 0$, the sets

$$E_\epsilon^+ = \left\{t \geqslant a : \int_a^t p(s)\, ds > \epsilon\right\} \quad \text{and} \quad E_\epsilon^- = \left\{t \geqslant a : \int_a^t p(s)\, ds < -\epsilon\right\}$$

have infinite measure. Then $P_f(a) \neq P_g(a)$ for f and g equal to the characteristic functions of E_ϵ^+ and E_ϵ^-, respectively. For the particular function $p(t) = \cos t$, we obtain $P_f(0) = 2/\pi$ and $P_g(0) = -2/\pi$ if f and g are taken to be the characteristic functions of E_0^+ and E_0^-, respectively. Hence, Theorem 5.1 implies that $y'' + y \cos t = 0$ oscillates.

Theorem 5.2. (Hartman [*28*; 1952]) *If* (5.2) *holds and if*

$$\sup_{0<v<\infty} \left| \int_u^{u+v} p(t)\,dt \right| \Big/ (1+v) \to 0, \qquad \text{as} \qquad u \to \infty,$$

then $y'' + py = 0$ *is oscillatory.*

Theorem 5.3. (Willett [*86*; 1968]) *Assume that* p *has a finite averaged integral* P *with respect to* $\mathfrak{F}_0$. *Then,* $y'' + py = 0$ *is disconjugate, if and only if, there exists a solution* $v \in C^1(a, \infty)$ *of*

$$v(t) = P(t) + \int_t^\infty v^2(s)\,ds. \tag{5.3}$$

Proof. Equation (5.3) implies that v satisfies $v' = -p - v^2$; hence, Theorem 2.1 implies that $y'' + py = 0$ is disconjugate. For the proof of the converse, which is more complicated, see Willett [*86*].

It is clear in the proof of Theorem 5.3 that (5.3) is sufficient for disconjugacy if P is any function such that $P' = -p$. Furthermore, (5.3) can be replaced in this instance by

$$v(t) \geqslant P(t) + \int_t^\infty v^2(s)\,ds \geqslant 0, \tag{5.4}$$

since $u = P + \int_t^\infty v^2(s)\,ds$ would then satisfy

$$u' = -p - v^2 \leqslant -p - u^2,$$

which also implies that $y'' + py = 0$ is disconjugate by Theorem 2.1.

For $P \in C[a, \infty)$, define

$$E_P \equiv E(t, s) = \exp\left(2 \int_t^s P(\tau)\,d\tau\right), \tag{5.5}$$

and

$$Q_P \equiv Q(t) = \int_t^\infty P^2(s)\,E(t, s)\,ds. \tag{5.6}$$

Clearly,

$$0 < E < \infty \qquad \text{and} \qquad 0 \leqslant Q \leqslant \infty.$$

Theorem 5.4. (Willett [86; 1968]) *Assume that p has a finite averaged integral P with respect to $\mathfrak{F}_0$. Then, $y'' + py = 0$ is disconjugate, if and only if, $Q \equiv Q_P$ is finite and there exists a solution $v \in C^1(a, \infty)$ of*

$$v(t) \geqslant Q(t) + \int_t^\infty E(t, s)\, v^2(s)\, ds \qquad (E \equiv E_P). \tag{5.7}$$

Proof. Condition (5.7) implies that the function

$$u(t) = P(t) + Q(t) + \int_t^\infty E(t, s)\, v^2(s)\, ds$$

satisfies $u' \leqslant -p - u^2$. Hence, Theorem 2.1 implies that $y'' + py = 0$ is disconjugate. Proof of the converse is more complicated. Theorem 5.3 implies that equation (5.3) has a solution. Let u be this solution. We can show next from (5.3) that the function

$$v(t) = \int_t^\infty u^2(s)\, ds,$$

satisfies

$$v(t) - Q(t) - \int_t^\infty E(t, s)\, v^2(s)\, ds = \lim_{\tau \to \infty} v(\tau) \exp\left(2 \int_t^\tau P(s)\, ds\right). \tag{5.8}$$

Hence, v satisfies (5.7).

We can actually show that the limit in (5.8) is zero. This means that Theorem 5.4 remains true if the inequality in (5.7) is replaced by equality. The proof of this fact, when $P(t) = \int_t^\infty p(s)\, ds$ exists and is finite, is due to Professor J. S. W. Wong. The proof for the general case when P is an averaged integral is similar and goes as follows:

Suppose the limit in (5.8) is positive for some value of t, $a \leqslant t < \infty$. Then there exist $\epsilon > 0$ and $b \geqslant a$ such that

$$v(s) \exp\left[2 \int_t^s P(\tau)\, d\tau\right] \geqslant \epsilon \qquad \text{for all} \quad s \geqslant b.$$

Hence, (5.8) implies that

$$v(t) \geqslant \int_t^\infty E(t, s)\, v^2(s)\, ds \geqslant \int_t^b E(t, s)\, v^2(s)\, ds + \epsilon^2 \int_b^\infty E^{-1}(t, s)\, ds$$

that is,

$$\int_b^\infty E^{-1}(t, s)\, ds = \int_b^\infty \exp\left[-2 \int_t^s P(\tau)\, d\tau\right] ds < \infty. \tag{5.9}$$

Condition (5.9) contradicts the following theorem, which is a generalization of results of Wintner [*91*; 1951] and Hartman [*28*; 1952]:

Theorem 5.5. *Assume that p has a finite averaged integral P with respect to $\mathfrak{F}_0$. If there exists a constant γ, $0 < \gamma \leqslant 4$, such that*

$$\int^{\infty} \exp\left[-\gamma \int^{t} P(s)\, ds\right] dt < \infty,$$

then $y'' + py = 0$ is oscillatory.

Proof. If $y'' + py = 0$ is nonoscillatory, then there exists a number $b \geqslant a$ and a function $v \in C[b, \infty)$ such that v is a solution to (5.3) for $b \leqslant t < \infty$. The classical proof can be carried from here to the usual contradiction.

Corollary 5.3. (Willett [*86*; 1968]) (i) *Let $P \in C^1[a, \infty)$ be such that $P' = -p$. If*

$$Q < \infty \qquad \text{and} \qquad \int_t^{\infty} Q^2(s)\, E(t, s)\, ds \leqslant Q(t)/4, \qquad a \leqslant t < \infty, \tag{5.10}$$

($E \equiv E_P$, $Q \equiv Q_P$) then $y'' + py = 0$ is disconjugate. (ii) *Assume that p has a finite averaged integral $P \equiv P_f$ with respect to $\mathfrak{F}_0$. If $Q(a) = \infty$ or if there exists $\epsilon > 0$ such that*

$$\int_t^{\infty} Q^2(s)\, E(t, s)\, ds \geqslant (1 + \epsilon)\, Q(t)/4 > 0, \qquad a \leqslant t < \infty, \tag{5.10$'$}$$

($Q \equiv Q_P$, $E \equiv E_P$, $P \equiv P_f$) then $y'' + py = 0$ is oscillatory.

Condition (5.10) holds if

$$\int_t^{\infty} P^2(s)\, ds \leqslant P(t)/4, \qquad a \leqslant t < \infty; \tag{5.11}$$

and condition (5.10′) holds for some $\epsilon > 0$, if there exists $\epsilon' > 0$ such that

$$\int_t^{\infty} P^2(s)\, ds \geqslant (1 + \epsilon')\, P(t)/4 > 0, \qquad a \leqslant t < \infty. \tag{5.11$'$}$$

For other simplifications of (5.10–5.10′), see Willett [*86*].

Under the assumption that $0 < P(t) \equiv \int_t^{\infty} p(s)\, ds < \infty$, Opial [*66*; 1958] originally proved that (5.11) and (5.11′) were sufficient for disconjugacy and oscillation, respectively. Wintner [*91*; 1951] had recognized earlier that the stronger hypothesis $P^2(t) \leqslant p(t)/4$ on $[a, \infty)$ implied disconjugacy.

Corollary 5.3 implies that the equation

$$y'' + (\mu t^{-1} \sin t)\, y = 0$$

is oscillatory when $|\mu| > 1/\sqrt{2}$ and nonoscillatory when $|\mu| < 1/\sqrt{2}$. See Willett [86] for other examples.

Theorem 5.6. *Assume that $y'' + p_1 y = 0$ is disconjugate and that p_1 has a finite averaged integral P_1 with respect to $\mathfrak{F}_0$. Assume that $P_2 \in C_1[a, \infty)$ satisfies $P_2' = -p_2$ and that $Q_2 \equiv Q_{P_2}$ is finite. If*

$$Q_1 \geqslant Q_2 \quad \textit{and} \quad Q_1 + P_1 \geqslant Q_2 + P_2 \quad (Q_1 \equiv Q_{P_1}), \tag{5.12}$$

then $y'' + p_2 y = 0$ is disconjugate.

Proof. Note first that the disconjugacy of $y'' + p_1 y = 0$ implies that Q_1 is finite. Theorem 5.4 implies that there exists $v \in C^1[a, \infty)$ such that

$$v(t) \geqslant Q_1(t) + \int_t^\infty E_1(t, s)\, v^2(s)\, ds, \qquad a \leqslant t < \infty \ (E_1 \equiv E_{P_1}). \tag{5.13}$$

Let

$$u(t) = P_2(t) + Q_2(t) + \int_t^\infty E_1(t, s)\, v^2(s)\, ds.$$

Because of (5.12) and (5.13), it is now an easy matter to show that $u' + p_2 + u^2 \leqslant 0$. Hence, Theorem 2.1 implies that $y'' + p_2 y = 0$ is disconjugate.

Corollary 5.4. (Taam [82; 1952]) *If $y'' + p_1 y = 0$ is disconjugate and*

$$\infty > \int_t^\infty p_1(s)\, ds \geqslant \left| \int_t^\infty p_2(s)\, ds \right|, \qquad a \leqslant t < \infty,$$

then $y'' + p_2 y = 0$ is disconjugate.

Theorem 5.6 for the case when

$$P_i(t) = \int_t^\infty p_i(s)\, ds < \infty, \qquad i = 1, 2,$$

was proven by Professor J. S. W. Wong in work not yet published. Corollary 5.4 with the additional assumption that $p_1 \geqslant 0$, $p_2 \geqslant 0$ was proven by Hille [31; 1948]. Corollary 5.4 has been rediscovered by

Kondratév [*40*; 1957], Wintner [*92*; 1957], Levin [*48*; 1960], and Drahlin [*15*; 1967].

Theorem 5.7. *Assume that p is not identically zero on any infinite subinterval of $[a, \infty)$ and that p has a finite nonnegative averaged integral P with respect to $\mathfrak{F}_0$. Then the equation $y'' + py = 0$ is disconjugate, if and only if, for each $b > a$, the smallest positive eigenvalue λ of the boundary value problem*

$$y'' + \lambda py = 0, \qquad y(a) = 0 = y'(b), \tag{5.14}$$

satisfies $\lambda > 1$.

Proof. If the eigenvalue condition is satisfied, it is obvious that no solution of $y'' + py = 0$ can have more than one zero.

In order to prove the converse, assume that $y'' + py = 0$ is disconjugate and that z is a positive solution on (a, ∞). Furthermore, suppose that there exist $b > a$ and $0 < \lambda \leqslant 1$ such that (5.14) has a nontrivial solution y on $[a, b]$. Let $w = zy' - yz'$. Then,

$$0 \leqslant (1 - \lambda)\,\lambda^{-1} \int_a^b [y'(s)]^2\, ds = (1 - \lambda) \int_a^b p(s)\, y^2(s)\, ds$$

$$= \int_a^b z^{-1}(s)\, y(s)\, w'(s)\, ds = -z^{-1}(b)\, y^2(b)\, z'(b) - \int_a^b z^{-2}(s)\, w^2(s)\, ds, \tag{5.15}$$

which implies that $z'(b) \leqslant 0$. Next, Theorem 5.3 implies that $v = z^{-1}z'$ satisfies

$$v(t) = P(t) + \int_t^\infty v^2(s)\, ds, \qquad a < t < \infty,$$

where $P \geqslant 0$ by assumption. Hence,

$$0 \geqslant v(b) = P(b) + \int_b^\infty v^2(s)\, ds \geqslant \int_b^\infty v^2(s)\, ds,$$

which can only occur if $v(t) = 0$ for all $b \leqslant t < \infty$. Therefore, z is constant in $[b, \infty)$, which implies that $p(t) = 0$ for $b \leqslant t < \infty$. This contradicts one of the hypothesis of the theorem.

Theorem 5.7 generalizes results of Nehari [*61*; 1957] and St. Mary

[*77*; 1968], who assume that $0 \leqslant P(t) \equiv \int_t^\infty p(s)\,ds < \infty$. Nehari also assumes that $p \geqslant 0$. St. Mary formulates his result as a necessary and sufficient condition for oscillation. We can also generalize this result to averaged integrals as follows:

Theorem 5.8. *Assume that p has a finite nonnegative averaged integral P with respect to $\mathfrak{F}_0$. Then, the equation $y'' + py = 0$ is oscillatory, if and only if, there exists a sequence of intervals $[a_n, b_n]$ with $a_n \uparrow \infty$ as $n \uparrow \infty$ such that the least positive eigenvalue λ_n of the system*

$$y'' + \lambda_n py = 0, \qquad y(a_n) = 0 = y'(b_n) \tag{5.16}$$

satisfies $\lambda_n \leqslant 1$, $n = 1, 2, \ldots$.

Proof. If $y'' + py = 0$ is oscillatory, then the eigenvalue condition is satisfied with $\lambda_n = 1$, $n = 1, 2, \ldots$.

Suppose the eigenvalue condition holds, and assume that $y'' + py = 0$ has a nonoscillatory solution z. Then, there exists $b \geqslant a$ such that z does not vanish in $[b, \infty)$. Let y denote the solution of (5.16) which corresponds to a_N, where $a_N \geqslant b$. The proof of Theorem 5.7 starting with (5.15) and applied to z and y of the present proof once again leads to the contradiction that $p(t) = 0$ for $t \geqslant a_N$. This contradicts the existence of the eigenvalues λ_n with $n > N$.

Other results involving eigenvalue conditions have been proven by Putnam [*70*; 1949], Barrett [*3*; 1959], and St. Mary [*77*; 1968]. Most of these results are of the comparison type.

Potter [*69*; 1953] lets $ry' = z$ in (1.1) to obtain the new equation

$$(q^{-1}z')' + r^{-1}z = 0. \tag{5.17}$$

Equation (5.17) is well defined on $[a, \infty)$ if $q > 0$. (We always assume that $r > 0$.) Furthermore, (5.17) and (1.1) have the same oscillatory behavior. Hence, with the additional hypothesis $q > 0$, we can apply most of the results mentioned in this paper to (5.17) to obtain new results for (1.1). Ráb [*73*; pp. 351–352] states specifically some of these results. Barrett [*3*; 1959] uses this transformation to obtain some new results of the eigenvalue type.

Bibliography

1. N. V. Adamov, On certain transformations not changing the integral curves of a differential equation of the first order. (Russian) *Mat. Sb.* **23** (65) (1948), no. 2, 187–228.

2. J. H. Barrett, Behavior of solutions of second order selfadjoint differential equations. *Proc. Amer. Math. Soc.* **6** (1955), 247–251.

3. J. H. Barrett, Disconjugacy of 2nd order linear differential equations with nonnegative coefficients. *Proc. Amer. Math. Soc.* **10** (1959), 552–561.

4. M. Bôcher, The theorems of oscillation of Sturm and Klein, I & II. *Bull. Am. Math. Soc.* **4** (1897–98), 295–313 and 365–376.

5. M. Bôcher, Non oscillatory linear differential equations of the second order. *Ibid* **7** (1900–01), 333–340.

6. M. Bôcher, Application of a method of d'Alembert to the proof of Sturm's theorem of comparison. *Trans. Amer. Math. Soc.* **1** (1900), 414–420.

7. M. Bôcher, An elementary proof of a theorem of Sturm. *Ibid* **2** (1901), 150–151.

8. P. Bohl, Über eine Differentialgleichung der Störungstheorie. *J. Reine Angew. Math.* **131** (1906), 268–321.

9. O. Borůvka, Théorie analytique et constructive des transformations différentielles linéaires du second ordre. *Bull. Math. Soc. Roum. Sci.* **1** (49), no. 2 (1957), 125–130.

10. O. Borůvka, Transformation of ordinary second-order linear differential equations." Differential Equations and their Applications (Conf. Proc., Prague, 1962), Acad. Pr., New York (1963), 27–38.

11. O. Borůvka, Lineare Differentialtransformationen 2. Ordnung, Veb Deutscher Verlag der Wissenschaften, Berlin (1967).

12. W. J. Coles, An oscillation criterion for second-order linear differential equations. *Proc. Amer. Math. Soc.* **19** (1968), 755–759.

13. W. J. Coles and D. Willett, Summability criteria for oscillation of second order linear differential equations. *Annali di mat. Pura ed App.*, **79** (1968), 391–398.

14. A. De Castro Brzezicki, Integrales oscilantes de las ecuaciones diferenciales lineales con segundo miembro. *Las Ciencias* **22** (1957), 5–28; **79** (1968), 391–398.

15. M. E. Drahlin, A comparison principle for differential equations of the second order on an infinite interval. (Russian) *Izv. Vysš. Učebn. Zaved. Mat.* (1967), no. 9 (64), 26–30.

16. M. I. El'šin, Qualitative problems on the linear differential equation of the second order. (Russian) *Dokl. Adad. Nauk SSSR* (N. S.) **68** (1949), no. 2, 221–224.

17. M. I. El'šin, The phase method and the classical method of comparison. (Russian) *Ibid* **68** (1949), 813–816.

18. M. I. El'šin, Qualitative investigation of a system of two linear homogeneous equations of the first order. (Russian) Ibid **94** (1954), 5–8.

19. M. I. El'šin, On a solution of a classical oscillation problem. (Russian) *Ibid* **147** (1962), 1013–1016; (*Soviet Math. Dokl.* **3** (1962), 1752–1755).

20. M. I. El'šin, Qualitative solution of a linear differential equation of the second order. (Russian) *Uspehi Mat. Nauk. (N. S.)* **5** (1950), no. 2 (36), 155–158.

21. A. M. Fink and D. F. St. Mary, A generalized Sturm Comparison theorem and oscillation coefficients. *Monat. Math.* **73** (1969), 207–212.

22. W. B. Fite, Concerning the zeros of the solutions of certain differential equations. *Trans. Amer. Math. Soc.* **19** (1918), 341–352.

23. T. Fort, Some theorems of comparison and oscillation. *Bull. Amer. Math. Soc.* **24** (1917–18), 330–334.
24. G. Fubini, Su un teorema di confronto per le equazioni del secondo ordine alle derivate ordinarie. *Ann. Scuola Norm. Super.* Pisa, S. II, 2 (1933), 283–284.
25. E. Gagliardo, Sui criteri di oscillazione per gli integrali di un'equazione differenziale lineare del secondo ordine," *Boll. Unione Mat. Ital.* **9** (1954), 177–189.
26. P. Hartman, On the linear logarithmico-exponential differential equation of the second order. *Amer. J. Math.* **70** (1948), 764–779.
27. P. Hartman, On linear second order differential equations with small coefficients. *Ibid* **73** (1951), 955–962.
28. P. Hartman, On non-oscillatory linear differential equations of second order. *Ibid* **74** (1952), 389–400.
29. P. Hartman and A. Wintner, On non-conservatory linear oscillators of low frequency· *Ibid* **70** (1948), 529–539.
30. P. Hartman and A. Wintner, On nonoscillatory linear differential equations. *Ibid* **75** (1953), 717–730.
31. E. Hille, Non-oscillation theorems. *Trans. Amer. Math. Soc.* **64** (1948), 234–252.
32. M. Ielchin, Sur le problème d'oscillation pour l'équation différentielle linéaire du deuxième ordre. *C. R. (Doklady) Acad. Sci.* URSS 18 (1938), no. 3, 141–145.
33. M. Jelchin, Sur une méthode d'évaluation de la phase d'une équation différentielle linéaire du second ordre. *Uchenye Zapiski Moskov. Gos. Univ. Mat.* **45** (1940), 97–108.
34. E. Kamke, A new proof of Sturm's comparison theorems. *Amer. Math. Monthly* *46* (1939), 417–421.
35. E. Kamke, Über Sturms Vergleichssätze fur homogene lineare Differentialgleichungen zweiter Ordnung und Systeme von zwei Differentialgleichungen erster Ordnung. *Math. Zeit.* **47** (1940), 788–795.
36. E. Kamke, Bemerkungen zu einigen Trennungssätzen von M. Nicolesco und S. Takahashi fur die Nullstellen der Lösungen von Differentialgleichungen. *Math. Cluj* **15** (1939), 201–203.
37. A. Kneser, Untersuchungen über die reelen Nullstellen der Integrale linearer Differentialgleichungen. *Math. Ann.* **42** (1893), 409–435.
38. Hans-Wilhelm, Knobloch, Wachstum und oszillatorisches Verhalten von Lösungen linearer Differentialgleichungen zweiter Ordnung. *Jber. Deutsch. Math. Verein.* **66** (1963–64), Abt. 1, 138–152.
39. V. A. Kondratév, Elementary derivation of a necessary and sufficient condition for nonoscillation of the solutions of a linear differential equation of second order. (Russian) *Uspehi Mat. Nauk (N.S.)* **12** (1957), no. 3 (75), 159–160.
40. V. A. Kondratév, Sufficient conditions for nonoscillatory or oscillatory nature of solutions of the second order equation $y'' + p(x)y = 0$. (Russian) *Dokl. Akad. Nauk SSSR* **113** (1957), 742–745.
41. E. E. Kummer, De generali quadam aequatione differentiali tertii ordinis. *J. Reine Angew. Math.* **100** (1887), 1–9 (Abdruck aus dem Programm des evangelischen Königl. und Stadtgym. in Liegnitz vom Jahre 1834.).
42. M. Laitoch, Sur une théorie des critères comparatifs sur l'oscillation des intégrales de l'équation différentielle $u'' = P(x)u$. *Spisy Přir. Fak. M U Brno* **365** (1955), 255–267.

43. G. LANDOLINO, Sul teorema di confronto di Sturm. *Boll. Un. Mat. Ital.*, Ser. III, 2 (1947), 16–19.

44. W. LEIGHTON, On self-adjoint differential equations of second order. *Proc. Nat. Acad. Sci.* **35** (1949), 656–657.

45. W. LEIGHTON, The detection of the oscillation of solutions of a second-order linear differential equation. *Duke J. Math.* **17** (1950), 57–62.

46. W. LEIGHTON, On self-adjoint differential equations of second-order. *J. London Math. Soc.* **27** (1952), 37–47.

47. W. LEIGHTON, Comparison theorems for linear differential equations of second order. *Proc. Amer. Math. Soc.* **13** (1962), 603–610.

48. A. JU. LEVIN, A comparison principle for second-order differential equations. (Russian) *Dokl. Akad. Nauk SSSR* **135** (1960), 783–786 (*Soviet Math. Dokl.* **1** (1961), 1313–1316).

49. A. JU. LEVIN, On linear second-order differential equations. (Russian) *Ibid* **153** (1963), 1257–1260 (*Soviet Math. Dokl.* **4** (1963), 1814–1817).

50. A. JU. LEVIN, Integral criteria for the equation $\ddot{x} + q(t)x = 0$ to be nonoscillatory. (Russian) *Uspehi Mat. Nauk* 20 (1965), no. 2 (122), 244–246.

51. S. LIE, Zur Theorie der Transformationsgruppen. *Leipziger Berichte* (1894), Heft III, 322–333 (*Gesammelte Abhandlungen*, V. 6, Part 1, No. XVII, 386–395, Teubner, Leipzig, 1927).

52. W. MAGNUS AND S. Winkler, "*Hill's Equation.*" Wiley, New York, 1966, pp. 56–78.

53. T. MANACORDA, Sul comportamento asintotico degli integrali dell'equazione: $y''(x) + p(x)y'(x) + q(x)y(x) = 0$ quando $\lim_{x\to\infty} q(x) = \infty$, I. *Atti Accad. Naz. Lincei Rend. Cl. Sci. Fis. Mat. Nat.* (8) 2 (1947), 537–541.

54. L. MARKUS AND R. A. MOORE, Oscillation and disconjugacy for linear differential equations with almost periodic coefficients. *Acta Math.* **96** (1956), 99–123.

55. J. C. P. MILLER, On a criterion for oscillatory solutions of a linear differential equation of the second order. *Proc. Cambridge Phil. Soc.* **36** (1940), 283–287.

56. R. MILOS, Kriterien für die Oszillation der Zösungen der Differentialgleichung $[p(x)y']' + q(x)y = 0$. *Casopis pro pestovani Mat.* **84** (1959), 3.

57. I. M. MIN, Sur les zéros des solutions des certaines équations différentielles. *Bull. Inst. Politehn. Bucuresti* **27** (1967), no. 6, 17–23.

58. R. A. MOORE, The behavior of solutions of a linear differential equation of second order. *Pacific J. Math.* **5** (1955), 125–145.

59. R. A. MOORE, The least eigenvalue of Hill's equation. *J. D'Analyse Math.* **5** (1956–57), 183–196.

60. H. NAKANO, Über die Verteilung der Nullstellen von dem Lösungen der Differentialgleichung $(d^2w)/(dz^2) + G(z)w = 0$ einer komplexen Veränderlichen. *Proc. Imp. Acad. Japan* **8** (1932), 337–339.

61. Z. NEHARI, Oscillation criteria for second-order linear differential equations. *Trans. Amer. Math. Soc.* **85** (1957), 428–445.

62. M. NICOLESCO, Sur le théorème de Sturm. *Math. Cluj* **1** (1929), 111–114.

63. L. D. NIKOLENKO, On oscillation of solutions of the differential equation $y'' + p(x)y = 0$. *Ukrain. Mat. Z* **7** (1955), 124–127.

64. C. O. OAKLEY, A note on the methods of Sturm. *Ann. Math.*, ser. II, **31** (1930), 660–662.

65. C. OLECH, Z. OPIAL, AND T. WAZEWSKI, "Sur le Problème d'oscillation des intégrales de l'équation $y'' + q(t)y = 0$," Bull. Acad. Polonaise Sci., cl. III, **5** (1957), 621–626.

66. Z. Opial, "Sur les intégrales oscillantes de l'équation différentielle $u'' + f(t)u = 0$," *Ann. Polon. Math.* **4** (1958), 308–313.
67. Z. Opial, Sur une Critère d'oscillation des Intégrales de l'équation Différentielle $(Q(t)x')' + f(t)x = 0$. *Ibid* **6** (1959–60), 99–104.
68. R. V. Petropavlovskaya, On oscillation of solutions of the equation $u'' + p(x)u = 0$, (Russian) *Dokl. Akad. Nauk SSSR* (N. S.) **105** (1955), 29–31.
69. R. L. Potter, On self-adjoint differential equations of 2nd order. *Pacific J. Math.* **3** (1953), 467–491.
70. C. R. Putnam, An oscillation criterion involving a minimum principle. *Duke Math. J.* **16** (1949), 633–636.
71. C. R. Putnam, Note on some oscillation criteria. *Proc. Amer. Math. Soc.* **6** (1955), 950–952.
72. M. Ráb, Eine Bemerkung zu der Frage über die oszillatorischen Eigenschaften der Lösungen der Differentialgleichungen $y'' + A(x)y = 0$. *Čas. Pěst. Mat.* **82** (1957), 342–348.
73. M. Ráb, Kriterien fur die Oszillation der Lösungen der Differentialgleichung $[p(x)y']' + q(x)y = 0$. *Ibid* **84** (1959), 335–370 (Erratum, *Ibid* **85** (1960), 91).
74. W. T. Reid, A comparison theorem for self-adjoint differential equations of second order. *Ann. Math.* **65** (1957), 197–202.
75. U. Richard, Sulla risoluzione asintoticonumerica dell'equazione differenziale $(py')' + qy = 0$ nel caso oscillante. *Atti Accad. Sci. Torino* **97** (1962–63), 857–890.
76. B. Riemann and H. Weber, "Die Partiellen Differentialgleichungen der Mathematischen Physik. vol. II, 5th ed., Braunschweig, 1912, pp. 53–72.
77. D. F. St. Mary, "Some oscillation and comparison theorems for $(r(t)y')' + p(t)y = 0$" Ph.D dissertation, U. of Nebraska, Lincoln, 1968.
78. G. Sansone, "Equazioni Differenziali nel Campo Reale." I, N. Zanichelli, Bologna 1948.
79. I. M. Sobol, Investigation with the aid of polar coordinates of the asymptotic behavior of a linear differential equation of the second order. (Russian) *Mat. Sb.* **28** (70) (1951), no. 3, 707–714.
80. P. Stäckel, Ueber Transformationen von Differentialgleichungen," *J. Reine Angew. Math.* **111** (1893), 290–302.
81. C. Sturm, "Sur les équations différentielles linéaires du second ordre," *J. Math. Pures Appl.* **1** (1836), 106–186.
82. Choy-tak, Taam, Non-oscillatory differential equations. *Duke Math. J.* **19** (1952), 493–497.
83. Shin-ichi, Takahashi, On the zero points of an integral of a linear differential equation. *Jap. J. Math.* **7** (1931), 335–346.
84. A. Tonolo, Sopra un teorema di confronto di Fubini. *Boll. Un. Mat. Italy* **14** (1935), 67–70.
85. E. Vessoit, *Encyklopedia Math. Wissenschaften*, IIa 4b, Leipzig (1899–1916), 271–285.
86. D. Willett, On the oscillatory behavior of the solutions of second order linear differential equations. *Ann. Polon. Math.* **21** (1969), 175–194.
87. A. Wintner, A norm criterion for non-oscillatory differential equations. *Quart. Appl. Math.* **6** (1948), 183–185.
88. A. Wintner, A criterion of oscillatory stability. *Ibid* **7** (1949), 115–117.
89. A. Wintner, Asymptotic integrations of the adiabatic oscillator in its hyperbolic range. *Duke Math. J.* **15** (1948), 55–67.

90. A. Wintner, On almost free linear motions. *Amer. J. Math.* **71** (1949), 595–602.
91. A. Wintner, On the non-existence of conjugate points. *Ibid* **73** (1951), 368–380.
92. A. Wintner, On the comparison theorem of Kneser-Hille. *Math. Scand.* **5** (1957), 2, 255–260.
93. M. Yelchin, Sur les conditions pour qu'une solution d'un système linéaire du second ordre possède deux zéros. *C. R. (Doklady) Acad. Sci. URSS* **51** (1946), no. 8, 573–576.
94. K. Yosida, On the distribution of α-points of solutions for linear differential equations of the second order. *Proc. Imp. Acad. Japan* **8** (1932), 335–336.
95. M. Zlámal, Oscillation criterions. *Cas. pěst. Mat. a Fis.* **75** (1950), 213–217.
96. A. F. Zubova, Concerning oscillation of the solutions of an equation of the second order. (Russian) *Vestnik Leningrad Univ.* **12** (211) (1957), no. 1, 168–174.
97. A. F. Zubova, Oscillations and stability of solutions of second-order equations, (Russian) *Sibirsk. Mat. Ž* **4** (1963), 1060–1070.

SUBJECT INDEX

T

V

W